Biofibers and Biopolymers for Biocomposites

Anish Khan · Sanjay Mavinkere Rangappa ·
Suchart Siengchin · Abdullah M. Asiri
Editors

Biofibers and Biopolymers for Biocomposites

Synthesis, Characterization and Properties

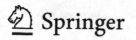 Springer

Editors
Anish Khan
Chemistry Department Faculty of Science
King Abdulaziz University
Jeddah, Saudi Arabia

Centre of Excellence for Advanced
Materials Research
King Abdulaziz University
Jeddah, Saudi Arabia

Suchart Siengchin
Department of Mechanical and Process
Engineering
The Sirindhorn International Thai—German
Graduate School of Engineering (TGGS)
President of King Mongkut's University
of Technology North Bangkok
Bangkok, Thailand

Sanjay Mavinkere Rangappa
Natural Composites Research Group Lab
Academic Enhancement Department
King Mongkut's University of Technology
North Bangkok
Bangkok, Thailand

Abdullah M. Asiri
Chemistry Department Faculty of Science
King Abdulaziz University
Jeddah, Saudi Arabia

Centre of Excellence for Advanced
Materials Research
King Abdulaziz University
Jeddah, Saudi Arabia

ISBN 978-3-030-40303-4   ISBN 978-3-030-40301-0   (eBook)
https://doi.org/10.1007/978-3-030-40301-0

This Springer imprint is published by the registered company Springer Nature Switzerland AG
The registered company address is: Gewerbestrasse 11, 6330 Cham, Switzerland

Editors are honored to dedicate this book to their parents

Preface

Biofibers and biopolymers for biocomposites is an emerging area in polymer science. The rise in ecological anxieties has forced scientists and researchers from all over the world to find new ecological materials. Therefore, it is necessary to expand knowledge about the properties of biofibers/biopolymers to expanding the range of their application. As the title indicates, the proposed books will emphasize new challenges for the synthesis, characterization and properties of biofibers, biopolymers, and biocomposites. This book presents the important areas of (synthesis, processing, characterization, and application) biofibers and biopolymers in a comprehensive manner. This book presents an extensive survey on recent improvements in the research and development of biofibers and biopolymers that are used to make biocomposites in various applications. Characterization techniques on the various aspects of biofibers and biopolymers are presented in this book. The structure and properties of biofibers and biopolymers were included along with surface modification techniques.

This book covers the void for the need for one-stop reference book for the researchers. Leading researchers from industry, academy, government, and private research institutions across the globe will contribute to this book. Academics, researchers, scientists, engineers, and students in the field of epoxy blends will benefit from this book which is highly application-oriented.

The proposed book will give a correlation of properties with biofibers and biopolymers and has become a point of great interest. Moreover, it will provide a cutting-edge research from around the globe in this field. Current status, trends, future directions, opportunities, etc., will be discussed in detail, making it friendly for the beginners and young researchers.

Jeddah, Saudi Arabia   Anish Khan
Bangkok, Thailand   Sanjay Mavinkere Rangappa
Bangkok, Thailand   Suchart Siengchin
Jeddah, Saudi Arabia   Abdullah M. Asiri

Contents

Editors and Contributors

About the Editors

Dr. Anish Khan Assistant Professor, Chemistry Department, Faculty of Science, King Abdulaziz University, Jeddah, Saudi Arabia.

Dr. Anish Khan received Ph.D. from Aligarh Muslim University, India in 2010. Research experience of working in the field of synthetic polymers, organic–inorganic electrically conducting nano-composites. He completed postdoctoral from the School of Chemical Sciences, Universiti Sains Malaysia (USM) electroanalytical chemistry in 2010–2011. He has research and teaching experience, published more than 100 research papers in reffered international journals. He participated in more than 20 international conferences/ workshops and 3 books have been published and 6 in progress and 12 book chapters. Around 20 research projects have been completed. He is the Managerial Editor of Chemical and Environmental Research (CER) Journal, Member of American Nano Society, and his fields of specialization are polymer nanocomposite/cation-exchanger/chemical sensor/microbiosensor/nanotechnology, application of nanomaterials in electroanalytical chemistry, material chemistry, ion-exchange chromatography, and electro-analytical chemistry, dealing with the synthesis, characterization (using different analytical techniques), and derivatization of inorganic ion-exchanger by the incorporation of electrically conducting polymers, preparation and characterization of hybrid nanocomposite materials and their applications, polymeric inorganic cation-exchange materials, electrically conducting polymeric, materials, composite material use as Sensors, green chemistry by remediation of pollution, heavy metal ion-selective membrane electrode, biosensor on neurotransmitter. e-mail: anishkhan97@gmail.com

Dr. Sanjay Mavinkere Rangappa Research Scientist, Department of Mechanical and Process Engineering, King Mongkut's University of Technology North Bangkok, 1518 Pracharaj 1, Wongsawang Road, Bangsue, Bangkok, Thailand.

Dr. Sanjay Mavinkere Rangappa, received B.E (Mechanical Engineering) from Visvesvaraya Technological University, Belagavi, India in the year 2010, M.Tech. (Computational Analysis in Mechanical Sciences) from VTU Extension Centre, GEC, Hassan, in the year 2013, Ph.D. (Faculty of Mechanical Engineering Science) from Visvesvaraya Technological University, Belagavi, India in the year 2017 and Postdoctorate from King Mongkut's University of Technology North Bangkok, Thailand, in the year 2019. He is a Life Member of Indian Society for Technical Education (ISTE) and Associate Member of Institute of Engineers (India). He is a reviewer for more than 40 international journals and international conferences (for Elsevier, Springer, Sage, Taylor & Francis, Wiley). In addition, he has published more than 70 articles in high-quality international peer-reviewed journals, 13 book chapters, 1 book, 11 books (Editor), and also presented research papers at national/international conferences. His current research areas include Natural Fiber Composites, Polymer Composites, and Advanced Material Technology. He is a recipient of DAAD Academic exchange-PPP Programme (Project-related Personnel Exchange) between Thailand and Germany to Institute of Composite Materials, University of Kaiserslautern, Germany. He has received the Top Peer Reviewer 2019 award, Global Peer Review Awards, Powdered by Publons, Web of Science Group. e-mail: mcemrs@gmail.com

Prof. Dr.-Ing. habil. Suchart Siengchin President of King Mongkut's University of Technology North Bangkok, Department of Materials and Production Engineering (MPE), The Sirindhorn International Thai-German Graduate School of Engineering (TGGS), King Mongkut's University of Technology North Bangkok, 1518 Pracharaj 1, Wongsawang Road, Bangsue, Bangkok, Thailand.

Prof. Dr.-Ing. habil. Suchart Siengchin, received his Dipl.-Ing. in Mechanical Engineering from University of Applied Sciences Giessen/Friedberg, Hessen, Germany in 1999, M.Sc. in Polymer Technology from University of Applied Sciences Aalen, Baden-Wuerttemberg, Germany in 2002, M.Sc. in Material Science at the Erlangen-Nürnberg University, Bayern, Germany in 2004, Doctor of Philosophy in Engineering (Dr.-Ing.) from Institute for Composite Materials, University of Kaiserslautern, Rheinland-Pfalz, Germany in 2008 and Postdoctoral Research from Kaiserslautern University and School of Materials Engineering, Purdue University, USA. In 2016, he received the habilitation at the Chemnitz University in Sachen, Germany. He worked as a Lecturer for Production and Material Engineering Department at The Sirindhorn International Thai-German Graduate School of Engineering (TGGS), KMUTNB. He has been Full Professor at KMUTNB and became the President of KMUTNB. He won the Outstanding Researcher Award in 2010, 2012 and 2013 at KMUTNB. His research interests are Polymer Processing and Composite Material. He is Editor-in-Chief: KMUTNB *International Journal of Applied Science and Technology* and the author of 150+ peer-reviewed Journal Articles. He has participated with presentations in more than 39 International and National Conferences with respect to Materials Science and Engineering topics. e-mail: suchart.s.pe@tggs-bangkok.org

Prof. Dr. Abdullah M. Asiri Director of the Center of Excellence for Advanced Materials Research (CEAMR), Chair, Chemistry Department, Faculty of Science, King Abdulaziz University, Jeddah, Saudi Arabia.

Prof. Dr. Abdullah M. Asiri, received Ph.D. (1995) from the University of Wales College, Cardiff, U.K. on tribochromic compounds and their applications. Currently, he is the Chairman of the Chemistry Department, King Abdulaziz University and also the Director of the Center of Excellence for Advanced Materials Research. Director of Education Affair Unit–Deanship of Community Services. Member of Advisory Committee for Advancing Materials, (National Technology Plan, King Abdul Aziz City of Science and Technology, Riyadh, Saudi Arabia). His area of interest includes: Color chemistry. Synthesis of novel photochromic and thermochromic systems, Synthesis of novel colorants and coloration of textiles and plastics, Molecular Modeling, Applications of organic materials into optics such as OEDS, High-performance organic dyes and pigments. New applications of organic photochromic compounds in new novelty. Organic synthesis of heterocyclic compounds as precursor for dyes. Synthesis of polymers functionalized with organic dyes. Preparation of some coating formulations for different applications. Photodynamic thereby using Organic Dyes and Pigments Virtual Labs and Experimental Simulations. He is member of Editorial board of Journal of Saudi Chemical Society, Journal of King Abdul Aziz University, Pigment and Resin Technology Journal, Organic Chemistry Insights, Libertas Academica, Recent Patents on Materials Science, Bentham Science Publishers Ltd. Besides he has professional membership of International and National Society and Professional bodies. e-mail: aasiri2@gmail.com; aasiri2@kau.edu.sa

Contributors

A. Ajithram Department of Mechanical Engineering, Kalasalingam Academy of Research and Education, Krishnankoil, Tamil Nadu, India

S. Basavarajappa Indian Institute of Information Technology, Dharwad, Dharwad, Karnataka, India

Sumayya Begum School of Physical Sciences, Swami Ramanand Teerth Marathwada University, Nanded, Maharashtra, India

K. N. Bharath Composite Materials and Engineering Center, Washington State University, Pullman, Washington State, United States of America;
Department of Mechanical Engineering, G.M. Institute of Technology, Davangere, Karnataka, India

Muhammad Bilal School of Life Science and Food Engineering, Huaiyin Institute of Technology, Huaian, China

J. S. Binoj Sree Vidyanikethan Engineering College (Autonomous), Tirupati, Andhra Pradesh, India

Kashinath A. Bogle School of Physical Sciences, Swami Ramanand Teerth Marathwada University, Nanded, Maharashtra, India

Luigi Calabrese Department of Engineering, University of Messina, Messina, Italy

Muthukumar Chandrasekar Department of Aerospace Engineering, Faculty of Engineering, University Putra Malaysia, Serdang, Selangor, Malaysia

C. Deepa Department of Computer Science and Engineering, KIT-Kalaignarkarunanidhi Institute of Technology, Coimbatore, Tamil Nadu, India

Vincenzo Fiore Department of Engineering, University of Palermo, Palermo, Italy

K. Gopalakrishna Center for Incubation, Innovation, Research and Consultancy, Jyothy Institute of Technology, Bengaluru, India

S. Indran Rohini College of Engineering & Technology, Palkulam, Tamil Nadu, India

Hafiz M. N. Iqbal School of Engineering and Sciences, Tecnologico de Monterrey, Monterrey, NL, Mexico

Mohammad Irfan Iqbal Department of Textile Engineering, Wuhan Textile University, Wuhan, Hubei, China;
Department of Textile Engineering, BGMEA University of Fashion & Technology, Dhaka, Bangladesh

Naman Jain Department of Mechanical Engineering, G. B. Pant University of Agriculture and Technology, Pantnagar, India

Aswathy Jayakumar School of Biosciences, Mahatma Gandhi University, Kottayam, Kerala, India

Jasila Karayil Department of Chemistry, Government Polytechnic Women's College, Calicut, Kerala, India

Rajendra S. Khairnar School of Physical Sciences, Swami Ramanand Teerth Marathwada University, Nanded, Maharashtra, India

Ayub Nabi Khan Department of Textile Engineering, BGMEA University of Fashion & Technology, Dhaka, Bangladesh

Rashed Al Mizan Department of Textile Engineering, BGMEA University of Fashion & Technology, Dhaka, Bangladesh

Nawshad Muhammad Interdisciplinary Research Centre in Biomedical Materials, COMSATS University Islamabad, Lahore Campus, Lahore, Pakistan

Lakshmanan Muthulakshmi Department of Materials Science, Madurai Kamaraj University, Madurai, Tamilnadu, India

Faran Nabeel School of Chemistry and Chemical Engineering, State Key Laboratory of Metal Matrix Composites, Shanghai Jiao Tong University, Shanghai, China

Vijaykiran N. Narwade School of Physical Sciences, Swami Ramanand Teerth Marathwada University, Nanded, Maharashtra, India;
Faculty of Natural Sciences, Department of Inorganic Chemistry, Comenius University, Bratislava, Slovakia

P. Navaneethakrishnan Faculty of Mechanical Engineering, School of Building and Mechanical Sciences, Kongu Engineering College, Perundurai, Erode, Tamilnadu, India

Sadia Naz Institute of Chemistry, University of the Punjab, Lahore, Pakistan

Jyotishkumar Parameswaranpillai Center of Innovation in Design and Engineering for Manufacturing, King Mongkut's University of Technology North Bangkok, Bangkok, Thailand

Avinash Parashar Department of Mechanical and Industrial Engineering, Indian Institute of Technology, Roorkee, India

Elpida Piperopoulos Department of Engineering, University of Messina, Messina, Italy

Yasir Beeran Pottathara Faculty of Mechanical Engineering, University of Maribor, Maribor, Slovenia

E. K. Radhakrishnan School of Biosciences, Mahatma Gandhi University, Kottayam, Kerala, India

Sabarish Radoor Department of Mechanical and Process Engineering, The Sirindhorn International Thai-German Graduate School of Engineering (TGGS), King Mongkut's University of Technology North Bangkok, Bangkok, Thailand

Nagarajan Rajini Centre for Composite Materials, Department of Mechanical Engineering, Kalasalingam Academy of Research and Education, Krishnankoil, Tamil Nadu, India

S. Ramakrishnan Faculty of Mechanical Engineering, School of Building and Mechanical Sciences, Kongu Engineering College, Perundurai, Erode, Tamilnadu, India

M. Ramesh Department of Mechanical Engineering, KIT-Kalaignarkarunanidhi Institute of Technology, Coimbatore, Tamil Nadu, India

Sanjay Mavinkere Rangappa Department of Mechanical and Process Engineering, The Sirindhorn International Thai-German Graduate School of Engineering (TGGS), King Mongkut's University of Technology North Bangkok, Bangkok, Thailand

Tahir Rasheed School of Chemistry and Chemical Engineering, State Key Laboratory of Metal Matrix Composites, Shanghai Jiao Tong University, Shanghai, China

Narendra Reddy Center for Incubation, Innovation, Research and Consultancy, Jyothy Institute of Technology, Bengaluru, India

T. P. Sathishkumar Faculty of Mechanical Engineering, School of Building and Mechanical Sciences, Kongu Engineering College, Perundurai, Erode, Tamilnadu, India

Thiagamani Senthil Muthu Kumar Department of Mechanical Engineering, Kalasalingam Academy of Research and Education, Krishnankoil, Tamil Nadu, India;
Department of Mechanical and Process Engineering, The Sirindhorn International Thai-German Graduate School of Engineering (TGGS), King Mongkut's University of Technology North Bangkok, Bangkok, Thailand

Krishnasamy Senthilkumar Department of Mechanical Engineering, Kalasalingam Academy of Research and Education, Krishnankoil, Tamil Nadu, India;
Department of Mechanical and Process Engineering, The Sirindhorn International Thai-German Graduate School of Engineering (TGGS), King Mongkut's University of Technology North Bangkok, Bangkok, Thailand

Faiza Sharif Interdisciplinary Research Centre in Biomedical Materials, COMSATS University Islamabad, Lahore Campus, Lahore, Pakistan

Suchart Siengchin Department of Mechanical and Process Engineering, The Sirindhorn International Thai-German Graduate School of Engineering (TGGS), King Mongkut's University of Technology North Bangkok, Bangkok, Thailand

Manoj Kumar Singh School of Engineering, Indian Institute of Technology (IIT), Mandi, Himachal Pradesh, India

V. K. Singh Department of Mechanical Engineering, G. B. Pant University of Agriculture and Technology, Pantnagar, India

Saravanasankar Subramaniam Department of Mechanical and Process Engineering, The Sirindhorn International Thai-German Graduate School of Engineering (TGGS), King Mongkut's University of Technology North Bangkok, Bangkok, Thailand

Maliha Uroos Institute of Chemistry, University of the Punjab, Lahore, Pakistan

Anumakonda Varada Rajulu Centre for Composite Materials, Department of Mechanical Engineering, Kalasalingam Academy of Research and Education, Krishnankoil, Tamil Nadu, India

Akarsh Verma Department of Mechanical and Industrial Engineering, Indian Institute of Technology, Roorkee, India

Nishant Verma School of Engineering, Indian Institute of Technology (IIT), Mandi, Himachal Pradesh, India

J. T. Winowlin Jappes Department of Mechanical Engineering, Kalasalingam Academy of Research and Education, Krishnankoil, Tamil Nadu, India

Sunny Zafar School of Engineering, Indian Institute of Technology (IIT), Mandi, Himachal Pradesh, India

Tahera Zafar Department of Science of Dental Materials, de'Montmorency College of Dentistry, Lahore, Pakistan

Yi Zhao Department of Nonwoven Materials and Engineering, College of Textiles, Donghua University, Shanghai, China

Surface Modification Techniques for the Preparation of Different Novel Biofibers for Composites

Akarsh Verma, Avinash Parashar, Naman Jain, V. K. Singh,
Sanjay Mavinkere Rangappa and Suchart Siengchin

Abstract This chapter reports on the various physical and chemical methods used in modifying the natural fibers properties for application in reinforcing composites. Low cost, low density and biodegradable nature of bio fibers have attracted composite industries to develop various useful products out of them. Nevertheless, associated disadvantages with these fibers are that they have poor compatibility with matrix, relative high water absorption capacity and sticking in bundles. For eradication of such unwanted characteristics, several physical and chemical treatments have been examined by the researchers. These treatments tend to alter the surface morphology and chemical structure for enhancing the adhesive strength between fiber and matrix. Mechanisms that are involved in this enhancement are the increase in fiber surface roughness and alteration in chemical polarity of natural fibers.

1 Introduction

Natural fibers (commonly known as bio/renewable fibers) reinforced polymeric composites have become the center of attention for various material scientists and engineering applications [1–3]. This is certainly because of the promising merits that these natural fibers possess, which includes low density and cost, renewable/biodegradable nature, adequate specific strength and modulus, available abundantly in nature, lessen tool wear for fabrication and enhanced energy recovery for the damping applications

A. Verma (✉) · A. Parashar
Department of Mechanical and Industrial Engineering, Indian Institute of Technology, Roorkee,
India
e-mail: akarshverma007@gmail.com

N. Jain · V. K. Singh
Department of Mechanical Engineering, G. B. Pant University of Agriculture and Technology,
Pantnagar, India

S. M. Rangappa · S. Siengchin
Department of Mechanical and Process Engineering, The Sirindhorn International Thai-German
Graduate School of Engineering (TGGS), King Mongkut's University of Technology North
Bangkok, Bangkok, Thailand

© Springer Nature Switzerland AG 2020   1
A. Khan et al. (eds.), *Biofibers and Biopolymers for Biocomposites*,
https://doi.org/10.1007/978-3-030-40301-0_1

[4–6]. Alternatively, some of their drawbacks like poor fiber-matrix interfacial adhesion, fiber surface irregularity and variability, fiber finite length and restricted thermal stability (accompanied by low processing temperature) have raised questions to their applications in extreme/harsh environment [7, 8]. From the performance point of view, interfacial bonding between the natural fibers-polymer matrix should be enhanced; so that a better transfer of load/stress from matrix to fiber is possible and we obtain amplified mechanical and thermal properties [9–11].

For fabrication of natural fiber based composites understanding of the interfacial bonding between matrix and fiber is required for which study of the chemical composition of fiber is a necessity [12–14]. On a chemical note, major components of natural fibers are cellulose (semi crystalline polysaccharide) in which D-glucopyranose units are linked together by b-(1-4)-glucosidic bonds; hemicelluloses (branched and fully amorphous polysaccharide) mostly consist of D-pentose sugars units strongly bonded with cellulose fibrils through hydrogen bonding; lignin (highly complex structure and amorphous) consisting of aromatic alcohol units such as coumaryl alcohol, coniferyl alcohol and sinapyl alcohol; pectin (a heterogeneous polysaccharide); waxes and other water soluble material. Cellulose fibrils are the major component that provides strength to the fiber and contain large amount of hydroxyl (–OH) functional groups. Since, the natural fibers retain hydroxyl and several other polar group sin their domain; this imparts hydrophilic nature to natural fibers and they tend to exhibit relatively high moisture absorption phenomenon [15]. This augmented moisture absorption ability of natural fibers results in initiation of biodegradation process and deteriorates the mechanical strength and dimensional stability. Furthermore, hydrophilic natural fibers and hydrophobic polymers matrix results in incompatibility and weak interfacial bonding amongst them. A better way to increase the adhesion of natural fibers in composites is through an appropriate surface treatment/modification [16–18]. Broadly, treatment of natural fibers is done to improve upon the interfacial adhesion bonding between fiber and matrix, increase the surface wettability of natural fiber in matrix, improve water resistance of fibers and increase in availability of active hydroxyl group to form bonding with matrix. Thus, to obtain optimum physical and mechanical properties of fabricated composites, performance of the reinforcing material is important and to improve upon the fiber performance their surface treatment is done. Also, degradation of natural fibers by microorganisms under the moist environment results in deterioration of cellulose fibrils that affect the overall performance of composites; so for that too we need a surface treatment process. Physically, presence of unwanted impurities (such as hemicelluloses, lignin, oil, dust and wax) result in reduction of availability of hydroxyl group of cellulose fibrils to interact with matrix polymer chain [19]. As the natural fibers are polar (due to presence of hydroxyl group on the surface of cellulose fibrils), this results in low compatibility with non-polar polymers (polyethene and polypropylene). The interaction between fibers and polymer occurs only in terms of hydrogen bonding, which results in agglomeration of fiber into bundles and unevenly distribution around the non-polar polymer matrix [20]. Moreover, wetting of fibers is insufficient by non-polar polymers due to which poor interfacial adhesion occurs between the fibers and polymer. This results in the reduction of load transfer efficiency from polymer to reinforcing fibers.

Literature has broadly classified these surface treatments into physical and chemical categories (refer Fig. 1).

Physical treatment techniques	Plasma treatment
	Corona treatment
	Ultrasound treatment
	Ultraviolet treatment

Chemical treatment techniques	Alkaline treatment
	Silane treatment
	Acetylation treatment
	Benzoylation treatment
	Acrylation and Acrylonitrile grafting
	Maleated coupling agents
	Permanganate treatment
	Peroxide treatment
	Isocyanate treatment
	Other treatments

Fig. 1 Classification of surface treatment of natural fibers

2 Physical Treatment Techniques

Physical treatment of fibers is performed basically for two reasons. Firstly, for separating the fiber bundles into individual filaments. This is done to avoid the agglomeration problem that leads to creation of stress concentration sites in the matrix region. Secondly, the modification of fiber surface to increase the compatibility for composite reinforcement application. Water and dew retting are the oldest method used for separation fiber bundle into filaments. In water retting bundles are submerged in the water for long period of time result in penetration of water to the stalk portion and swallow the inner wall which busts the outermost layer. In dew retting fermentation of stem material around the fiber bundle by the action of sun, air, bacteria and dew is take place. Major disadvantages of retting are water absorption, water pollution, decay of fiber by bacterial and low quality of fiber. Steam explosion process is another method to disintegrate the fiber bundle by treating plant stalks with hot steam (120–180 °C) under certain pressure and then rupture of rigid structure biomass fiber by explosion [21]. Thermo-mechanical process is good alternative in which the thermal treatment of fiber is done around glass transition temperature of lignin (142 °C). Prasad and Sain thermally treated the hemp fiber from 160 to 240 °C (above glass transition temperature of lignin) for 30 min under inner and air environment [22]. On thermally heating the hemp fiber, depolarization of lignin and hemicelluloses into lower molecules occurs. Tensile strength of treated fiber under inner environment show higher strength as compared to air environment. In air environment, oxidation of fiber surface takes place that lead to lower strength. Other physical treatments to modify the fiber surface and leading to an increase in the effectiveness of fiber are presented in subsequent passages.

3 Plasma Treatment

Plasma is an ionized gas having equal positive and negative charge density. Plasma treatment only affects the top layer of fiber and does not alter the bulk properties of fiber [23]. Plasma treatment can be done in vacuum pressure and atmospheric pressure as well. Further, the low temperature plasma is preferred over high temperature plasma; as they can burn the material. In cold plasma (low temperature plasma) treatment, temperature of ions is near atmospheric; but has sufficient high energy to break the covalent bond. Plasma treatment is done to:

(a) clean the fiber surface and improve adhesion with polymer or coupling agent. For cotton fiber, O_2 plasma treatment is done under low pressure to remove polyvinyl alcohol (PVA) particles [24]. Figure 2 represent the cleaning effect of plasma treatment on flax fibers and improving the interfacial bonding with matrix [25];

(b) increase the surface roughness by etching the material from surface and promote interlocking between fiber and matrix. Sun and co-authors increased the surface

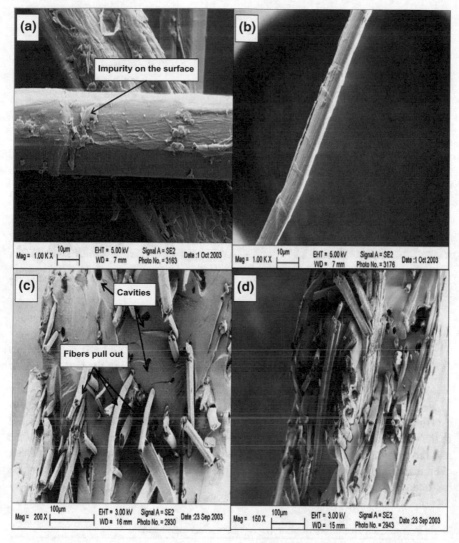

Fig. 2 SEM images of (**a**) and (**b**) untreated and treated flax fiber, and (**c**) and (**d**) untreated and treated flax fiber/unsaturated polyester composites, respectively (reproduced with permission from the Ref. [25])

roughness of polyimide fiber by oxygen plasma treatment, which further resulted in about 15% improvement in interlaminar shear strength [26]; and

(c) functionalizing the fiber surface to promote chemical bonding with polymer as shown in Fig. 3. Free radicals on the surface of fibers are created by direct attachment of ions and gas phase free radicals. In free radicalization, breaking of surface bonds occurs when the gas/ions radicals with sufficient energy are

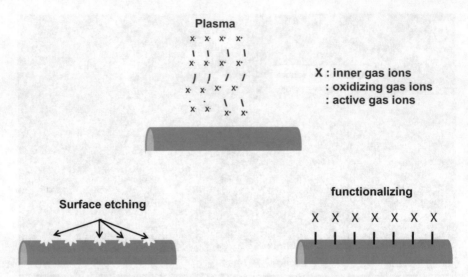

Fig. 3 Effect of plasma treatment on the surface modification of fibers by etching and functionalizing

bombarded on the surface. This result in absorption of atoms on the fiber surface. When wool fibers are plasma treated with H_2, N_2 and air; this result in increase in the oxygen atomic concentration and decrease in the carbon atomic concentration [27].

Furthermore, Wang and Qiu reported that after the plasma treatment, water absorption time for both sides of the fabrics was reduced [28]. This decrement was attributed to the destruction of scale structure due to plasma etching on the wool fiber surface and the introduction of more polar groups such as hydroxyl groups due to plasma chemical modification. Table 1 represents the various plasma gases used for surface treatment purpose.

Table 1 Application of different plasma gases for surface modification

Gas	Application
Oxidizing gas (H_2O, O_2 or N_2O)	Cleaning the surface by oxidation of organic compound and leaving oxygen species
Reducing gas (H_2 or N_2)	Remove organic compound that sensitive to oxidation and replace fluorine or oxygen atoms on the surface
Noble gas (Ar or He)	Promote free radicals on the surface which can cross-linked or leave active site
Active gas (NH_3)	Surface modification by leaving amino groups on the surface which can form covalent bond with polymer matrix
Polymerizing gas	Act as coating layer on the surface of fiber which form cross-linked bond with polymer matrix

4 Corona Treatment

Corona treatment is a non-thermal plasma treatment done at atmospheric pressure [29]. Corona is a luminous discharge that occurs, when the breakdown of air between two electrode take places because of excessive localization electric field. When the high electric field is applied in air, the breakdown of air into ions or excited molecules occur, which act as reactive species for free radicals. Belgacem et al. [30] improved the mechanical properties of cellulose fiber and polypropylene composite by corona treatment. Quantitatively, the yield strength, elastic modulus and energy to break the composite of treated cellulose fiber was 24.5, 42.4 and 25.9% higher than the untreated fiber. Amirou et al. [31] observed that the untreated algerian date palm fibers (ADPF) resulted in interfacial gaping between ADPF and polylactic acid as compared to the corona treated fibers as shown in Fig. 4. Corona treatment increased the surface roughness, due to which the interfacial bonding between fiber and polymer matrix increased. As per the research of Gassan and team [32, 33], corona treatment leads to an increase in the polar components (such as hydroxyl and carboxyl functional groups) at the surface of jute fiber, which then act as free radicals to form covalent bonds with the polymer matrix. Dong et al. [34] reported a decrement in the apparent melt viscosity of cellulose-filled PE composites, if one (or both) of the constituents was corona treated. Also, Bataille et al. [35] used the corona treatment for activating cellulose, in order to improve upon the efficiency of grafting process.

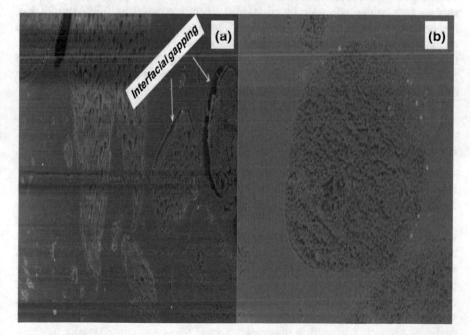

Fig. 4 SEM images of algerian date palm fibers and polylactic acid composite (**a**) before treatment (**b**) after treatment (reproduced with permission from the Ref. [31])

5 Ultrasound Treatment

Ultrasound refers as sound with frequency higher than the audible range (above 25 Hz) and is used in various industries [36]. Propagation of ultrasound wave through liquid bath results in formation of longitudinal wave in water molecules to further generate rarefaction and compression wave. Compression wave forms bubbles of microscopic size which collapse violently to form shock wave. Thus, cleaning of fiber surface occurs in two ways that is micro-jetting and micro-streaming. In micro-jetting, the bubbles collide with fiber surface resulting in tearing of intermolecules. Whereas in micro-streaming, oscillation of bubbles take place that scrubber the fiber surface. Effect of ultrasound treatment in water and alkali medium on oil palm empty fruit fiber (EFB) was studied by Alam et al. [37]. Results showed that the tensile strength, tensile modulus and impact strength of ultrasound treated of EFB in alkali medium was higher by 31.25%, 109.5% and 46.1% respectively, as compared to the untreated fibers. Fourier-transform infrared spectroscopy (FTIR) shows a reduction in C=O bond peak in ultrasound treated UEF fiber, which confirm the removal of fatty acid from the fiber surface. However, alkali bath ultrasound treated fibers shows an additional peak that can be attributed to possible reaction between EFB fibers and alkali as represented in Fig. 5. Kadam et al. [38] performed the ultrasound treatment on wool fiber, which resulted in reduction of residual grease content from the raw wool fibers. Moreover, the improvement in whiteness was greater than 160% and decrease in yellowness was about 44%. Study of Liu et al. [39] showed that the surface free energy and polarity of aramid fibers increased by 6.3% and 23.5%, respectively, when treated with ultrasound frequency of 20 Hz without affecting the mechanical properties. Laine and Goring [40] worked on the influence of ultrasonic irradiation on physical and chemical properties of pulp fibers; thereby, reporting an increase in the fiber wall porosity and a slight increase in the carbonyl group content of the fibers, due to the oxidation of carbohydrate hydroxyl groups.

6 Ultraviolet Treatment

Ultraviolet light is an electromagnetic radiation having wavelength longer that the X-rays but shorter than visible light ranges from 10–40 nm. Photons energy of ultraviolet radiation is high enough to break down C–C, C–O, C–F, C–Si or C–H bonds which may help in cleaning of fatty acids from the fibers surfaces. When treatment is done in the presence of air, photons react with O_2 to form reactive species such as O, O_3 and O_2^* that can act as the free radicals. This can improve compatibility between the hydrophobic matrix and hydrophilic fiber. Kato et al. [41] treated the cellulose fiber with ultraviolet irradiation in vacuum for surface oxidation. FTIR spectroscopy shows that a new peak at 1720 cm^{-1} corresponding to C=O bond to carbonyl group is appeared in ultraviolet treated fibers without changing mechanical properties and physical appearance. Benedetto et al. [42] studied the effect of

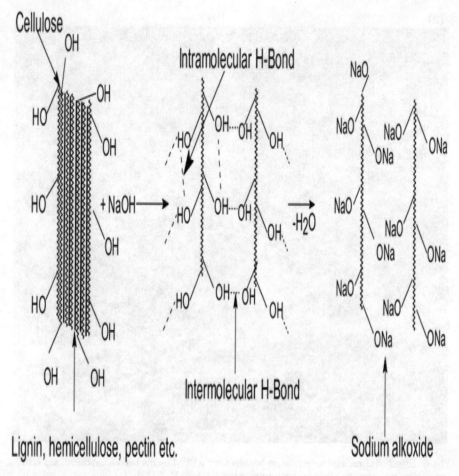

Fig. 5 Possible chemical reaction between EFB fibers and alkali during alkali bath ultrasound treatment (reproduced with permission from the Ref. [37])

ultraviolet radiation on physico-chemical and mechanical properties of banana fiber. Under ultraviolet radiation, inter-crosslinking of celluloses molecules occurs due to which the tensile strength of banana fiber increased by 28.5%. FTIR spectroscopy shows that peaks at 1731 cm^{-1} (C=O bond in hemicellulose) and 1620 cm^{-1} (C=O bond in lignin) are absent in the treated fibers representing removal of hemicellulose and lignin. This is also confirmed by scanning electron microscopy (SEM) images in which treated banana fibers have longitudinal cracks and higher surface roughness as shown in Fig. 6. In a study by Gassan and Gutowski [33], tossa jute fibers were treated by ultraviolet treatment, which lead to higher polarities of the fiber surface. Consequently, the wettability of fibers and composite strength were improved. Oosterom et al. [43] in a bid to improve the polarity of ultra-high molecular weight polyethylene, used different techniques including an ultraviolet source for surface

(a)   (b)

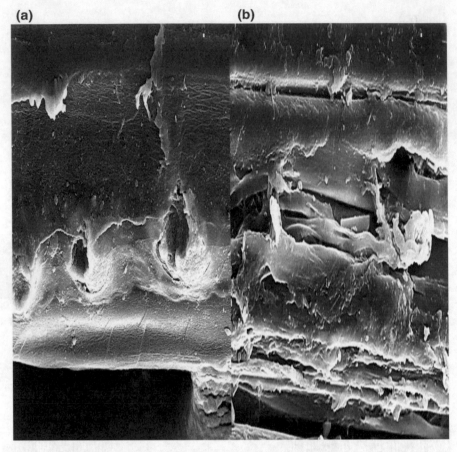

Fig. 6 Effect of ultraviolet treatment on banana fiber **a** SEM image before treatment and **b** SEM image after treatment (reproduced with permission from the Ref. [42])

modification and improvement of polarity of polyethylene (PE). They have commented that corona and glow discharge treatments were found to activate the surface by an increase in surface energy of over 100% in an order of less than a minute, corresponding to an increase in ultimate shear stress from 0.12 to 0.40 MPa. In contrast, ultravoilet/ozone required exposure times in the order of minutes to have an effect that was still incomparable to the other gas-phase treatments examined.

7 Chemical Treatment Techniques

Natural fibers are hydrophilic in nature as derived from lignocelluloses, when interact with hydrophobic matrix polymer result in poor interfacial interaction. Hydrophilic nature also results in moisture absorption and rotting of fiber by micro-organism attack. Poor wetting of polar fiber with non-polar polymer matrix also result in weak interface. Therefore, chemical modification of fiber is required to improve the compatibility of forming interfacial bonds between reinforcement and matrix phase. Chemical treatment such as alkali treatment is use to remove fatty acids, max and unwanted impurity from the fiber surface and increase the availability of active –OH group to interlock with polymer [44]. Application coupling agent on fiber surface is another method to overcome the hydrophilic nature of fiber and fungi deterioration. Coupling agent act as the bridge between polymer and fiber which improves the load transfer ability. Thus, chemical modification of fiber surface cleans up the unwanted impurity from fiber surface; improve wettability of fiber; form the covalent bond between fiber and matrix to improve interfacial adhesion and produce tough and flexible layer by forming intermolecular bonds. Different chemical treatments applied to modify the fiber surface are elaborated in detail.

8 Alkaline Treatment

Alkalization (also known as mercerization) is the widely used chemical treatment process for natural fibers when applying to fabricate the composites [45]. Here, the fibers are first washed several times by fresh water to clean the initial dirt and mud. After that, the fibers are soaked in an alkali solution for a certain time period at room temperature. Before reinforcing with matrix, the alkali treated fibers are dipped in an acetic acid solution to neutralize any effect of alkali (such as potassium hydroxide (KOII) and sodium hydroxide (NaOH)) that is present on the fibers surface [46–48]. Then the fibers are again washed thoroughly several times in distilled water before drying them. Since, the natural fibers have a distinctive property of lacking strong interfacial adhesion with the polymers. Therefore, the alkaline treatment is used to boost up the interfacial adhesion between matrices and fibers by disrupting the hydrogen bonds in their network structure [49–53]. This increases the surface roughness and hence, we get more contact points/surface contact area between the fiber and matrix that would result in high mechanical interlocking phenomenon. More contact points would mean more gripping sites among fiber and matrix that would lead to an increase in the stress transfer phenomenon and thus, the composites mechanical strength and toughness properties will get enhanced. In addition, this treatment boosts up the number of reaction sites by increasing the cellulose amount exposed on the fiber surface. The major compositions of natural fibers are cellulose (main building block in natural fibers structure and provides strength to it), hemi-cellulose, pectin and lignin. The alkaline treatment eliminates certain amount of lignin, wax and oils from the natural fiber external surface, depolymerizes cellulose and exposes the short length crystallites [54, 55]. When we add aqueous form of an alkali (for example NaOH) to natural fiber, the transformation of hydroxyl (on fibers surface) to alkoxide group multiplies [56]. Refer to Fig. 7 for the chemical reaction.

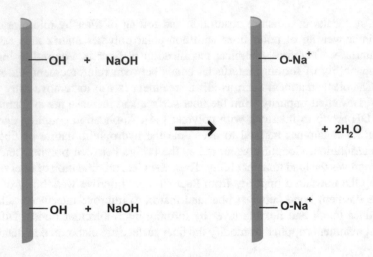

Fig. 7 Reaction utilized in alkaline treatment of natural fibers

Chemical and physical bonds are formed between the matrix and hydroxyl group of cellulose that act as an interface to transfer the load from matrix to fiber. Presence of unwanted impurity (such as oil, dust and wax) results in reduction of availability of hydroxyl group to interact with the matrix polymer chain. Presence of these impurities result in surface separation debonding and void formation, due to which the effectiveness of the fiber reduces. To overcome these limitations, the alkaline treatment has been done; so that the impurities (coating the exterior surface of fiber), lignin and hemicellulose will be removed as shown in Fig. 8.

Initially, the interfibrillar region of fiber is less dense after alkali treatment (due to dissolution of hemicellulose). Therefore, the fiber can rearrange their orientation along the loading direction; thus, resulting in better load transfer and improved the mechanical properties of composites. Moreover, splitting of fibers take place that increases the effective surface area and improve the interfacial bonding between fiber and matrix. Cai et al. [57] studied the effect of alkali treatment on surface morphology of abaca fiber and predicted an improvement in the interfacial adhesion between composite constituents. Surface of raw abaca fiber bundle consist of hemicelluloses, pectin, lignin, oil, wax and other unwanted materials that resulted in glossy surface as shown in Fig. 9a. But, after treatment with 5 weight percentage (wt.%) NaOH solution, there is about 17.7% weight loss in the abaca fiber. Weight loss signifies the dissolution of hemicelluloses, lignin and removal of unwanted impurity, that further results in rough surface of abaca fiber (refer to Fig. 9b).

Alkali treatment results in an increase in the availability of hydroxyl group (–OH) to form chemical and physical bonds with the polymer chains. In physical bonding, the hydroxyl group of cellulose fiber form hydrogen bonding with the hydroxyl group of polymer chain; whereas, in chemical bonding, the cellulose fiber–OH react with alkali (NaOH) to form fiber–O–Na^+ that further make bond with polymer chains as shown in Fig. 8. Also, alkali treatment increases the surface roughness, which results

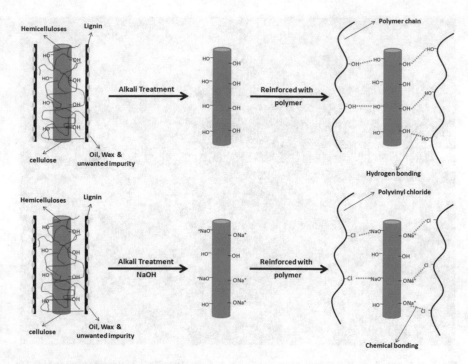

Fig. 8 Alkaline treatment of natural fiber

in higher mechanical interlocking of fiber with polymer. When basalt and glass fibers were treated with NaOH solution flake type structures appeared on the surface of fibers that increases the surface roughness as shown in Fig. 9c, d [58].

Researchers have treated natural fibers with alkali with different wt.% and time periods. Garcia et al. [59] observed that 2% alkali at 200 °C and 1.5 MPa (for 90 s) was an optimum parametrization for degumming and defibrillation of individual fibers. Morrison et al. [60], Mishra et al. [61] and Ray et al. [62] treated the jute, sisal and flax fibers, respectively with NaOH solution (5%) for 2–72 h at room temperature. They also reported that alkali leads to an increase in the amorphous cellulose content at the cost of crystalline cellulose. High amount of alkali concentration should also be not used as it results in excess delignification of natural fiber and resulting in a weaker/damaged fiber. Table 2 showcases the effect of alkali treatment on the mechanical properties of various composites.

9   Silane Treatment

The interaction of natural or inorganic (such as glass fiber) fibers with polymer matrix occurs in term of hydrogen bonding (The effectiveness/bond strength of hydrogen

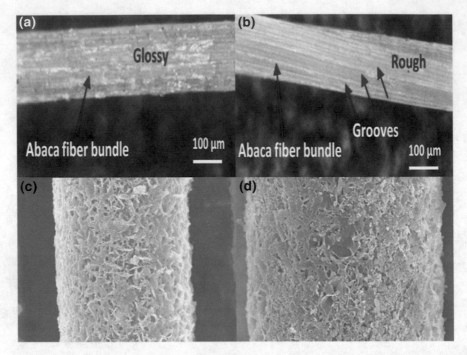

Fig. 9 **a** Abaca fiber before alkali treatment, **b** Abaca fiber after alkali treatment (reproduced with permission from the Ref. [57]), **c** and **d** increase in surface roughness of basalt and glass fiber after alkali treatment (reproduced with permission from the Ref. [58])

bonding is less as compared to the covalent bond) only. To improve upon the load transfer efficiency from polymer matrix to fiber can be done by developing a chemical bridge. Silane (SiH_4) is a coupling agent/stabalizer used to improve the adhesion between fiber and matrix interface by building a cross-linked interface between polymer and matrix, which act as chemical bridge to transfer the load from polymer to fiber. Interestingly, these silane coupling agents has the tendency to reduce cellulose hydroxyl groups at the interface.

As per the study of Sang et al. [68], SEM image represents the clear interphase gapping between the untreated polybutylene succinate (PBS) and basalt fiber (BF) composites as compared to silane modified basalt fiber (MBF) as shown in Fig. 10a, b. Tensile strength of PBS/MBF composites also increased by 48.7% as compared to the untreated fibers. Tan et al. [69] studied the effect of silane treatment on interfacial adhesion of unsaturated polyester resin (UPS) and empty fruit bunch fibers (EFB). SEM image (Fig. 10c, d) of fractured surface represent an improver bonding between EFB and UPS as compared to treated EFB fibers; resulting in 21% improvement in the tensile strength. Rong et al. [70] treated the sisal fibers in a solution of 2% aminosilane (in 95% alcohol) for 5 min followed by 30 min air drying for hydrolysing the coupling agent. Valadez-Gonzalez et al. [71] treated the henequen fibers with silane solutions in water and ethanol mixture (with concentration of 0.033 and 1%). Their team also

Table 2 Effect of alkali treatment on the mechanical properties of composites

Fiber	Matrix	Alkali treatment	Improvement
Sugar palm [63]	Epoxy	Sugar palm fiber were dipped into NaOH solution for 1, 4 and 8 h	The flexural strength and modulus of fabricated composite improved by 16.7% and 66.8%, respectively
Coir [64]	Cement	Coir Fibers were dipped for 30 min into 5 wt.% NaOH solution at 20 °C	The natural frequency, tensile strength, tensile modulus, flexural strength and flexural modulus of composite increased by 4.2%, 17.8%, 6.9%, 16.7% and 7.4%, respectively
Flax [65]	Polyester	Flax fiber were dipped into 2.5–7 wt.% alkali solution (NaOH) at different temperatures (room temperature, 35 and 45 °C)	Crystallinity of flax fiber increased by 13.8% when treated with 2.5 wt.% NaOH solution at 35 °C
Flax [66]	Epoxy	Treatment with alkali was carried out	40% improvement in the tensile (longitudinal) properties was observed
Sisal [67]	Natural rubber	Treatment with NaOH (0.5, 1, 2, 4, 10%) was carried out	Maximum strength of composite was found to be with 4% NaOH treated alkali
Human hair [47]	Epoxy	Human hairs were dipped into 5 wt.% KOH solution at room temperature	Improvement in the mechanical properties was observed

predicted that silane treated fibers results in higher tensile strength than alkali treated ones.

Silane coupling agents are consisting of bifunctional groups with generic chemical structure $R_{(4-n)}-Si-(R'X)_n$ (where R-alkoxy, X-organofuntionality, R'-alkyl and $n = 1, 2$). One functional group of silane coupling agent react with hydroxyl group of fibers and other with matrix results in formation of bridge by chemical interlocking in between them. Many silane agents were been used by different researchers to provide coupling between inorganic fibers and organic polymers whereas there are limited silane agents used for natural fiber reinforced composites. The organofunctionality reactive group of the silane agent results in formation of covalent bond with polymer depending upon polymer functionality reactivity. Different organofunctionality of silane agents are $-NH_2$ (amino), $-HS$ (mercapto), $-C_3H_5O_2$ (glycidoxy), $-CH=CH_2$ (vinyl), or $-(CH_2)_3-OOC(CH_3)C=CH_2$ (methacryloxy) groups. On the other hand, compatibility between non-polar matrix and non-reactive alkyl group of silane increases because of similar polarity. Different silane coupling agents used for fabrication of composite are presented in Table 3. Most commonly used silane

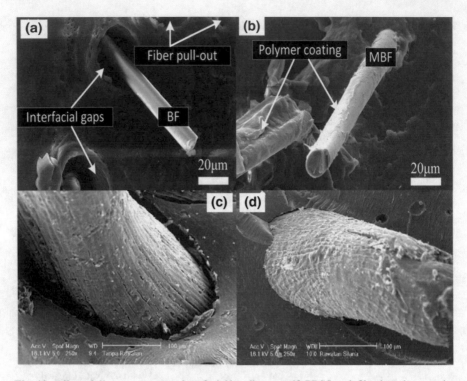

Fig. 10 Effect of silane treatment on interfacial bonding. **a** and **b** PBS/basalt fiber based composites (reproduced with permission from the Ref. [68]), and **c** and **d** polyester/fruit bunch fibers composites (reproduced with permission from the Ref. [69])

Table 3 Different types of silane coupling agents used

Name	Structure
(3-Aminopropyl)triethoxysilane (APS)	$(CH_3O)_3Si–(CH_2)_3–NH_2$
Vinyltrimethoxyysilane (VTS)	$(CH_3O)_3Si–CH=CH_2$
3-(Trimethoxysilyl)propylmethacrylate (MPS)	$(CH_3O)_3Si–(CH_2)_3–OOC(CH_3)C=CH_2$
	$(RO)_3Si–(CH_2)_3–SH$
(3-Glycidyloxypropyl)trimethoxysilane (GPS)	$(CH_3O)_3Si–(CH_2)_3–O–CH_2CHCH_2O$
Dichlorosilane (DCS)	$H_2–Si–Cl_2$
	$(RO)_3Si–(CH_2)_{15}CH_3$

coupling agents by researchers for natural fibers are (3-Aminopropyl) triethoxysilane (APS), Vinyltrimethoxyysilane (VTS) agent that are able to develop covalent bonds with polymer matrix, and 3-(Trimethoxysilyl)propylmethacrylate (MPS) that has high reactivity for unsaturated polyester matrices.

Mechanism of coupling of silane agents with natural fibers mainly occurs in four steps:

(a) Hydrolysis: In this process, reactive silanol groups are obtained by hydrolyzing the silane monomers in the presence of catalyst (normally acid or base) as shown in Fig. 11. Cellulose fibrils hydroxyl group do not react with silane because of low acidity of hydroxyl group of cellulose even at elevated temperature. Therefore, activation of alkoxysilane is done by hydrolyzing the alkoxy group that further leads to development of silanol reactive groups. Thus, in presence of moisture (water) hydrolysable alkoxy group leads to the formation of silanols.

(b) Self-condensation: Silanol reactive group obtained from hydrolysis react with each other to form silonals oligomers known as self-condensation as shown in Fig. 11. Condensation results in decrease in the availability of silanol group to react with hydroxyl group of cellulose fiber. Moreover, it also results in higher molecular size oligomers, which do not diffuse easily with cellulose cell walls. Therefore, to overcome these problems hydrolysis process should be controlled by maintaining solution acidic, which accelerate the hydrolysis process and slow down the condensation.

(c) Absorption: In this process, monomers/oligomers of silonal obtained after hydrolysis and self-condensation form hydrogen bonding with –OH group present on the surface of fiber. Moreover, free silanol react with each other to form rigid –Si–O–Si– polysiloxane structure as shown in Fig. 11.

(d) Grafting: In this process, hydrogen bond formed between –OH group and silanols are converted to stable –Si–O–C– covalent bonds at higher temperature. This creates a crosslinked network because of covalent bonding between the matrix and the fiber. Further, self-condensation of unbounded silonals group also takes place with each other as show in Fig. 11. This chemisorption completes the silane treatment process.

10 Acetylation Treatment

Major disadvantage of the natural fibers is their hydrophilic nature that results in dimensional instability in the fabricated composite, low interfacial bonding with hydrophobic polymer matrix and micro-organism degradation of fibers. Hydrophilic nature of fibers occurs due to the presence of hydroxyl group in lignocelluloses fibers. To overcome this delinquent, acetylation of natural fibers is done. Acetylation is an esterification reaction in which an acetyl functional group (CH_3COO^-) is introduced onto the surface of fiber, resulting in plasticization of cellulosic fibrils. In acetylation reaction, hydroxyl group of cellulose fibers react with acetic anhydride (CH_3–C(=O)–O–C(=O)–CH_3) resulting in formation of acetyl groups on the cell wall of fiber, while generating acetic acid (CH_3COOH) as a by-product (as shown in Fig. 12). Since, the hydroxyl groups are now replaced by acetyl groups on cell wall, this converts it into hydrophobic nature [72]. This acetic acid must be detached from the lignocellulosic fiber before utilizing it for the fabrication purpose. Sometimes reaction between acetic anhydride and fiber is not proper and moreover to accelerate the reaction many

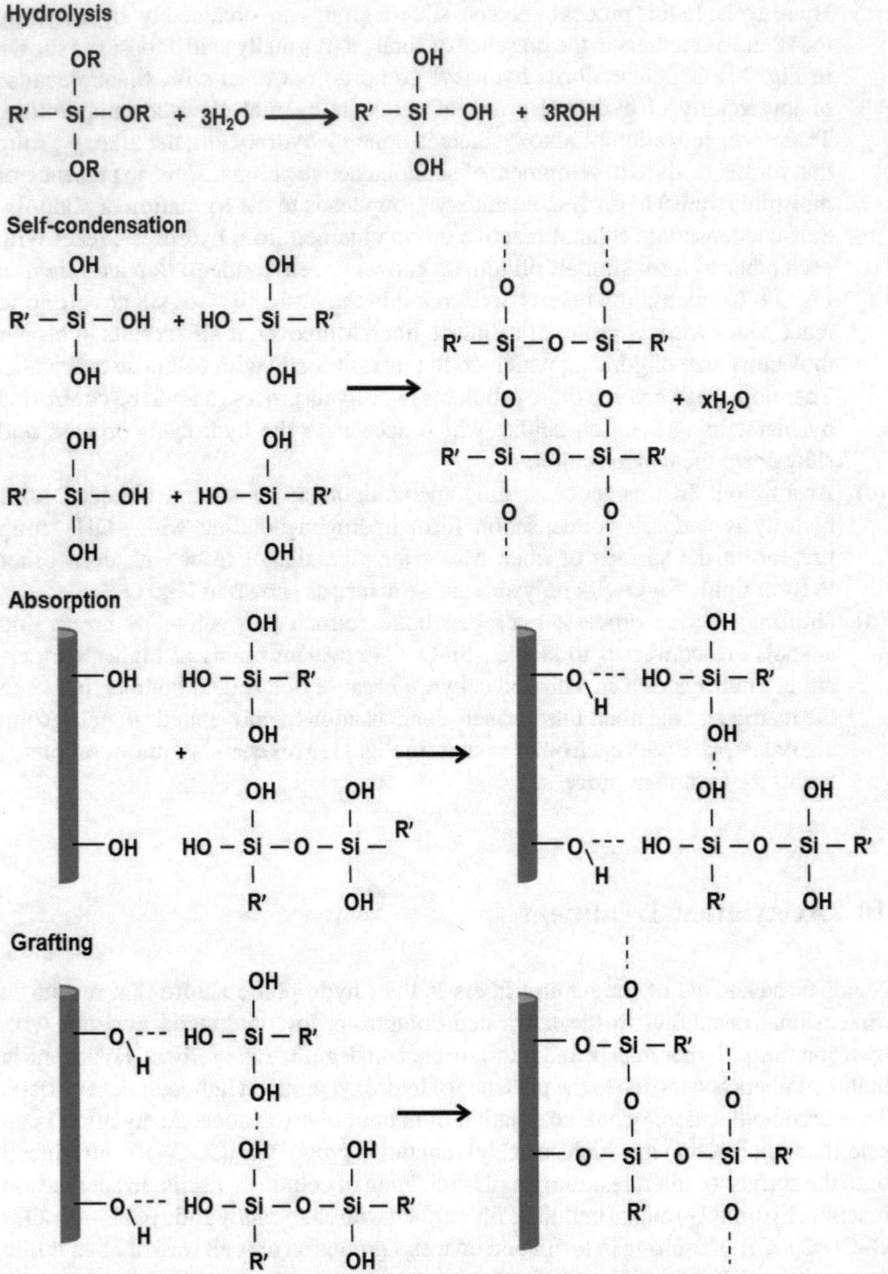

Hydrolysis

$$\begin{array}{c} OR \\ | \\ R' - Si - OR \\ | \\ OR \end{array} + 3H_2O \longrightarrow \begin{array}{c} OH \\ | \\ R' - Si - OH \\ | \\ OH \end{array} + 3ROH$$

Self-condensation

Absorption

Grafting

Fig. 11 Coupling mechanism of silane agent with natural fiber

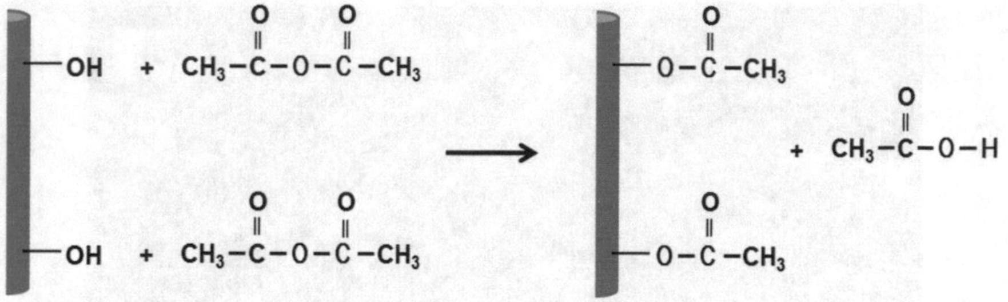

Fig. 12 Reaction involved in acetylation process

researchers have also used catalyst. Acetylation reduces the hygroscopic nature of natural fibers, thereby increasing the dimensional stability of composites.

Bledzki et al. [73] studied the effect of acetylation on cellulose content and water absorption capacity of flax fiber. For acetylation process, flax fibers were dipped in an acetylating solution (consisted of acetic anhydride 125 mL, toluene 250 mL and perchoric acid as a catalyst). Process temperature was maintained at 60 °C for 1–3 h. Before investigation, the unwanted acid was removed by washing it in distilled water. Presence of catalyst accelerates the reaction process and also improves the degree of acetylation. At 65% relative humidity (RH) and 95% RH, water absorption of flax fibers reduces to 50% and 45%, respectively. SEM image of untreated flax fiber surface consisted of wax and protruding part; whereas, the treated fiber surface is smoother due to fibrillation, removal of wax and lignin. On increase in degree of acetylation beyond 12% degradation of the fiber starts as shown in Fig. 13. Mishra et al. [74] performed the acetylation on sisal fibers to improve the tensile, impact and flexural strength sisal/glass polyester composites. Before acetylation process, dewaxed fibers were alkali treated for 1 h. The alkali treated fibers were soaked in acetic acid at 30 °C for 1 h and then decanted in acetic anhydride solution (50 mL) for 5 min. To accelerated the reaction, one drop of concentrated sulphuric acid (H_2SO_4) was also added. Improvement in tensile, impact and flexural strength of treated sisal fiber hybrid composites were about 18.5%, 14% and 6%, respectively. Khalil and Ismail [75] reported that the acetylated treated natural fiber-reinforced polyester composites had higher bio-resistance and less tensile strength loss relative to the composites with silane treated fiber in the biological tests.

11 Benzoylation Treatment

Benzoylation treatment of fiber surface is another method to overcome the hydrophilicity of natural fibers. Benzoylation reaction is done to introduce benzoyl ($C_6H_5C=O$) group onto the surface of fiber, which improves the interaction of fiber with hydrophobic polymer matrix. Benzoyl chloride is most commonly used for benzoylation treatment. In benzoylation reaction, hydroxyl group of cellulose

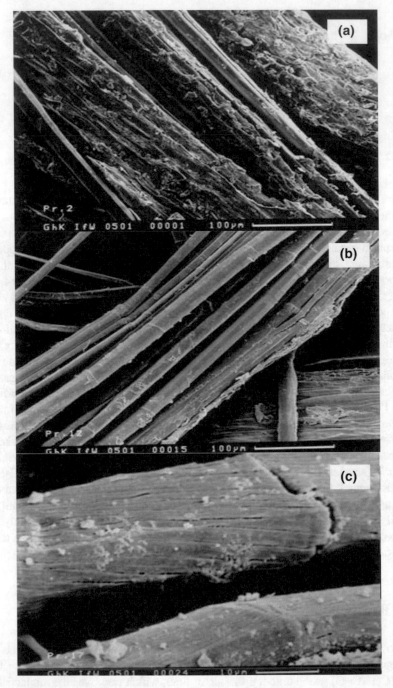

Fig. 13 SEM images of (**a**) untreated flax fiber, (**b**) treated flax fiber and (**c**) degradation of fiber at higher degree of acetylation (reproduced with permission from the Ref. [73])

fibers react with benzoyl chloride that results in addition of benzoyl groups on the cell wall of fiber (see Fig. 14 for the reactions involved [76]). Nair et al. [77] studied the effect of benzoylation of sisal fiber on the dynamic mechanical properties of fabricated composite. Fibers were first cleaned and chopped into 6 mm length. Then the fibers were soaked in 18% NaOH solution for 30 min. After that the treated fibers were dipped into 10% NaOH and benzoyl chloride solution for 15 min. Also, the treated fibers were soaked in ethanol solution to remove any unreacted benzoyl chloride. Benzoylation of sisal fiber results in improvement of storage modulus and activation energy at glass transition temperature of fabricated composites. Kalia et al. [78] modified the sisal with benzoyl chloride to improve the physical, mechanical and thermal properties of fiber. Sisal fibers cut in definite size were soaked into 5% NaOH solution for 30 min and then treated with 15% and 30% benzoyl chloride for

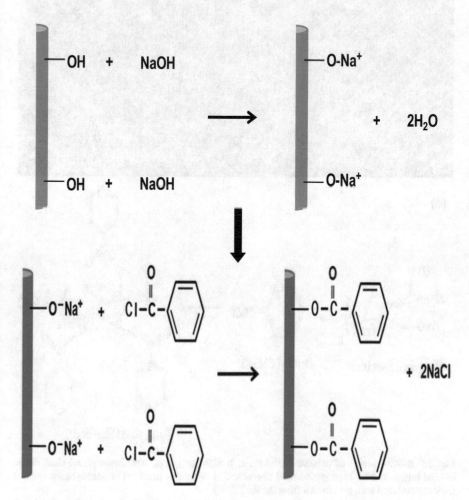

Fig. 14 Reactions involve in benzoylation process

another 30 min; after that they were dried at 80 °C. Figure 15 represents the SEM
images of untreated sisal fiber, treated sisal fiber and chemical reaction of sisal fiber
with benzoyl chloride. SEM image of untreated fibers surface was smoother than
treated due to the presence of waxy and oily material; whereas, the benzoylation
results in improvement of surface roughness. Moreover, crystallinity of benzoylated
sisal fiber increased by 2–4%.

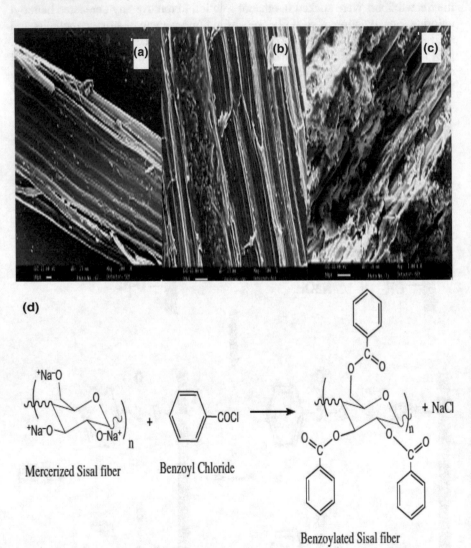

Fig. 15 **a** SEM image of untreated sisal fiber, **b** SEM image of 10% benzoylated sisal fibers,
c SEM image of 10% benzoylated sisal fibers, and **d** Reactions involved in benzoylation of sisal
fiber (reproduced with permission from the Ref. [78])

12 Acrylation and Acrylonitrile Grafting

Grafting of natural fiber with acrylic acid ($CH_2=CHCOOH$) and acrylonitrile ($CH_2=CH–CN$) is another method to alter the physical and chemical modification of fibers (refer reactions presented in Figs. 16 and 17). Before acrylate and acrylonitrile grafting, free radicals are initiated on the surface of cellulose fibrils accompanied by the chain scission; for that the cellulose fibers are pretreated with high energy radiation [79]. In grafting process, free radical sites are generated by radiation technique (such as ceric ion), which activate the oxygen atom of hydroxyl group of cellulose fiber through electron transfer (as shown in Fig. 16). These active oxygen atoms react with acrylic acid or acrylonitrile to modify the surface of fibers. Singha and Rana [80] modified the cellulose fiber by grafting with acrylonitrile (AN) and ethyl acrylate (EA). Additional peaks at 2241.7 and 736.1 cm^{-1} representing –CN and –C=O bonds were observed in FTIR spectroscopy, which confirm the presence of AN and EA; whereas, the crystallinity (and crystallinity index) of fiber decreased due to the disorientation of cellulose crystals. Moisture absorption of fiber also decreased, due to the unavailability of reactive site to water molecules. Sreekala et al. [81] modified the oil palm fiber by acrylic acid and acrylonitrile solutions. In acrylation, fibers were soaked in 10% NaOH solution for 30 min at room temperature. After that the

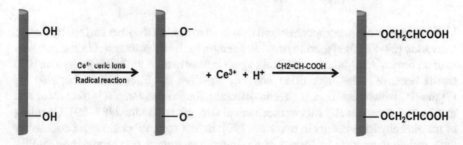

Fig. 16 Reaction involved in acrylation process

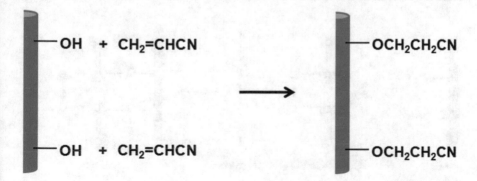

Fig. 17 Reaction involved in acrylonitrile process

fibers were treated with various concentrations of acrylic acid solution at 50 °C for 1 h; whereas, for acrylonitrile grafting the fibers were soaked in 2% NaOH solution for 30 min, and further oxidized with $KMnO_4$ and then put in 1% H_2SO_4 containing acrylonitrile. The water uptake of modified fiber reduced, whereas the Young's modulus increased slightly. Mishra et al. [82] modified the surface of sisal fiber with acrylonitrile grafting through a combination of $NaIO_4$ and $CuSo_4$ initiators. Before grafting, the sisal fibers were dewaxed by treating with alkali solution and then soaked with AN for 10 min in the reaction vessel. Then known concentration of $CuSO_4$ was added, after that required amount of $NaIO_4$ solution was mixed. The reaction was carried out for a desired time at 50, 60, and 70 °C. Water absorption affinity of AN grafted sisal fiber highly reduced from 64.39% to 11.11% for 25% grafting concentration. Tensile strength and modulus of modified fiber increased by about 6% and 37.5% for 5% grafting, which also had the maximum elongation before fracture. This may be attributed to the orderly arrangement of polyacrylonnitrile units at lower concentration. Li et al. [83] showed that the tensile strength of acrylic acid treated flax fiber–high density PE composites improved and the water absorption capacity decreased.

13 Maleated Coupling Agents

Maleated coupling is another chemical treatment to modify the fiber surface by maleic anhydride [84–86]. It is used to provide strength to the filler material in the polymer matrix composites. Maleated coupling agent is mostly used with polypropylene (PP) matrix because maleic anhydride not only modifies the fiber surface but also the PP matrix, which results in higher interfacial adhesion between reinforcement and matrix, and consequently higher mechanical strength is obtained [87–89]. Coupling of maleic anhydride occurs in two steps [90]: in first step, PP chain react cohesively with maleic anhydride to form maleic anhydride grafted polypropylene (MaPP); and in the second step, hot MaPP copolymers matrix react with cellulose fibers resulting in formation of covalent bonds between the interfaces (as shown in Fig. 18).

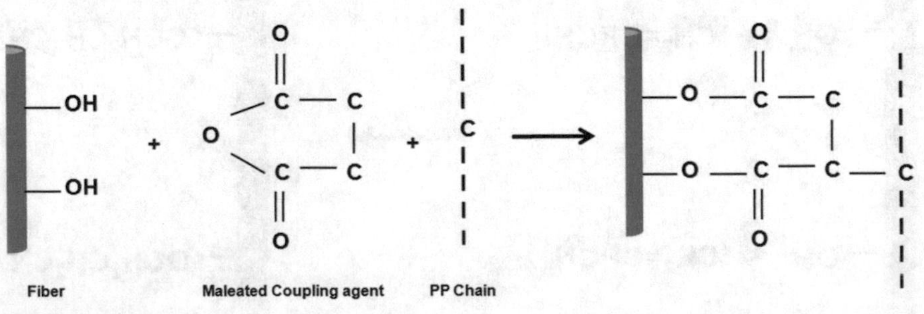

Fig. 18 Reaction involved in Maleated coupling process

Prior to esterification of cellulose fibers, activation of MaPP copolymer is done by heating it to ~170 °C. After this treatment, the surface energy of cellulose fibers get amplified closer to the surface energy of polymer matrix; thereby, resulting in in better wettability and higher interfacial adhesion of the fiber. Keener et al. [86] enhanced the mechanical properties of agro fiber (lax/jute) and polypropylene composites by optimizing the molecular weight and maleic anhydride content. Moreover, efficiency of newly developed MaPE couplers for wood/PE was also studied. Coupling resulted in improvement of interfacial adhesion between agro fiber/polypropylene, as the composites with coupler shows reduction in void formation and pulling out of fibers. Also, an enhancement in mechanical properties of composites by approximately 60% was observed. Mohanty et al. [91] utilized the MAPP as a coupling agent for the surface modification of jute fibers. From their results, it was found that 30% fiber loading with 0.5% MAPP concentration in toluene and 5 min impregnation time (with 6 mm average fiber lengths) gave the optimum results. A 72.3% increase in flexural strength was also observed from the treated composites. In addition to PP matrix, Mishra et al. [92] reported that the maleic anhydride treatment reduced water absorption capacity in hemp, banana and sisal fiber-reinforced novolac composites. Mechanical properties of plant fiber-reinforced composites increased after maleic anhydride treatment.

14 Permanganate Treatment

Permanganate (compound having MnO_4^- functional group) treatment is another method to overcome the hydrophilic nature of natural fiber by introducing MnO_3^- ions on the surface of cellulose fiber. Most commonly used compound is the potassium permanganate ($KMnO_4$) solution (in acetone solution) at different concentrations for a certain time period. Highly reactive Mn^{+3} ions tend to initiate the grafting polymerization process through mentioned mechanism reactions (see Fig. 19 for the operative reactions) [93].

Paul et al. [94] treated the alkali soaked sisal fiber with $KMnO_4$ solution in acetone at different concentration for 1 min. As a result, the water absorption of composites decreases due to the reduction in hydrophilic tendency of sisal fiber. Khan et al. [95]

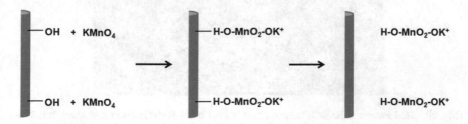

Fig. 19 Reactions involved in permanganate treatment

treated the jute fiber with $KMnO_4$ in oxalic acid medium. Jute fibers were soaked
for 2 min at room temperature. Tensile, impact and bending strength of treated jute
fiber based composite were increased by 34.2%, 27.7% and 27.3%, respectively.
Moreover, thermal stability of treated composite also improved and water up take of
treated composites reduced from 13% to 10.2%. Arsyad and Soenoko [96] studied
the effect of $KMnO_4$ on the surface roughness of coconut fibers. Coconut fibers were
treated with different concentration of $KMnO_4$ as shown in Fig. 20. The surface

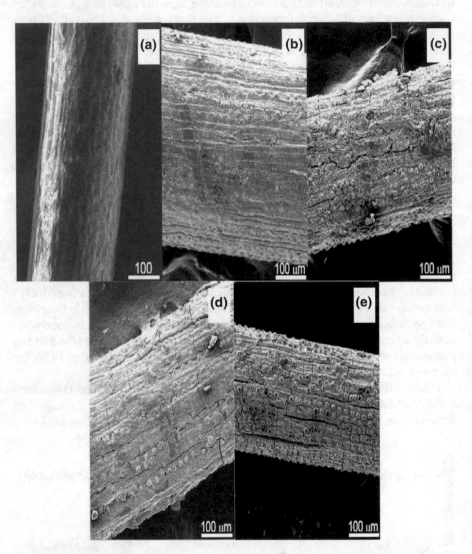

Fig. 20 SEM images of $KMnO_4$ treated coconut fiber **a** 0% $KMnO_4$ (untreated fiber), **b** 0.25%
$KMnO_4$, **c** 0.5% $KMnO_4$, **d** 0.75% $KMnO_4$, and **e** 1% $KMnO_4$ (reproduced with permission from
the Ref. [96])

roughness of untreated fiber is 1.62 μm, but after KMnO$_4$ treatment the surface roughness of coconut fiber increased to 3.17 μm.

15 Peroxide Treatment

Organic peroxides (functional group having chemical formula ROOR and containing the divalent ion O–O) have the tendency to dissociate into free radicals (RO·), which then reacts with hydrogen molecules present in cellulose fibers and matrix. Major peroxide chemicals used for the natural fiber treatment are benzoyl peroxide and dicumyl peroxide. Mechanism that is followed (in particular with polyethylene (PE)) are presented in the reactions 1–4 [94].

$$RO-OR \rightarrow 2RO^{·} \tag{1}$$

$$RO^{·} + PE-H \rightarrow ROH + PE \tag{2}$$

$$RO^{·} + Cellulose - H \rightarrow ROH + Cellulose^{·} \tag{3}$$

$$PE^{·} + Cellulose \rightarrow PE-Cellulose^{·} \tag{4}$$

After alkali pretreatment, fibers are coated with saturated solutiosns of peroxide in acetone. Sreekala et al. [97] reported that the peroxide treated oil palm fiber-reinforced phenol formaldehyde composites have more capability to withstand tensile stress to a higher strain level, due to the decreased hydrophilicity of fibers. Joseph et al. [98] investigated the effect of peroxide treatment on short sisal fiber reinforced PE composites; and predicted that the tensile strength of composites increased with the increase in concentration of peroxide up to a certain level (4% for dicumyl peroxide and 6% for benzoyl peroxide).

16 Isocyanate Treatment

Due to its property of reacting rigorously with fiber hydroxyl group, Isocyanate functional group (–N=C=O) has been utilized as a coupling agent in fiber-reinforced composites [94, 99]. Reaction followed in this treatment is represented in the Fig. 21.

Joseph et al. [98] examined that the cardanol derivative of toluene diisocyanate treatment increased the tensile strength of PE-sisal fiber composite, by reducing the hydrophilic nature of sisal fiber. Wu et al. [100] inspected the grafting of isocyanate-terminated elastomers onto the surfaces of carbon fibers. George et al.

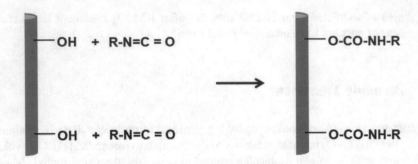

Fig. 21 Reaction involved in isocyanate treatment

[101] treated the pineapple leaf fiber with polymethylene-polyphenyl-isocyanate solution to improve interfacial adhesion between the fiber–matrix.

17 Other Treatments

Zafeiropoulos et al. [102] used the stearic acid in ethyl alcohol solution for flax fiber treatment and reported that this treatment removed the non-crystalline constituents of fibers; thus, altering its surface topography. They also examined that the treated flax fibers were more crystalline as compared to the untreated ones and stearation decreased the fiber surface free energy. Mishra et al. [82] dipped the untreated sisal fiber (for use in sisal-polystyrene bio composites) in sodium chlorite solution. Tensile strength of bleached sisal fiber composite was less than other chemical treated fiber composites because of the delignification of the fiber. But, the flexural strength was better for bleached fiber composite because of lower stiffness and more flexible character of fibers after delignification. After delignification, the polymer replaces the role of lignin in fibers and makes composites more hydrophobic and tougher. It has also been described that the triazine ($C_3H_3N_3$) derivatives (for example $C_3N_3Cl_3$) could form covalent bonds with the cellulose fibers. Hence, these can be used in the modification of vegetable fibers to reduce the number of cellulose hydroxyl groups that are available for water uptake in the composites [90].

18 Concluding Remarks

Natural fibers are a potential reinforcement for the composite industry because of their low cost and density as compared to the non-biodegradable man-made fibers. On the other hand, its application is limited due to their strong polar character that creates incompatibility with most polymer matrices. To overcome this problem and enhance the interfacial adhesion, physical and chemical surface treatments of natural fibers have been considered as a solution. Several discussed physical and chemical treatments have been utilized to separate the natural fibers from bundles, chemically couple the fiber to matrix and decrease the water absorption in composites. Most of the treatments have achieved success in enhancing the mechanical and thermal properties of composites. Major mechanisms that govern this enhancement are the increase in fiber surface roughness and alteration in chemical polarity of natural fibers. Overall the physical and chemical treatments are an effective means to improve the fiber-matrix interaction in composites.

References

1. Mukhopadhyay, S., & Fangueiro, R. (2009). Physical modification of natural fibers and thermoplastic films for composites—A review. *Journal of Thermoplastic Composite Materials, 22*(2), 135–162.
2. Oksman, K., Skrifvars, M., & Selin, J. F. (2003). Natural fibres as reinforcement in polylactic acid (PLA) composites. *Composites Science and Technology, 63*(9), 1317–1324.
3. Pickering, K. L., Efendy, M. A., & Le, T. M. (2016). A review of recent developments in natural fibre composites and their mechanical performance. *Composites Part A: Applied Science and Manufacturing, 83*, 98–112.
4. Verma, A., Singh, V. K., Verma, S. K., & Sharma, A. (2016). Human hair: A biodegradable composite fiber–a review. *International Journal of Waste Resources, 6*, 1–4.
5. Verma, A., & Singh, V. K. (2016). Experimental investigations on thermal properties of coconut shell particles in DAP solution for use in green composite applications. *Journal of Material Science and Engineering, 5*(3), 1–5.
6. Verma, A., Singh, C., Singh, V. K., & Jain, N. (2019). Fabrication and characterization of chitosan-coated soy protein-based green composite. *Journal of Composite Materials, 53*(18), 2481–2504.
7. Rana, A. K., Mandal, A., & Bandyopadhyay, S. (2003). Short jute fiber reinforced polypropylene composites: Effect of compatibiliser, impact modifier and fiber loading. *Composites Science and Technology, 63*(6), 801–806.
8. Singleton, A. C. N., Baillie, C. A., Beaumont, P. W. R., & Peijs, T. (2003). On the mechanical properties, deformation and fracture of a natural fibre/recycled polymer composite. *Composites Part B: Engineering, 34*(6), 519–526.
9. Verma, A., Singh, V. K., & Arif, M. (2016). Study of flame retardant and mechanical properties of coconut shell particles filled composite. *Research and Reviews: Journal of Material Sciences, 4*(3), 1–5.
10. Verma, A., Parashar, A., & Packirisamy, M. (2018). Atomistic modeling of graphene/hexagonal boron nitride polymer nanocomposites: A review. *Wiley Interdisciplinary Reviews: Computational Molecular Science, 8*(3), e1346.
11. Verma, A., Parashar, A., & Packirisamy, M. (2019). Effect of grain boundaries on the interfacial behaviour of graphene-polyethylene nanocomposite. *Applied Surface Science, 470*, 1085–1092.
12. Keller, A. (2003). Compounding and mechanical properties of biodegradable hemp fibre composites. *Composites Science and Technology, 63*(9), 1307–1316.
13. Jain, N., Singh, V. K., & Chauhan, S. (2017). Review on effect of chemical, thermal, additive treatment on mechanical properties of basalt fiber and their composites. *Journal of the Mechanical Behavior of Materials, 26*(5–6), 205–211.
14. Jain, N., Verma, A., & Singh, V. K. (2019). Dynamic mechanical analysis and creep-recovery behaviour of polyvinyl alcohol based cross-linked biocomposite reinforced with basalt fiber. *Materials Research Express, 6*(10), 105373.

15. Jain, N., Singh, V. K., & Chauhan, S. (2017). A review on mechanical and water absorption properties of polyvinyl alcohol based composites/films. *Journal of the Mechanical Behavior of Materials, 26*(5–6), 213–222.

16. Cruz, J., & Fangueiro, R. (2016). Surface modification of natural fibers: A review. *Procedia Engineering, 155*, 285–288.

17. Jain, N., Singh, V. K., & Chauhan, S. (2019). Dynamic and creep analysis of polyvinyl alcohol based films blended with starch and protein. *Journal of Polymer Engineering, 39*(1), 35–47.

18. Chaurasia, A., Verma, A., Parashar, A., & Mulik, R. S. (2019). Experimental and computational studies to analyze the effect of h-BN nanosheets on mechanical behavior of h-BN/Polyethylene nanocomposites. *The Journal of Physical Chemistry C, 123*(32), 20059–20070.

19. Deepmala, K., Jain, N., Singh, V. K., & Chauhan, S. (2018). Fabrication and characterization of chitosan coated human hair reinforced phytagel modified soy protein-based green composite. *Journal of the Mechanical Behavior of Materials, 27*(1–2), 1–8.

20. Wielage, B., Lampke, T., Utschick, H., & Soergel, F. (2003). Processing of natural-fibre reinforced polymers and the resulting dynamic–mechanical properties. *Journal of Materials Processing Technology, 139*(1–3), 140–146.

21. Kessler, R. W., Becker, U., Kohler, R., & Goth, B. (1998). Steam explosion of flax—a superior technique for upgrading fibre value. *Biomass and Bioenergy, 14*(3), 237–249.

22. Prasad, B. M., & Sain, M. M. (2003). Mechanical properties of thermally treated hemp fibers in inert atmosphere for potential composite reinforcement. *Materials Research Innovations, 7*(4), 231–238.

23. Mittal, K. L. (2014). *Plasma surface modification of polymers: Relevance to adhesion.* CRC Press.

24. Cai, Z., Qiu, Y., Zhang, C., Hwang, Y. J., & Mccord, M. (2003). Effect of atmospheric plasma treatment on desizing of PVA on cotton. *Textile Research Journal, 73*(8), 670–674.

25. Marais, S., Gouanvé, F., Bonnesoeur, A., Grenet, J., Poncin-Epaillard, F., Morvan, C., et al. (2005). Unsaturated polyester composites reinforced with flax fibers: effect of cold plasma and autoclave treatments on mechanical and permeation properties. *Composites Part A: Applied Science and Manufacturing, 36*(7), 975–986.

26. Sun, X., Bu, J., Liu, W., Niu, H., Qi, S., Tian, G., et al. (2017). Surface modification of polyimide fibers by oxygen plasma treatment and interfacial adhesion behavior of a polyimide fiber/epoxy composite. *Science and Engineering of Composite Materials, 24*(4), 477–484.

27. Jovančić, P., Jocić, D., Radetić, M., Topalović, T., & Petrović, Z. L. (2005). The influence of surface modification on related functional properties of wool and hemp. *Materials Science Forum, 494*, 283–290.

28. Wang, C. X., & Qiu, Y. P. (2007). Two sided modification of wool fabrics by atmospheric pressure plasma jet: Influence of processing parameters on plasma penetration. *Surface and Coatings Technology, 201*(14), 6273–6277.

29. Uehara, T. (1999). Corona discharge treatment of polymers. In K. L. Mittal & A. Pizzi (Eds.), *Adhesion promotion techniques* (pp. 139–174). New York: Marcel Dekker.

30. Belgacem, M. N., Bataille, P., & Sapieha, S. (1994). Effect of corona modification on the mechanical properties of polypropylene/cellulose composites. *Journal of Applied Polymer Science, 53*(4), 379–385.

31. Amirou, S., Zerizer, A., Haddadou, I., & Merlin, A. (2013). Effects of corona discharge treatment on the mechanical properties of biocomposites from polylactic acid and Algerian date palm fibres. *Scientific Research and Essays, 8*(21), 946–952.

32. Gassan, J., Gutowski, V. S., & Bledzki, A. K. (2000). About the surface characteristics of natural fibres. *Macromolecular Materials and Engineering, 283*(1), 132–139.

33. Gassan, J., & Gutowski, V. S. (2000). Effects of corona discharge and UV treatment on the properties of jute-fibre epoxy composites. *Composites Science and Technology, 60*(15), 2857–2863.

34. Dong, S., Sapieha, S., & Schreiber, H. P. (1992). Effect of corona discharge on cellulose polyethylene composites. *Polymer Engineering Science, 32*(22), 1737–1741.

35. Bataille, P., Dufourd, M., & Sapieha, S. (1994). Copolymerization of styrene on to cellulose activated by corona. *Polymer International, 34*(4), 387–391.
36. Willems, P. (1962). Kinematic high-frequency and ultrasonic treatment of pulp. *Pulp & Paper Magazine Canada, 63*, T455–T462.
37. Alam, A. M., Beg, M. D. H., Prasad, D. R., Khan, M. R., & Mina, M. F. (2012). Structures and performances of simultaneous ultrasound and alkali treated oil palm empty fruit bunch fiber reinforced poly (lactic acid) composites. *Composites Part A: Applied Science and Manufacturing, 43*(11), 1921–1929.
38. Kadam, V. V., Goud, V., & Shakyawar, D. B. (2013). Ultrasound scouring of wool and its effects on fibre quality. *Indian Journal of Fibre & Textile Research, 38*, 410–414.
39. Liu, L., Huang, Y. D., Zhang, Z. Q., Jiang, Z. X., & Wu, L. N. (2008). Ultrasonic treatment of aramid fiber surface and its effect on the interface of aramid/epoxy composites. *Applied Surface Science, 254*(9), 2594–2599.
40. Laine, J. E., & Goring, D. A. I. (1977). Influence of ultrasonic irradiation on the properties of cellulosic fibres. *Cellulose Chemistry and Technology, 11*(5), 561–567.
41. Kato, K., Vasilets, V. N., Fursa, M. N., Meguro, M., Ikada, Y., & Nakamae, K. (1999). Surface oxidation of cellulose fibers by vacuum ultraviolet irradiation. *Journal of Polymer Science Part A: Polymer Chemistry, 37*(3), 357–361.
42. Benedetto, R. M. D., Gelfuso, M. V., & Thomazini, D. (2015). Influence of UV radiation on the physical-chemical and mechanical properties of banana fiber. *Materials Research, 18*, 265–272.
43. Oosterom, R., Ahmed, T. J., Poulis, J. A., & Bersee, H. E. N. (2006). Adhesion performance of UHMWPE after different surface modification techniques. *Medical Engineering & Physics, 28*(4), 323–330.
44. Benyahia, A., Merrouche, A., Rokbi, M., & Kouadri, Z. (2013). Study the effect of alkali treatment of natural fibers on the mechanical behavior of the composite unsaturated Polyester-fiber Alfa. 21ème Congrès Français de Mécanique Bordeaux (pp. 1–6).
45. Li, X., Tabil, L. G., & Panigrahi, S. (2007). Chemical treatments of natural fiber for use in natural fiber-reinforced composites: A review. *Journal of Polymers and the Environment, 15*(1), 25–33.
46. Verma, A., Gaur, A., & Singh, V. K. (2017). Mechanical properties and microstructure of starch and sisal fiber biocomposite modified with epoxy resin. *Materials Performance and Characterization, 6*(1), 500–520.
47. Verma, A., & Singh, V. K. (2019). Mechanical, microstructural and thermal characterization of epoxy-based human hair-reinforced composites. *Journal of Testing and Evaluation, 47*(2), 1193–1215.
48. Verma, A., Negi, P., & Singh, V. K. (2019). Experimental analysis on carbon residuum transformed epoxy resin: Chicken feather fiber hybrid composite. *Polymer Composites, 40*(7), 2690–2699.
49. Verma, A., Negi, P., & Singh, V. K. (2018). Physical and thermal characterization of chicken feather fiber and crumb rubber reformed epoxy resin hybrid composite. *Advances in Civil Engineering Materials, 7*(1), 538–557.
50. Verma, A., Negi, P., & Singh, V. K. (2018). Experimental investigation of chicken feather fiber and crumb rubber reformed epoxy resin hybrid composite: mechanical and microstructural characterization. *Journal of the Mechanical Behavior of Materials, 27*(3–4), 1–24.
51. Verma, A., Joshi, K., Gaur, A., & Singh, V. K. (2018). Starch-jute fiber hybrid biocomposite modified with an epoxy resin coating: Fabrication and experimental characterization. *Journal of the Mechanical Behavior of Materials, 27*(5–6), 1–16.
52. Verma, A., Budiyal, L., Sanjay, M. R., & Siengchin, S. (2019). Processing and characterization analysis of pyrolyzed oil rubber (from waste tires)-epoxy polymer blend composite for lightweight structures and coatings applications. *Polymer Engineering & Science, 59*(10), 2041–2051.
53. Verma, A., Kumar, R., & Parashar, A. (2019). Enhanced thermal transport across a bi-crystalline graphene–polymer interface: An atomistic approach. *Physical Chemistry Chemical Physics, 21*, 6229–6237.

54. Mohanty, A. K., Misra, M., & Drzal, L. T. (2001). Surface modifications of natural fibers and performance of the resulting biocomposites: An overview. *Composite Interfaces, 8*(5), 313–343.

55. Jähn, A., Schröder, M. W., Füting, M., Schenzel, K., & Diepenbrock, W. (2002). Characterization of alkali treated flax fibres by means of FT Raman spectroscopy and environmental scanning electron microscopy. *Spectrochimica Acta Part A: Molecular and Biomolecular Spectroscopy, 58*(10), 2271–2279.

56. Agrawal, R., Saxena, N. S., Sharma, K. B., Thomas, S., & Sreekala, M. S. (2000). Activation energy and crystallization kinetics of untreated and treated oil palm fibre reinforced phenol formaldehyde composites. *Materials Science and Engineering: A, 277*(1–2), 77–82.

57. Cai, M., Takagi, H., Nakagaito, A. N., Li, Y., & Waterhouse, G. I. (2016). Effect of alkali treatment on interfacial bonding in abaca fiber-reinforced composites. *Composites Part A: Applied Science and Manufacturing, 90*, 589–597.

58. Wei, B., Cao, H., & Song, S. (2010). Tensile behavior contrast of basalt and glass fibers after chemical treatment. *Materials & Design, 31*(9), 4244–4250.

59. Garcia-Jaldon, C., Dupeyre, D., & Vignon, M. R. (1998). Fibres from semi-retted hemp bundles by steam explosion treatment. *Biomass and Bioenergy, 14*(3), 251–260.

60. Morrison Iii, W. H., Archibald, D. D., Sharma, H. S. S., & Akin, D. E. (2000). Chemical and physical characterization of water-and dew-retted flax fibers. *Industrial Crops and Products, 12*(1), 39–46.

61. Mishra, S., Misra, M., Tripathy, S. S., Nayak, S. K., & Mohanty, A. K. (2001). Graft copolymerization of acrylonitrile on chemically modified sisal fibers. *Macromolecular Materials and Engineering, 286*(2), 107–113.

62. Ray, D., Sarkar, B. K., Rana, A. K., & Bose, N. R. (2001). Effect of alkali treated jute fibres on composite properties. *Bulletin of Materials Science, 24*(2), 129–135.

63. Bachtiar, D., Sapuan, S. M., & Hamdan, M. M. (2010). Flexural properties of alkaline treated sugar palm fibre reinforced epoxy composites. *International Journal of Automotive and Mechanical Engineering, 1*(1), 79–90.

64. Yan, L., Chouw, N., Huang, L., & Kasal, B. (2016). Effect of alkali treatment on microstructure and mechanical properties of coir fibres, coir fibre reinforced-polymer composites and reinforced-cementitious composites. *Construction and Building Materials, 112*, 168–182.

65. Hosur, M., Maroju, H., & Jeelani, S. (2015). Comparison of effects of alkali treatment on flax fibre reinforced polyester and polyester-biopolymer blend resins. *Polymers and Polymer Composites, 23*(4), 229–242.

66. Van de Weyenberg, I., Ivens, J., De Coster, A., Kino, B., Baetens, E., & Verpoest, I. (2003). Influence of processing and chemical treatment of flax fibres on their composites. *Composites Science and Technology, 63*(9), 1241–1246.

67. Jacob, M., Thomas, S., & Varughese, K. T. (2004). Mechanical properties of sisal/oil palm hybrid fiber reinforced natural rubber composites. *Composites Science and Technology, 64*(7–8), 955–965.

68. Sang, L., Zhao, M., Liang, Q., & Wei, Z. (2017). Silane-treated basalt fiber-reinforced poly (butylene succinate) biocomposites: Interfacial crystallization and tensile properties. *Polymers, 9*(8), 351.

69. Tan, C., Ahmad, I., & Heng, M. (2011). Characterization of polyester composites from recycled polyethylene terephthalate reinforced with empty fruit bunch fibers. *Materials & Design, 32*(8–9), 4493–4501.

70. Rong, M. Z., Zhang, M. Q., Liu, Y., Yang, G. C., & Zeng, H. M. (2001). The effect of fiber treatment on the mechanical properties of unidirectional sisal-reinforced epoxy composites. *Composites Science and technology, 61*(10), 1437–1447.

71. Valadez-Gonzalez, A., Cervantes-Uc, J. M., Olayo, R., & Herrera-Franco, P. J. (1999). Chemical modification of henequen fibers with an organosilane coupling agent. *Composites Part B: Engineering, 30*(3), 321–331.

72. Hill, C. A., Khalil, H. A., & Hale, M. D. (1998). A study of the potential of acetylation to improve the properties of plant fibres. *Industrial Crops and Products, 8*(1), 53–63.

73. Bledzki, A. K., Mamun, A. A., Lucka-Gabor, M., & Gutowski, V. S. (2008). The effects of acetylation on properties of flax fibre and its polypropylene composites. *Express Polymer Letters, 2*(6), 413–422.

74. Mishra, S., Mohanty, A. K., Drzal, L. T., Misra, M., Parija, S., Nayak, S. K., et al. (2003). Studies on mechanical performance of biofibre/glass reinforced polyester hybrid composites. *Composites Science and Technology, 63*(10), 1377–1385.

75. Khalil, H. A., & Ismail, H. (2000). Effect of acetylation and coupling agent treatments upon biological degradation of plant fibre reinforced polyester composites. *Polymer Testing, 20*(1), 65–75.

76. Kalia, S., Kaith, B. S., & Kaur, I. (2009). Pretreatments of natural fibers and their application as reinforcing material in polymer composites—a review. *Polymer Engineering & Science, 49*(7), 1253–1272.

77. Nair, K. M., Thomas, S., & Groeninckx, G. (2001). Thermal and dynamic mechanical analysis of polystyrene composites reinforced with short sisal fibres. *Composites Science and Technology, 61*(16), 2519–2529.

78. Kalia, S., Kaushik, V. K., & Sharma, R. K. (2011). Effect of benzoylation and graft copolymerization on morphology, thermal stability, and crystallinity of sisal fibers. *Journal of Natural Fibers, 8*(1), 27–38.

79. Bledzki, A. K., & Gassan, J. (1999). Composites reinforced with cellulose based fibres. *Progress in Polymer Science, 24*(2), 221–274.

80. Singha, A. S., & Rana, R. K. (2012). Functionalization of cellulosic fibers by graft copolymerization of acrylonitrile and ethyl acrylate from their binary mixtures. *Carbohydrate Polymers, 87*(1), 500–511.

81. Sreekala, M. S., Kumaran, M. G., & Thomas, S. (2002). Water sorption in oil palm fiber reinforced phenol formaldehyde composites. *Composites Part A: Applied Science and Manufacturing, 33*(6), 763–777.

82. Mishra, S., Mishra, M., Tripathy, S. S., Nayak, S. K., & Mohanty, A. K. (2002). The influence of chemical surface modification on the performance of sisal-polyester biocomposites. *Polymer Composites, 23*(2), 164–170.

83. Li, X., Panigrahi, S. A., Tabil, L. G., & Crerar, W. J. (2004). Flax fiber-reinforced composites and the effect of chemical treatments on their properties. In *North central ASAE/CSAE Annual Intersectional Meeting*, Winnipeg, Canada.

84. Van de Velde, K., & Kiekens, P. (2003). Effect of material and process parameters on the mechanical properties of unidirectional and multidirectional flax/polypropylene composites. *Composite Structures, 62*(3–4), 443–448.

85. Cantero, G., Arbelaiz, A., Llano-Ponte, R., & Mondragon, I. (2003). Effects of fibre treatment on wettability and mechanical behaviour of flax/polypropylene composites. *Composites Science and Technology, 63*(9), 1247–1254.

86. Keener, T. J., Stuart, R. K., & Brown, T. K. (2004). Maleated coupling agents for natural fibre composites. *Composites Part A: Applied Science and Manufacturing, 35*(3), 357–362.

87. Gassan, J., & Bledzki, A. K. (1997). The influence of fiber-surface treatment on the mechanical properties of jute-polypropylene composites. *Composites Part A: Applied Science and Manufacturing, 28*(12), 1001–1005.

88. Van den Oever, M., & Peijs, T. (1998). Continuous-glass-fibre-reinforced polypropylene composites II. Influence of maleic-anhydride modified polypropylene on fatigue behaviour. *Composites Part A: Applied Science and Manufacturing, 29*(3), 227–239.

89. Joseph, P. V., Joseph, K., Thomas, S., Pillai, C. K. S., Prasad, V. S., Groeninckx, G., et al. (2003). The thermal and crystallisation studies of short sisal fibre reinforced polypropylene composites. *Composites Part A: Applied Science and Manufacturing, 34*(3), 253–266.

90. Bledzki, A. K., Reihmane, S., & Gassan, J. (1996). Properties and modification methods for vegetable fibers for natural fiber composites. *Journal of Applied Polymer Science, 59*(8), 1329–1336.

91. Mohanty, S., Nayak, S. K., Verma, S. K., & Tripathy, S. S. (2004). Effect of MAPP as a coupling agent on the performance of jute–PP composites. *Journal of Reinforced Plastics and Composites, 23*(6), 625–637.

92. Mishra, S., Naik, J. B., & Patil, Y. P. (2000). The compatibilising effect of maleic anhydride on swelling and mechanical properties of plant-fiber-reinforced novolac composites. *Composites Science and Technology, 60*(9), 1729–1735.

93. Frederick, T. W., & Norman, W. (2004). *Natural fibers plastics and composites*. New York: EUA: Kluwer Academic Publishers.

94. Paul, A., Joseph, K., & Thomas, S. (1997). Effect of surface treatments on the electrical properties of low-density polyethylene composites reinforced with short sisal fibers. *Composites Science and Technology, 57*(1), 67–79.

95. Khan, J. A., Khan, M. A., & Islam, R. (2012). Effect of potassium permanganate on mechanical, thermal and degradation characteristics of jute fabric-reinforced polypropylene composite. *Journal of Reinforced Plastics and Composites, 31*(24), 1725–1736.

96. Arsyad, M., & Soenoko, R. (2018). The effects of sodium hydroxide and potassium permanganate treatment on roughness of coconut fiber surface. In *MATEC Web of Conferences (EDP Sciences), 204*, 05004.

97. Sreekala, M. S., Kumaran, M. G., Joseph, S., Jacob, M., & Thomas, S. (2000). Oil palm fibre reinforced phenol formaldehyde composites: influence of fibre surface modifications on the mechanical performance. *Applied Composite Materials, 7*(5–6), 295–329.

98. Joseph, K., Thomas, S., & Pavithran, C. (1996). Effect of chemical treatment on the tensile properties of short sisal fibre-reinforced polyethylene composites. *Polymer, 37*(23), 5139–5149.

99. Sreekala, M. S., & Thomas, S. (2003). Effect of fibre surface modification on water-sorption characteristics of oil palm fibres. *Composites Science and Technology, 63*(6), 861–869.

100. Wu, Z., Pittman, C. U., Jr., & Gardner, S. D. (1996). Grafting isocyanate-terminated elastomers onto the surfaces of carbon fibers: Reaction of isocyanate with acidic surface functions. *Carbon, 34*(1), 59–67.

101. George, J., Janardhan, R., Anand, J. S., Bhagawan, S. S., & Thomas, S. (1996). Melt rheological behaviour of short pineapple fibre reinforced low density polyethylene composites. *Polymer, 37*(24), 5421–5431.

102. Zafeiropoulos, N. E., Williams, D. R., Baillie, C. A., & Matthews, F. L. (2002). Engineering and characterisation of the interface in flax fibre/polypropylene composite materials. Part I. Development and investigation of surface treatments. *Composites Part A: Applied Science and Manufacturing, 33*(8), 1083–1093.

Structure and Surface Morphology Techniques for Biopolymers

Sabarish Radoor, Jasila Karayil, Aswathy Jayakumar,
E. K. Radhakrishnan, Lakshmanan Muthulakshmi,
Sanjay Mavinkere Rangappa, Suchart Siengchin
and Jyotishkumar Parameswaranpillai

Abstract Different techniques such as optical microscopy, scanning electron microscopy, transmission electron microscopy, atomic force microscopy, nuclear magnetic resonance, X-ray diffraction, and Fourier-transform infrared spectroscopy are used for the examination of biopolymer-based materials. This chapter discusses the characterisation of structure and surface morphology of the biopolymers, their blends, and composites by these techniques. A careful examination of biopolymers, their blends and composites are essential for the fruitful application of these materials.

1 Introduction

Biopolymers are naturally occurring polymers having superior features than other polymers which makes them potential candidate in research as well as in other sectors. They are biocompatible, biodegradable, non-toxic, highly stable and safe to use. They can be blended with other synthetic polymers for improving the mechanical as well as the degradation properties. The development of polymers from natural sources is

S. Radoor · S. M. Rangappa · S. Siengchin
Department of Mechanical and Process Engineering, The Sirindhorn International Thai-German Graduate School of Engineering (TGGS), King Mongkut's University of Technology North Bangkok, 1518 Wongsawang Road, Bangsue, Bangkok 10800, Thailand

J. Karayil
Department of Chemistry, Government Polytechnic Women's College, Calicut, Kerala, India

A. Jayakumar · E. K. Radhakrishnan
School of Biosciences, Mahatma Gandhi University, P.D. Hills (P.O.), Kottayam, Kerala 686560, India

L. Muthulakshmi
Department of Materials Science, Madurai Kamaraj University, Madurai 625021, Tamilnadu, India

J. Parameswaranpillai (✉)
Center of Innovation in Design and Engineering for Manufacturing, King Mongkut's University of Technology North Bangkok, 1518 Wongsawang Road, Bangsue, Bangkok 10800, Thailand
e-mail: jyotishkumarp@gmail.com

© Springer Nature Switzerland AG 2020
A. Khan et al. (eds.), *Biofibers and Biopolymers for Biocomposites*,
https://doi.org/10.1007/978-3-030-40301-0_2

one of the fastest-growing areas of material research. Biopolymers find application in food packaging, cosmetics, food additives, fabrics, medical, biosensors, etc. [1–3].

The structural changes occurring in biopolymers can be analysed by various techniques like optical microscopy, electron microscopy, atomic force microscopy (AFM), etc. To analysis, the crystallinity and chemical structure of biopolymer, common techniques such as nuclear magnetic resonance (NMR), X-ray diffraction (XRD), and Fourier-transform infrared spectroscopy (FT-IR) are used. The morphology, structure, and composition of the biopolymers can be analysed by using these techniques. The emergence of highly sophisticated and integrated techniques has immense promises in biopolymer research as well as in applications. Apart from chemical and structural analysis, some of these techniques can determine the crystallinity, molecular orientation, as well as mechanical behaviours. Hence this chapter deals with some of the surface and structural characterization techniques of biopolymers [4, 5].

2   X-Ray Diffraction (XRD)

XRD is one of the important tools to determine the structural features of materials such as particle sizes, interplanar distance, percentage crystallinity, etc. In this technique, a monochromatic X-ray (CuKα or MoKα source) is allowed to fall on the specimens produce constructive interference. For constructive interference, the lattice planes should obey Bragg's law which is represented as

$$n\lambda = 2d \sin\theta \tag{1}$$

where, λ, d, and θ are the wavelength of the X-rays, the distance between successive lattice planes, and diffraction angle, respectively, and n is an integer which is generally known as the order of reflection [6].

The width of the diffraction lines can be calculated using Debye–Scherrer formula:

$$d_{hkl} = k\lambda/\beta \cos\theta \tag{2}$$

where d_{hkl}, λ, β and θ are the averaged crystallite size, wavelength of the X-ray, peak width at half maximum and diffraction angle, respectively, and k is a constant, regularly taken as 1 [7].

Researchers successfully employed the XRD technique to study the crystallinity and crystallinity index of biopolymers. Trivedi et al. [8] employed XRD to evaluate the crystalline nature of untreated and treated cellulose and cellulose acetate. The XRD results revealed that the crystallinity of biopolymer is retained even after biofield treatment (Fig. 1b, d). The peak appears at $2\theta = 16.27°$ and $22.53°$ are assigned to the characteristic crystalline peak of cellulose (Fig. 1) sample. On the other hand, cellulose acetate is more semi-crystalline with peaks at $10.50°$ and $13.13°$.

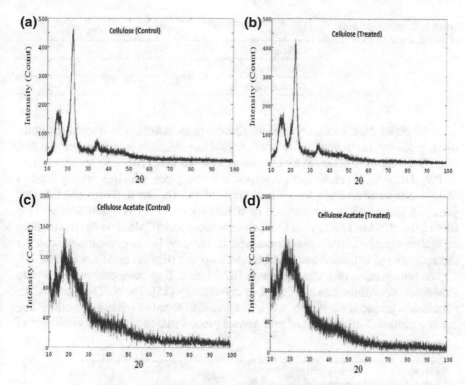

Fig. 1 XRD diffractogram of: **a** cellulose (control), **b** cellulose (treated), **c** cellulose acetate (control), **d** cellulose acetate (treated) [8]

Wulandari and co-workers developed nanocellulose by acid treatment. Based on XRD studies (Fig. 2) they reported that the nanocellulose obtained by the acid hydrolysis of cellulose using 50% acid (nanocellulose B) possess greater crystallinity and crystallnity index than native cellulose (Table 1). However, on increasing the strength

Fig. 2 XRD pattern of
a cellulose; **b** nanocellulose
A; **c** nanocellulose B [9]

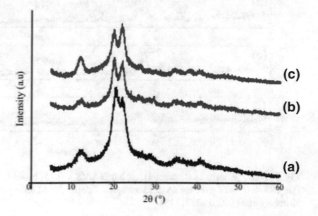

Table 1 Crystallinity index of cellulose and nanocellulose [9]

Samples	Crystallinity index (%)
Cellulose	70.62
Nanocellulose A	67.83
Nanocellulose B	76.01

of acid (>60%), the crystallinity of nanocellulose (nanocellulose A) is found to diminishes, probably due to destruction of the crystalline structure by acid hydrolysis [9]. Similar results were observed by Cohoua and co-workers [10].

Poly(lactic acid) (PLA) is an amorphous biopolymer, which is widely used to make biodegradable films for various application. The crystallinity of PLA was generally improved by the addition of nanoparticles [11–13]. Shuhua et al. [14] studied the effect of cellulose-based carbon microsphere (CMSs) on the crystallinity of PLA composites. The authors observed an increase in the crystallinity with the incorporation of cellulose-based carbon microsphere (CMSs) into PLA matrix.

The introduction of sodium alginate (SA) into n-Hap composite was found to enhance its crystalline size and degree of crystallinity [15]. The XRD patterns of the composites prepared are shown in Fig. 3. From XRD analysis, it can be noticed that the concentration of sodium alginate plays a pivotal role to tune the crystallinity of

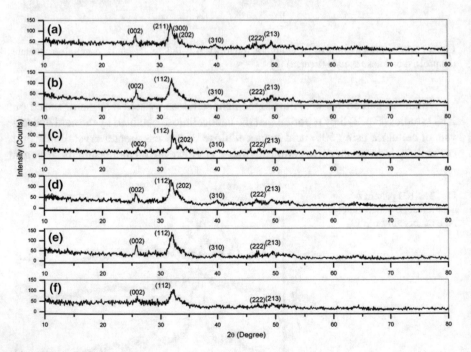

Fig. 3 XRD patterns of **a** pure n-HAp, **b** n-HAp/SA-0.75, **c** n-HAp/SA-1.5, **d** n HAp/SA-2.25, **e** n-HAp/SA-3.0 and **f** n-HAp/SA-3.75 samples. [Reproduced with permission from Elsevier, License Number: 4671140615253] [15]

n-HAp. The enhancement of crystallinity and crystalline size of n-HAp at low wt% of SA could be due to SA induced crystal growth of n-HAP. However, the higher percentage of SA was found to inhibits the crystallinity of n-HAP, causing a reduction in the intensity of crystalline peak and crystal size.

Usha et al. [16] were the pioneer to report the self-assembly of collagen without mineralization by XRD technique. It is evident from Fig. 4 that the glycated collagen nano-fibrils in presence of aminoguanidine (AG) exhibit high crystalline peak than

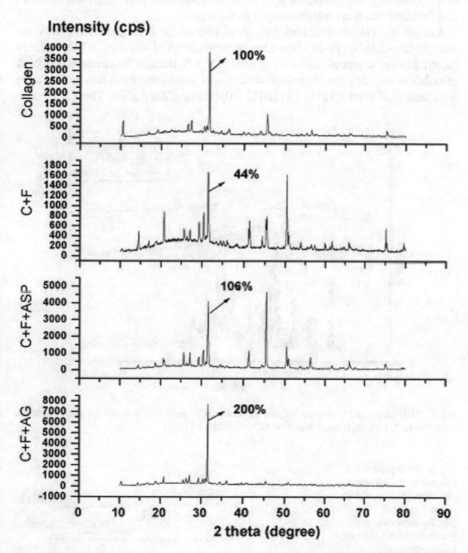

Fig. 4 XRD patterns of collagen, collagen + fructose, collagen + fructose + aspirin and collagen + fructose + AG treated reconstituted collagen fibrils. [Reproduced with permission from Elsevier, License Number: 4671141403394] [16]

native collagen and aspirin-treated glycated collagen nano-fibrils. The authors thus suggested that AG enhance the crystallinity and self-assembly processes of collagen nanofibril than aspirin.

There is ample work in literature were XRD is used to assess the physical inter-action. For instance, Chauhan et al. [17] employed XRD to understand the physical interaction between gelatin and curcumin. It is clear from the XRD spectrum (Fig. 5) that curcumin loaded gelatin film possess only the characteristic peak of gelatin. A drop-in intensity and prominent shift in the characteristic peak of gelatin indicates that the curcumin is completed trapped in the film.

Lin et al. [18] synthesised cellulose/chitosan and cellulose/chitosan/silver nanoparticles (AgNPs) composite films for antimicrobial activities. XRD profile of the synthesised composite films is depicted in Fig. 6. Besides the characteristic peak of cellulose and chitosan, the composite exhibits an addition reflection peak at $2\theta = 34.0°$ and $41.0°$ ascribed to (1 1 1) and (2 0 0) planes of the AgNPs. The XRD results

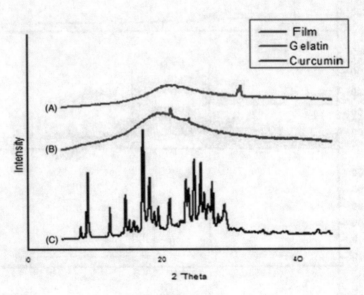

Fig. 5 XRD pattern of (A) optimized film, (B) gelatin; (C) curcumin. [Reproduced with permission from Taylor & Francis, License Number: 4671650408306] [17]

Fig. 6 XRD pattern of chitosan/cellulose and chitosan/cellulose-AgNPs composite films. [Reproduced with permission from Elsevier, License Number: 4671150961169] [18]

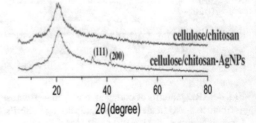

Fig. 7 X-ray diffraction patterns of the pure CS film and CS composite films with various ESP concentrations [19]

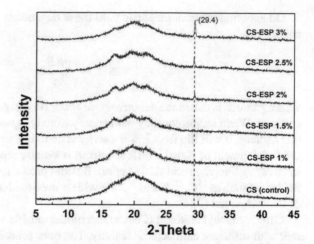

thus confirm the successful incorporation of AgNPs in cellulose/chitosan composite film.

Jiang et al. [19] studied the X-ray diffraction patterns of the pure CS (natural corn starch) film and CS composite films with varying eggshell powder concentrations as shown in Fig. 7. This result indicates that the eggshell powder (ESP), were uniformly distributed in the matrix, therefore the diffraction peak appeared at $2\theta = 29.4°$, especially with high concentration of eggshell powder which leads to semicrystalline nature of the composite films. This is due to calcite crystal and the agglomeration of eggshell powder.

3 Nuclear Magnetic Resonance Spectroscopy (NMR)

NMR (nuclear magnetic resonance) is an excellent analytical method used by chemists to identify the structure of organic molecules. The basic principle of NMR is that the nuclei of many elemental isotopes have either integral spins (e.g. $I = 1, 2, 3$ ....) or fractional spins (e.g. $I = 1/2, 3/2, 5/2$ ....). For nuclei with spin $I = 1/2$ the two possible orientation are $+1/2$ and $-1/2$ [20].

Energy levels for a nucleus with spin quantum number 1/2

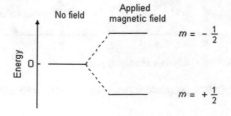

On applying external magnetic field these degenerate energy levels split with an energy gap

$$\Delta E \; = \; \frac{\gamma h B}{2\pi} \tag{3}$$

where gamma, is called the *magnetogyric ratio*, B is magnetic field and h is Planck's constant. When radiofrequency with energy exactly equals to delta E is irradiated on this nucleus, it will flip from lower energy state to higher spin state thus resulting in the development of signal. NMR spectrum is usually represented by plotting signal intensity against chemical shift (δ ppm). Besides proton (proton NMR), several NMR nucleuses such as ^{13}C, ^{15}N and ^{31}P are widely used to identify the chemical structure of the compound [7].

Carboxymethyl chitosan (CMCs) is a biodegradable and biocompatible biopolymer with sufficient antibacterial activity. The poor solubility of CMCs at neutral pH urges researchers to modify CMCs by using various chemical treatment. Quaternized carboxymethyl chitosan (QCMCS) is a modified form of CMCs with sufficient solubility in a wide range of pH. Yin et al. [21] developed quaternized carboxymethyl chitosan/poly(vinyl alcohol)/Cu blend film by solution casting method for biomedical and packaging applications. The chemical structure of QCMCS was studied by proton NMR analysis as shown in Fig. 8.

Sun et al. [22] grafted water-soluble starch with polyacrylonitrile by electrospinning technique. The synthesized St-g-PAN (starch-graft-polyacrylonitrile) nanofiber

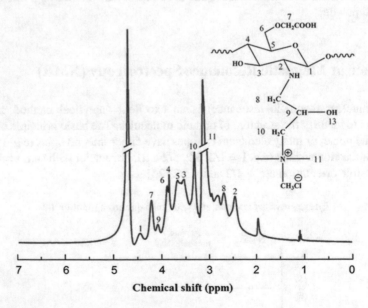

Fig. 8 ^{1}H NMR spectra of QCMCS. [Reproduced with permission from Elsevier, License Number: 4671160088980] [21]

Fig. 9 ^{13}C NMR spectrum of **a** St and **b** synthesized St-g-PAN. [Reproduced with permission from Elsevier, License Number: 4671160259922] [22]

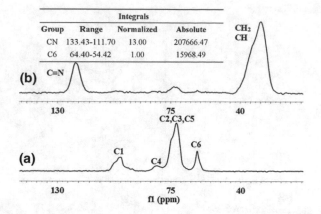

Integrals				
Group	Range	Normalized	Absolute	
CN	133.43–111.70	13.00	207666.47	
C6	64.40–54.42	1.00	15968.49	

exhibits good biocompatibility, and tensile intensity. The water hydrophilicity of St (starch) decreases with grafting AN. Figure 9 shows the ^{13}C NMR spectrum of (a) St and (b) St-g-PAN. ^{13}C NMR of starch displays characteristic peaks at $\delta = 60$–110 ppm, corresponding to the carbon present in starch. In addition to the characteristic peak of starch, the St-g-PAN display another sharp peak at $\delta = 121.18$ ppm and 29.75 ppm due to the carbon in –CN and –CH (–CH$_2$) group respectively. The degree of substitution of St and the weight fraction of the grafted acrylonitrile of St-g-PAN were calculated from NMR and found to be 13.0 and 81.0%.

The successful grafting of polyacrylic acid (PAA) on (CMS carboxymethyl starch) was confirmed by proton NMR. NMR spectra of CMS and three different types of CMS-g-PAA namely CMS-g-PAA1, CMS-g-PAA3 and CMS-g-PAA4 are shown in Fig. 10. Since the chemically equivalent protons of C-1a, C-1b and C-1c carbon atoms of the anhydroglucose unit (AGU) are directly attached to electronegative oxygen atom it appears downfield at $\delta = 5.30$–5.82 ppm. The peak at $\delta = 3.36$–4.26 ppm corresponds to protons attached to C-2 to C-6 carbon atoms of AGU. In the case of CMC grafted PAA, peak in the range of 2.36–2.60 ppm and 1.58–2.07 ppm (C-8) appears and is attributed to the methane proton (C-9 carbon atom) and methylene proton (C-9 carbon atom) of the grafted PAA. The additional peaks in the PAA grafted CMS spectrum confirmed the grafting of PAA on CMS [23].

The superior features of starch nanocrystals (SNCs) enables it to act as a reinforcing agent for polymeric composites materials. Hao et al. [24] enhance the hydrophobicity of potato starch nanocrystals (SNCs) by modifying it with octenyl succinic anhydride (OSA). They used proton NMR to assess the structure and degree of OSA substitution (Fig. 11). The peak observed at $\delta = 5.12$ and 4.78 ppm were assigned to 1 and 1′ respectively and the peak at $\delta = 3.23$–3.86 ppm was assigned to the other protons (2–6). It is evident from NMR spectra that with an increase in the degree of OSA substitution, the characteristic peak of anomeric proton of α-1,4-linkages ($\delta = 5.10$–5.28 ppm) becomes broader. Also, the OSA-SNCS sample displays a new peak at 0.86 ppm due to the protons of the terminal methyl group of OS chains. The enhancement of the intensity of this peak with OSA percentage indicates that OSA

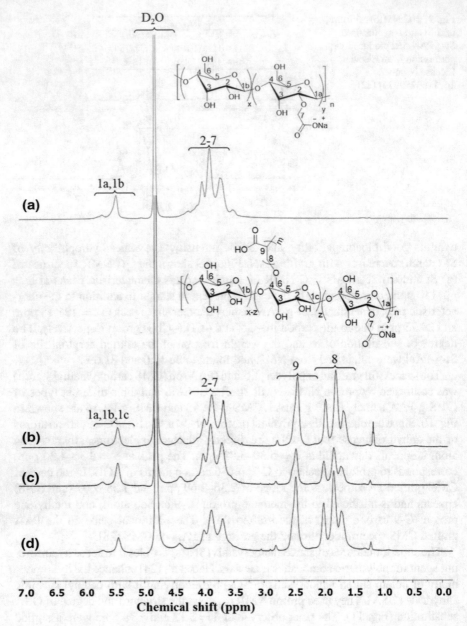

Fig. 10 ^{1}H NMR spectra of **a** CMS, **b** CMS-g-PAA1, **c** CMS-g-PAA3 and **d** CMS-g-PAA4. [Reproduced with permission from Elsevier, License Number: 4671160600793] [23]

Fig. 11 ¹H NMR spectra of (A) SNCs and (B) OSA-SNCs with different OSA amount (1) 12.5%, (2) 25%, (3) 50%, and (4) 75%. [Reproduced with permission from Elsevier, License Number: 4671160850165] [24]

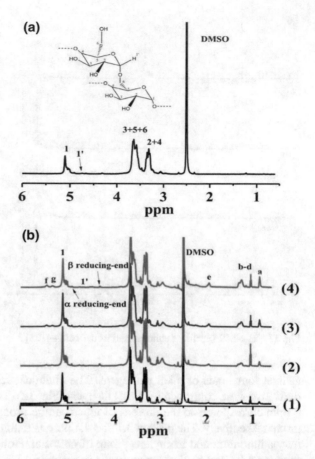

group is linked with SNCs. The degree of substitution of OSA-SNCs with different amount of OSA was calculated and is presented in Table 2. The increase in DS with OSA amount is attributed to the greater interaction between OSA and hydroxyl group of SNCs.

Quaternary phosphonium salts exhibit superior antiseptics property than quaternary ammonium salt. Therefore, researchers are focussing to develop quaternary phosphonium based materials as broad spectrum antiseptics. Tan et al. [25] introduce quaternary phosphonium into chitosan and investigated their antifungal activity

Table 2 Degree of substitution of OSA in SNC with different OSA addition amount. [Reproduced with permission from Elsevier, License Number: 4671160850165] [24]

Samples	OSA addition amount (%)	DS
OSA-SNC1	12.5	0.0575 ± 0.0015
OSA-SNC2	25	0.0686 ± 0.0024
OSA-SNC3	50	0.0926 ± 0.0040
OSA-SNC4	75	0.1271 ± 0.0021

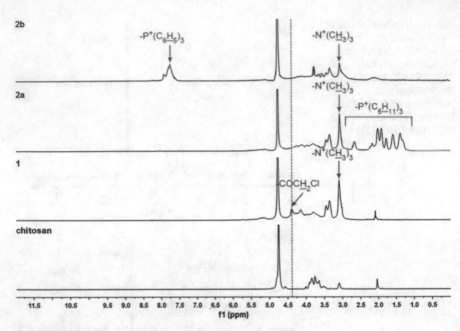

Fig. 12 ^{1}H-NMR spectra of chitosan and its derivatives [25]

against four kinds of plant pathogens. The chemical structure of chitosan and its derivative were confirmed by NMR analysis (Fig. 12). The peak appears at 3–5 and 2 ppm corresponds to the protons of glucosamine group of chitosan and N-acetyl group respectively. The peak at 3.1 and 4.4 ppm is assigned to the proton of quaternary ammonium and methylene group of chitosan. However, the chitosan derivate (compound a and b) shows new peaks in the range 1.2–2.7 and 7.5–5 ppm, corresponding to protons of methyl/methylene and phenyl groups respectively. In addition to proton NMR, ^{31}P spectra of the samples were also employed to understand the structure. The peaks at 33.9 and 20.1 ppm of chitosan derivative sample is due to the presence of phosphorus atoms in the quaternary phosphonium groups (Fig. 13).

4 Atomic Force Microscopy (AFM)

AFM (atomic force microscopy) is a versatile microscopic technique invented jointly by Gerd Binning, Calvin Quate and Christoph Gerber in 1986 [26]. Besides visualizing the surface of the sample in nanometer scale this technique will also provide various types of surface measurements [27]. One of the main attractive features of AFM technique is that it can be used for the surface analysis of almost any sample ranging from hard ceramic/metallic article to soft biomolecules/flexible polymers. Moreover, it has the advantage of probing the surface of the sample without special

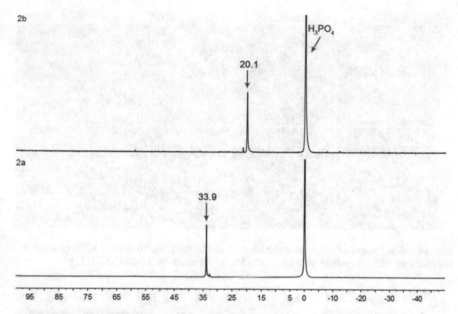

Fig. 13 ^{31}P-NMR spectra of chitosan derivatives [25]

treatment as require in complementary techniques like SEM and TEM [28]. AFM consists of the following components: a cantilever with sharp tip/probe, laser beam, PSPD (position sensitive photodiode) and a scanner. The working of AFM is similar to an SPM (scanning probe microscope), and utilize a sharp tip/probe to scan the surface of the sample [29]. As the probe travels across the sample in a raster pattern, it interacts with the sample (electrostatic, magnetic, capillary, Van der Waals) and causes a slight bending/deflection in the cantilever. This defection is detected by a laser beam and is recorded precisely by the PSPD (position sensitive photodiode) and 3D surface profile of the sample is generated [30].

AFM technique is one of the essential tools to visualize the surface of biopolymers. Bonardd et al. [31] designed a chitosan composite with improved dielectric property by introducing CN-CNC (cyanoethylated cellulose nanocrystals). AFM images of chitosan and nanocomposite with different percentage of CN-CNC (10, 30 and 50 wt%) is depicted in Fig. 14. It can be seen from AFM images that CN-CNC is responsible for the surface roughness of nanocomposites: an enhancement in surface roughness from 7.19 to 24.8 nm was observed on increasing the CN-CNC weight % from 10 to 50. It can be also noted that even at high wt%, the CN-CNC is properly dispersed in the chitosan matrix.

A structure tunable film was generated by coupling method from chitosan (CTS) and TEMPO-oxidized nano fibrillated cellulose (TONCs). The AFM images of CTS-TONCs, chitosan and TONCs are shown in Fig. 15a–c. Adsorption of chitosan onto the TONC's surfaces can be identified from AFM analysis. It can be also observed

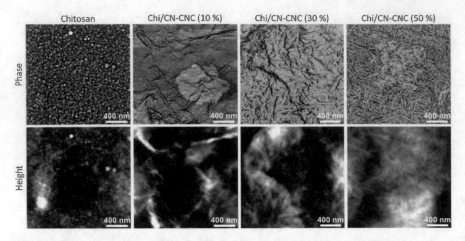

Fig. 14 AFM images of chitosan and nanocomposites with 10, 30 and 50 wt% of CN-CNC. [Reproduced with permission from Elsevier, License Number: 4671170435007] [31]

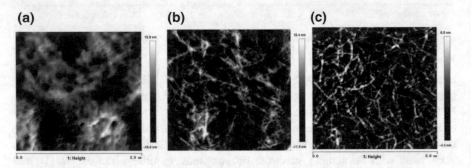

Fig. 15 AFM images of **a** chitosan, **b** CTS-TONCs, and **c** TONCs. [Reproduced with permission from Elsevier, License Number: 4671180478653] [32]

that the fiber diameter of CTS-TONCs (2.36 ± 0.26 nm) was wider than TONCs (2.96 ± 0.66 nm), probably due to the effect of chitosan on the TONC [32].

Sodium alginate is a cheap and easily available biopolymer with excellent film-forming ability [33]. However, poor water resistance restricted its application in various fields [34]. Recently Yang et al. [35] modify sodium alginate (SA) by grafting it with hydrophobic acrylonitrile. AFM analysis strongly supports that the introduction of acrylonitrile (AN) improves the water resistance property of the membrane. It is evident from the figure (Fig. 16) that the sodium alginate film has a smooth and flat surface with low roughness (0.487 nm) while the AN treated sample possess sufficient roughness ranging 0.87 to 4.2 nm. The result thus indicates that AN grafted SA membrane possesses better hydrophobic character than pure SA membrane.

Li and co-workers developed a novel (poly (lactic acid)) PLA/Au nanocomposite for the targeted delivery of an anticancer drug, daunorubicin. Self-assembly of drug

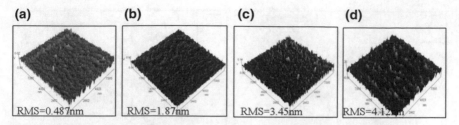

Fig. 16 AFM images of the SA-g-AN film with various contents of acrylonitrile: **a** SA/AN ¼ 1/0; **b** SA/AN ¼ 1/8; **c** SA/AN ¼ 1/11 and **d** SA/AN ¼ 1/14 [35]

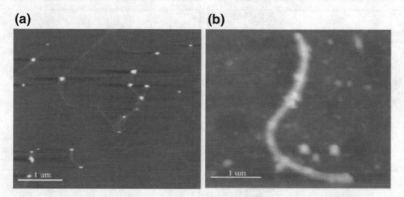

Fig. 17 AFM images of **a** nano PLA/Au polymer nanofibers and **b** PLA/Au nanocomposites conjugated with daunorubicin [36]

and Au onto PLA nanofiber results in the generation of the large spherical particle around the PLA nanofiber and is confirmed by AFM analysis (Fig. 17). This is attributed to electrostatic interaction between positively charged drug with negative PLA/Au nanoparticle. Their studies thus claim that PLA/Au nanoparticle is a potential candidate for the encapsulating anticancer drug, daunorubicin [36].

The aqueous solution of NaOH and polyethylene glycol (PEG) is reported to be an alternative solvent system for urea and thiourea for cellulose. In their recent work, Cernencu et al. [37] adapted this green solvent to develop a novel cellulose/alginate films with different feed ratio (1:1 (CA11) and 3:1 (CA31)) for biomedical engineering. Three different chain length of PEG such as P200, P1000, and P3000 were used to develop cellulose/alginate film and its effect on the morphological properties of the cellulose/alginate film is also studied. 3D and 2D AFM images (Fig. 18) indicates that the longer the chain length of PEG, higher is its compatibility with CA31 films. Consequently, the roughness of CA31 films is found to be low (21 ± 2, 21 ± 3, and 35 ± 4 nm). However, the CA11 film exhibit high roughness (57 ± 4 nm) probably due to its lower compatibility with the films.

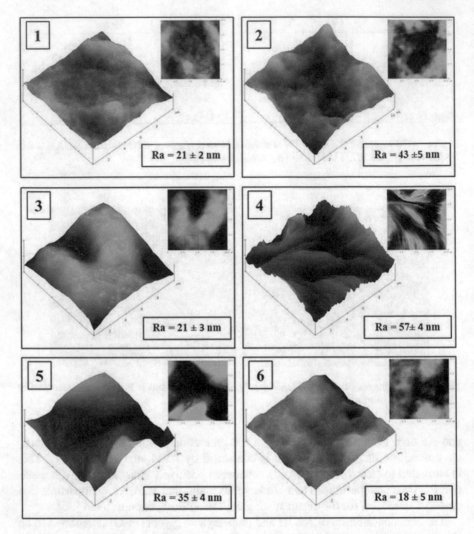

Fig. 18 AFM 3D images and 2D surface topography (inset) of CA films: (1) CA31P3000, (2) CA11P3000, (3) CA31P1000, (4) CA11P1000, (5) CA31P200 and (6) CA11P200. [Reproduced with permission from Springer, License Number: 4671640123406] [37]

5   Transmission Electron Microscopy (TEM)

TEM is one of the most powerful microscopic technique which is widely exploited in research fields such as chemical, biological, nanotechnology, medical and material research. It employed a high energy electron to visualize the specimens and gather information about the composition, morphology, and crystallography of the specimens. The main advantage of TEM is the high magnification of the micrographs.

However, the requirement of vacuum condition restricts its application for the study of living cells [38].

TEM image of a novel bio-nanocomposite chitosan/PVA (polyvinyl alcohol)/ZnO (zinc oxide) is shown in Fig. 19. Absence of agglomerated particle in TEM can be taken as a strong evidence for the proper mixing of the constituent in the nanocomposite. The uniform distribution of zinc oxide (ZnO) nanoparticles in the composite (Fig. 20a, b) allow it to act as an efficient adsorbent against AB 1 (Acid Black 1) [39]. A similar observation was reported by Upadhyaya et al. [40], based on TEM analysis they proved that ZnO nanoparticle are homogeneously distributed in ZnO/CMC (carboxymethyl cellulose) and curcumin loaded ZnO/CMC nanocomposites (Fig. 20a, b). It can also be noted that nanoparticle diameter is almost same (~15 nm) even after encapsulating the nanoparticle with curcumin.

Perotti et al. [41] also employed TEM to confirm the proper dispersion of clay particle in cassava starch-clay nanocomposites. The clay particles can be seen as small dark fringes in the TEM images (Fig. 21). It can also be visualised from figure

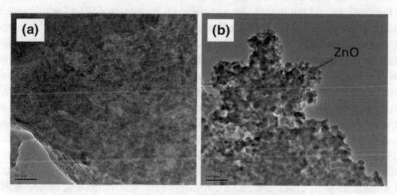

Fig. 19 TEM images of CS/PVA/ZnO samples. [Reproduced with permission from Elsevier, License Number: 4671640401412] [39]

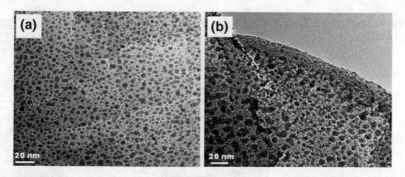

Fig. 20 TEM micrograph of **a** ZnO/carboxymethyl cellulose nanocomposites, **b** curcumin/ZnO/carboxymethyl cellulose nanocomposites. [Reproduced with permission from Springer, License Number: 4671640593301] [40]

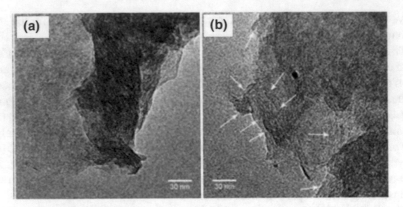

Fig. 21 TEM images of the starch-clay nanocomposite [41]

that even at high clay content, there is sufficient compatibility between biopolymer and clay particles, which makes the membrane a homogeneous and stable one.

Rath and co-workers [42] enhance the strength of chitosan composite by introducing it with multiwalled carbon nanotube (MWCT). They used TEM analysis (Fig. 22) to confirm the attachment of chitosan to CNT. Successful bonding of MWCT to the chitosan surface was visible in the TEM micrograph.

Photodynamic therapy (PDT) is one of the emerging areas of phototherapy which utilize a photosensitiser and light to destroy bacteria. Agel et al. [43] in their recent work encapsulated curcumin in poly(lactic-co-glycolic acid) (PLGA) and found it to be efficient for gram-positive and gram-negative bacteria. The effect of chitosan on the antibacterial efficiency of curcumin/PLGA nanoparticles was also presented in their work. The morphological change in the bacteria (S. saprophyticus and E. coli) incubated with curcumin/PLGA nanoparticle and chitosan/curcumin/PLGA nanoparticle with and without radiation is shown in the Figs. 23 and 24. It can be seen from Fig. 23, that the chitosan modified curcumin loaded nanoparticles (CS CUR

Fig. 22 TEM of MWCNT-chitosan [42]

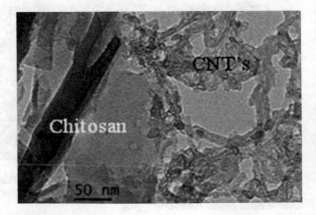

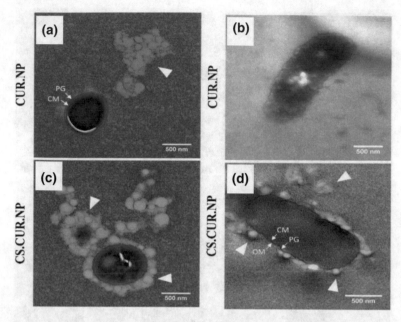

Fig. 23 TEM micrographs of S. staphylococcus (**a** and **c**) and E. coli alpha (**b** and **d**) incubated with CUR.NP CS.CUR.NP before irradiation. White arrowheads represent nanoparticles attached or in close proximity to the bacteria. [Reproduced with permission from Elsevier, License Number: 4671281408293] [43]

NP) show better attachment to the bacterial cell wall than curcumin nanoparticle (CU NP). This is attributed to greater electrostatic interaction between the negative charged bacterial cell wall and positively charged CS CUR NP.

Their studies also revealed that nanoparticle has a prominent effect on the cellular structure of bacteria. Concurrent presence of curcumin nanoparticles (CUR.NP) and chitosan modified curcumin loaded nanoparticles (CS.CUR.NP) and light cause cellular leakage and drastically affect the cellular morphology (Fig. 24).

Tiwari et al. [44] enhance the hydrophilicity and biocompatibility of mesoporous silica nanoparticles (MSNs) by functionalizing it with amino group as well as by grafting it with carboxymethyl cellulose. The nano-dimension of mesoporous silica nanoparticles was confirmed by TEM image (Fig. 25). On introducing amine functionalization, the particle becomes porous (Fig. 25a, b). However, the porosity was reduced in the case of CMC grafted sample.

Ni et al. [45] fabricated a promising tissue regeneration material: polyvinyl alcohol (PVA)/sodium alginate (SA)/nano-hydroxyapatite (nHAP) by electrospinning method. The TEM images were complementary to EDX (energy-dispersive X-ray) and demonstrate the successful incorporation of nHAP particles into the fibers (Fig. 26). Furthermore, (002) lattice planes of hexagonal HAP can also be seen in the TEM micrograph.

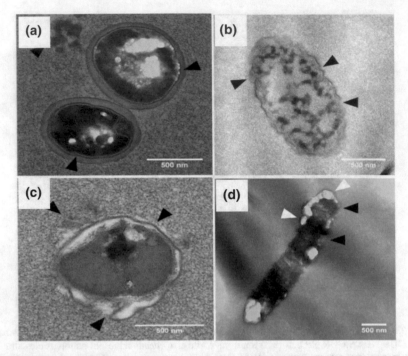

Fig. 24 TEM micrographs of S. staphylococcus (**a** and **c**) and E. coli alpha (**b** and **d**) incubated with CUR.NP or CS.CUR.NP after irradiation. Black arrowheads show the ultrastructural changes of the bacterial membrane. [Reproduced with permission from Elsevier, License Number: 4671281408293] [43]

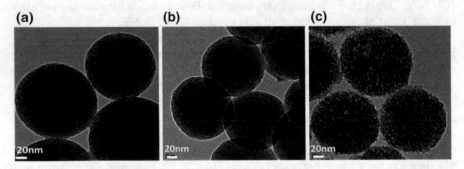

Fig. 25 Transmission electron microscopy (TEM) images of **a** MSN, **b** MSN-NH$_2$ and **c** MSNCMC [44]

6 Optical Microscopy

Optical microscopy (or light microscopy) is the traditional techniques which magnify the object 1500 times greater than human eyes [46]. As this technique utilizes visible

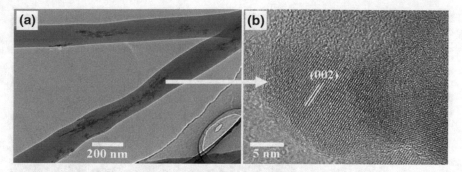

Fig. 26 TEM observation of electrospun PVA/SA/nHAP fiber membrane. [Reproduced with permission from Elsevier, License Number: 4671640730793] [45]

light, the observer can visualize the specimen by direct eyes. The magnification in optical microscope is done using powerful objective lens and eyepiece lens. The light sensitive camera captures this magnified image and generates optical micrograph. Polarizing optical microscope (POM) is an advanced form of optical microscope which uses polarized light to visualize birefringent materials [47, 48].

Optical microscopy has been used to investigate the microstructural information of biopolymer. Ali et al. [49] reported that corn and wheat hulls are potential candidates to reinforce starch film. The optical microscopic images of starch composite with hulls are displayed in Fig. 27. It can be seen that both corn and wheat hulls are uniformly distributed in the starch matrix, without losing their characteristic geometries, crystallinity and particle size.

Ashok et al. [50] utilized optical microscopic technique to visualize the distribution and nature of plant fiber, Thespesia lampas microfibers (TLMFs) in the cellulose matrix. Polarised optical microscopic images revealed that the TLMFs are uniformly distributed in the cellulose composite. It can be also visualized from Fig. 28 that on increasing the filler loading the distribution of TLMF in the matrix also increases.

A novel scaffold was developed from chitosan and f-MWCNT (f-multiwalled carbon nanotube) for orthopaedic application. Based on the optical microscopic analysis (Fig. 29) authors reported that the scaffold has porous nature and the f-MWCNT is homogeneously distributed in the composite [51].

The mechanical properties of PLA were improved by incorporating a biodegradable aliphatic polyester, PBS (polybutylene succinate). PLA/PBS blends with different weight ratios such as 90/10, 70/30, 50/50, and 30/70 were developed and their optical images are shown in Fig. 30. It can be noted that in both PLA (Fig. 30a) and PBS (Fig. 30f) tiny crystals are observed. The PLA crystals are found to be uniformly distributed while the PBS has nonuniform distribution. Furthermore, the crystal size of PLA/PBS blend is found to increases with PBS content, ascribed to the enhancement in the crystallisation of PLA by PBS [52].

Core-shell microspheres are one of the ideal drug carriers which are being exploited in the bioengineering field for the controlled release of the drug. Xu and co-workers in their recent project fabricated double layered alginate microsphere by

Fig. 27 (**A, a**) starch matrix containing wheat hull and (**B, b**) corn hull particles observed under an OM with normal and polarized lights. [Reproduced with permission from Springer, License Number: 4671641032865] [49]

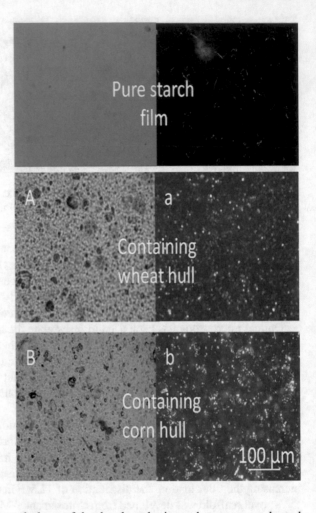

electrospray method. The morphology of the developed microspheres was evaluated by optical microscope and fluorescent microscopy. A uniformly dispersed double layer particles were observed in the optical microscopic image (Fig. 31). The successful coating of three layers of chitosan into the alginate microsphere is observed [53].

7 Scanning Electron Microscopy (SEM)

Scanning electron microscopy (SEM) is a topographic technique which uses a beam of electrons to provide the surface morphology of specimens. The interaction of the electron with samples provide the structural features of the materials. Here the beam of electron hit the sample surface results in the emission of high energy backscattered

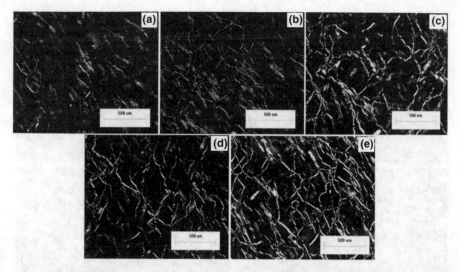

Fig. 28 Polarized OM of cellulose/TLMFs composite films with **a** 1 wt%; **b** 2 wt%; **c** 3 wt%; **d** 4 wt% and **e** 5 wt% TLMFs content. [Reproduced with permission from Elsevier, License Number: 4671290507653] [50]

electrons, low energy secondary electrons, and Auger electrons from the specimen surface [54]. SEM uses back scattered electrons and secondary electrons for sample analysis. The coating of samples with conducting layers is preferred in traditional SEM. However, recent SEM techniques use low energy electron beams which do not require the coating of samples.

The surface morphology of biopolymers can be analysed successfully by the use of SEM. Yang et al. reported the SEM micrographs of whole wheat starch and their A-type and B-type granules (Fig. 32) [55]. They observed that the morphology of both A-type and B-type granules were close to that of wheat starch regardless of their freezing storage. A-type granules showed a disc shape, on the other hand, B-type showed ellipsoid shape. With freezing storage, the surface of A-type granules got a slightly rough appearance, while the B-type granules have irregular edges with the incompleteness of particles were observed. A similar observation was observed for whole wheat starch.

Li et al. reported the SEM images of biodegradable scaffold of chitosan –alginate and chitosan having application in tissue engineering (Fig. 33) [56]. They observed the porous scaffold with a pore size of around 100–300 μm which favours the cell attachment and bone tissue ingrowth. The SEM micrographs reveal the porous interconnectivity at higher magnification.

Suika et al. reported the SEM micrograph of starch granules (potato, corn, rice, and wheat) before and after the modification in ethanol and water using ultrasounds (Fig. 34) [57]. From the images, most of the corn and rice starch granules have an angular shape with an uneven surface, with few granules are entirely smooth ones. But after the ultrasound treatment in water, some granules showed small fissures and

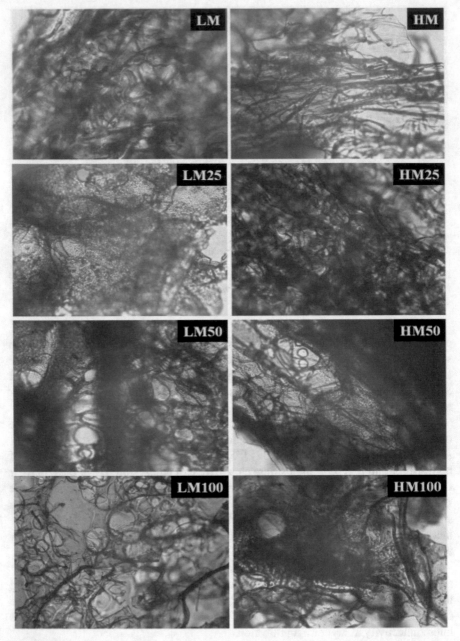

Fig. 29 OM images of chitosan and chitosan/f-MWCNT composite scaffolds (magnification = 40×). [Reproduced with permission from Elsevier, License Number: 4671290793368] [51]

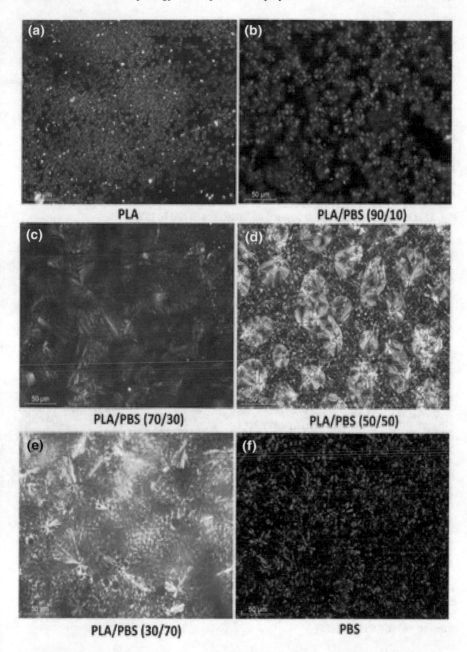

Fig. 30 Optical microscopy images for PLA/PBS blends: **a** PLA; **b** PLA/ PBS (90/10); **c** PLA/PBS (70/30), **d** PLA/PBS (50/50); **e** PLA/PBS (30/70); **f** PBS [52]

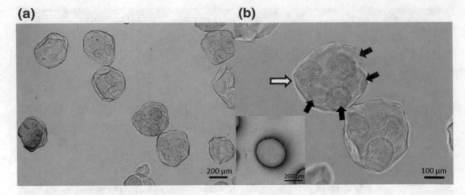

Fig. 31 OM images of the double-layered microspheres (**a** and **b**) [53]

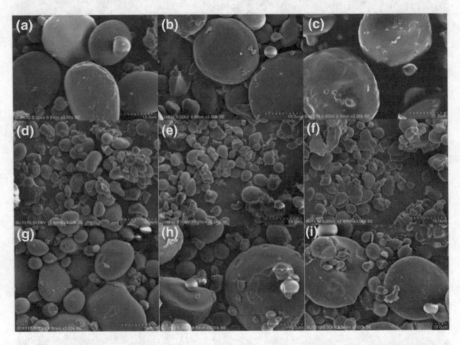

Fig. 32 SEM images of the **a** 0 week A type starch **b** 4 week A type starch, **c** 8 week A type starch, **d** 0 week B type starch, **e** 4 week B type starch, **f** 8 week B type starch, **g** 0 week wheat starch, **h** 4 week wheat starch, **i** 8 week wheat starch. [Reproduced with permission from Elsevier, License Number: 4671640911100] [55]

depressions on the surface. The potato granules were smooth, while wheat granules with numerous small depressions can be observed. After treatment with water, potato granules show cracks and scratches, while after ethanol treatment few fissures are observed. Similar results were observed for wheat granules.

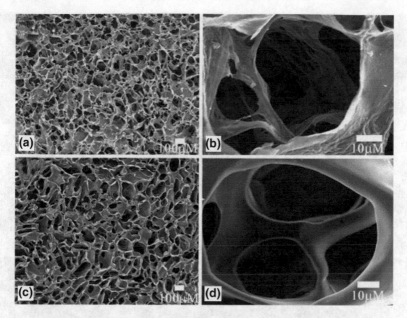

Fig. 33 SEM images of cross section view of chitosan alginate scaffold (**a** and **b**), chitosan scaffold (**c** and **d**). [Reproduced with permission from Elsevier, License Number: 4671291378372] [56]

Moshaverinia et al. reported the SEM images showing the morphology of alginate hydrogel with the application as a scaffold for dental derived stem cells [58]. Figure 35a represents the morphology of alginate hydrogel. The spheroidal stem cells positioned inside the alginate microbeads can be observed from the SEM images (Fig. 35b). The mineral formation by the stem cells after culturing in osteogenic induction medium can be observed from Fig. 35c.

Liu et al. investigated the morphology of pomelo peel cellulose and pomelo peel microcrystalline cellulose [59]. The surface of pomelo peel was smooth and the cellulose bundles can be observed from the Fig. 36a. The appearance of the rough surface of pomelo peel cellulose was imaged in Fig. 36b. In Fig. 36c, shows the surface morphology of pomelo peel cellulose after hydrolysis with acid. Here, the penetration of acid causes the cleavage of β-1,4-linkage between the cellulose and hence shorter pomelo peel microcrystalline cellulose was obtained. The appearance of an ordered crystalline arrangement may be due to the formation of inter and intramolecular H bonding between the hydroxyl groups. The commercially available microcrystalline cellulose has rod-like structures and was longer and smoother than pomelo peel microcrystalline cellulose (Fig. 36d).

Wasserman et al. reported the SEM images of granules of maize starch at different degree of hydrolysis by glucoamylase (Fig. 37). They observed the irregular polygonal shapes of starch granules with sizes of wider distribution and a relatively small portion of granules were oval in shape [60].

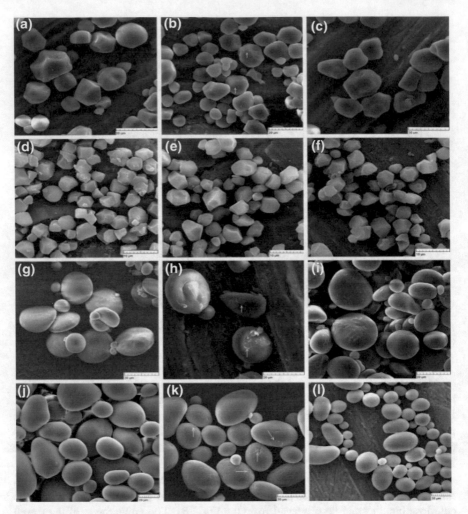

Fig. 34 SEM images of native and modified starch granules. **a**, **d**, **g**, and **j** represents the granules of native corn, rice, wheat and potato starch. **b**, **e**, **h**, and **k** represents the modified granules (water) **c**, **f**, **i**, and **l** modified granules (ethanol). [Reproduced with permission from Elsevier, License Number: 4671641296401] [57]

8 Fourier Transform Infrared Spectroscopy

Fourier transform infrared spectroscopy (FTIR) is a powerful technique for analysing the functional groups presents in organic molecules. The presence or absence of functional groups, protonation states, and the newly formed interactions can be monitored as a function of the intensity of bands and its position [61]. It collects high spectral resolution data that covers over a wide range of spectra. Most of the biopolymers have been successfully studied by using this technique. Several integrated approaches have

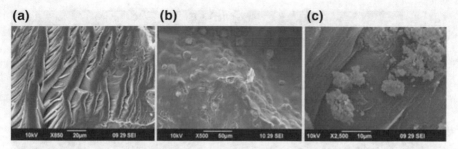

Fig. 35 SEM micrographs of alginate microbeads with encapsulated stem cells. [Reproduced with permission from Springer, License Number: 4671641474458] [58]

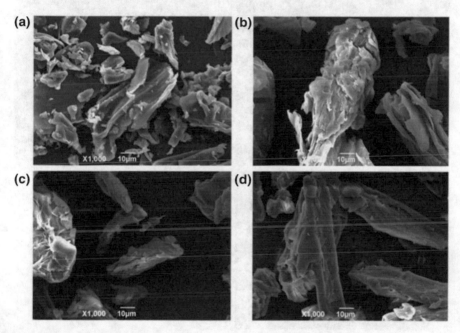

Fig. 36 SEM images of pomelo peel (**a**), pomelo peel cellulose (**b**), pomelo peel micro crystalline cellulose (**c**), and commercially available micro crystalline cellulose (**d**). [Reproduced with permission from Elsevier, License Number: 4671300171825] [59]

been developed in recent years. For example, Nano-FTIR is an integrated approach that utilizes FTIR and scattering-type scanning near field optical microscopy. All the biological samples, polymers, and other soft matters can be analysed and can be compared with standard FTIR databases. The operation is similar to that of non-contact mode in AFM that benefits biological samples [62]. The integration of thermogravimetric analysis (TGA) with FTIR provides the quantitative assessment of samples via thermogram and the identification of the chemical composition of evolved gases by IR spectra [63].

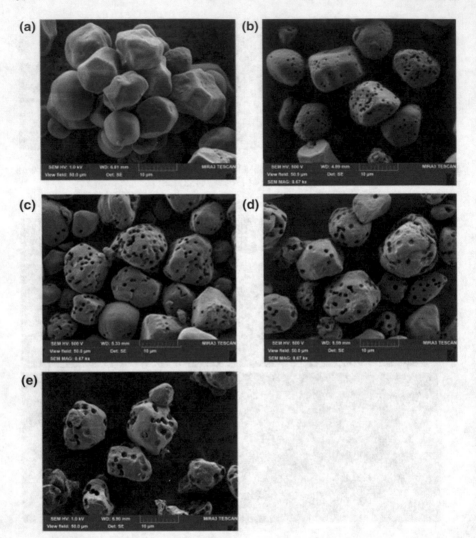

Fig. 37 SEM images of native (**a**) and hydrolysed granules of starch (**b, c, d, e** with degree of hydrolysis at 19.3, 30.4, 40.4 and 52.0%). [Reproduced with permission from Elsevier, License Number: 4671650111568] [60]

Mendes et al. reported the FTIR spectra of thermoplastic starch (TPS), thermo-plastic chitosan (TPC) and its blends (Fig. 38) [64]. For thermoplastic starch absorption band at 920, 1022 and 1148 cm^{-1} corresponds to CO stretching. The band at 1648 cm^{-1} corresponds to bound water. The band at 3277 cm^{-1} is due to the presence of –OH groups, while the band at 2914 cm^{-1} corresponds to CH stretching. For the thermoplastic chitosan films, the band at 3300 cm^{-1} indicate the –OH stretching which overlaps with –NH stretching. The band at 1647 cm^{-1} corresponds to

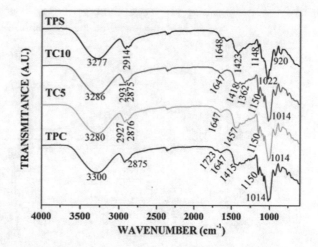

Fig. 38 FTIR spectra of TPS, TPC and TPS/TPC blends [64]

C=O (amide I) stretching. The band at 1717 cm^{-1} shows the presence of carbonyl groups. The bands at 2875, 1415 and 1150–1014 cm^{-1} indicated the presence of –CH stretching, carboxyl and CO groups. For the blends, the bands identical to TPS was observed this is expected since the amount of TPC is blend is small.

Kulig et al. reported the FTIR spectra of chitosan, sodium alginate, and their blends. The FTIR spectra is given in Fig. 39 [65]. The band at 3232 cm^{-1} corresponds to the stretching vibrations of O–H groups in chitosan. The asymmetric or symmetric stretching of CH$_2$ can be observed at 2973, 2933, 2881, 1413, 1310, and 1230 cm^{-1}. The characteristic absorption band of chitosan was observed at 156 m cm^{-1}. The spectra of sodium alginate showed a broad band at 3293 cm^{-1} (O–H stretching) and a weak band at 2926 cm^{-1} (C–H stretching). The authors also discussed the formation of polyelectrolyte complex by blending chitosan, sodium alginate which is indicated by the characteristic peak at 1730 cm^{-1}.

Behera et al. reported the FTIR spectra of pure chitosan and chitosan/titanium dioxide membranes (Fig. 40) [66]. From the FTIR spectra, the immobilization of TiO$_2$ in the chitosan matrix was confirmed by the presence of characteristic peak of TiO$_2$ at 400–700 cm^{-1}.

9 Summary

In recent years, there is a tremendous increase in biopolymer-based research due to environmental awareness and the demand for alternatives to petroleum-based resources. The non-toxic nature, biocompatibility, biodegradability makes them superior over others and have been used in several applications. The extensive research on structural and surface analysis by various microscopic techniques enables a better understanding of biopolymers. Most of the features such as morphology,

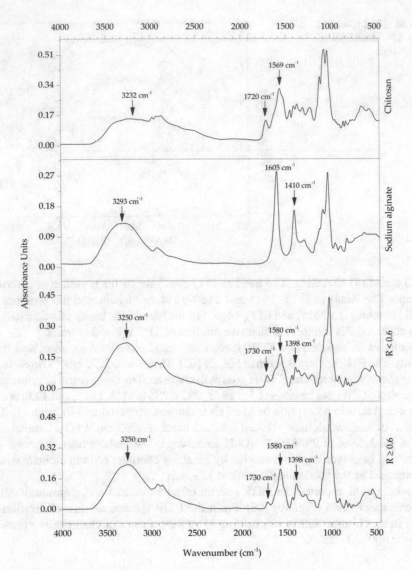

Fig. 39 FTIR spectra of chitosan, sodium alginate and their polyelectrolyte blend [65]

topology, structure, and composition can be analysed by using XRD, NMR, TEM, SEM, OM, AFM and FTIR. Majority of the application studies is based on the structural interpretation which is always in correlation with other properties. Hence understanding the surface and structure of biopolymers by microscopic and others can have immense promises in every field.

Fig. 40 FTIR spectra of titanium dioxide (TiO₂), pure chitosan (CS) and chitosan/titanium dioxide (CS/TiO₂) composite membranes. [Reproduced with permission from Elsevier, License Number: 4671300657703] [66]

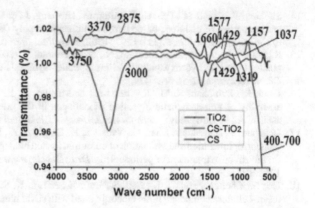

References

1. Grumezescu, A. M. (2017). *Food packaging*. Amsterdam; Boston: AP/Elsevier, 768 pp.
2. Poole-Warren, L., Martens, P., & Green, R. (2015). *Biosynthetic polymers for medical applications*. Elsevier.
3. Grumezescu, A. M. (2016). *Nanobiomaterials in Galenic formulations and cosmetics: Applications of nanobiomaterials*. Amsterdam; Boston: Elsevier/WA, William Andrew is an imprint of Elsevier, 433 pp.
4. Bajpai, P. (2019). *Biobased polymers: Properties and applications* (1st ed.). Cambridge: Elsevier, 250 pp.
5. Das, T. K., & Prusty, S. (2017). Biopolymer composites in field-effect transistors. In *Biopolymer composites in electronics* (pp. 219–229). Elsevier.
6. Epp, J. (2016). X-ray diffraction (XRD) techniques for materials characterization. In *Materials characterization using nondestructive evaluation (NDE) methods* (pp. 81–124). Woodhead Publishing.
7. Chatterjee, A. K. (2000). X-ray diffraction. In *Handbook of analytical techniques in concrete science and technology* (pp. 275–332).
8. Trivedi, M. K., Nayak, G., Patil, S., Tallapragada, R. M., & Mishra, R. (2015). Impact of biofield treatment on chemical and thermal properties of cellulose and cellulose acetate. *Journal of Bioengineering and Biomedical Sciences*. https://doi.org/10.4172/2155-9538.1000162.
9. Wulandari, W. T., Rochliadi, A., & Arcana, I. M. (2016). Nanocellulose prepared by acid hydrolysis of isolated cellulose from sugarcane bagasse. *IOP Conference Series: Materials Science and Engineering,107*, 012045. https://doi.org/10.1088/1757-899X/107/1/012045.
10. Cohuo, S. C. P., Escamilla, G. C., González, A. V., Escamilla, V. V. A. F., & Calderon, J. U. (2018). Production and modification of cellulose nanocrystals from Agave tequilana weber waste and is effect on the melt rheology of PLA. *International Journal of Polymer Science*. https://doi.org/10.1155/2018/3567901.
11. Gupta, K. K., Mishra, P. K., Srivastava, P., Gangwar, M., Nath, G., & Maiti, P. (2013). Hydrothermal in situ preparation of TiO₂ particles onto poly(lactic acid) electrospun nanofibers. *Applied Surface Science,264*, 375–382.
12. Chu, Z., Zhao, T., Li, L., Fan, J., & Qin, Y. (2017). Characterization of antimicrobial poly (lactic acid)/nano-composite films with silver and zinc oxide nanoparticles. *Materials,10*, 659. https://doi.org/10.3390/ma10060659.
13. Shameli, K., Ahmad, M. B., Yunus, W. M. Z. W., Ibrahim, N. A., Rahman, R. A., Jokar, M., et al. (2010). Silver/poly (lactic acid) nanocomposites: preparation, characterization, and antibacterial activity. *International Journal of Nanomedicine,5*, 573–579.

14. Shuhua, W., Qiaoli, X., Fen, L., Jinming, D., Husheng, J., & Bingshe, X. (2014). Preparation and properties of cellulose-based carbon microsphere/poly (lactic acid) composites. *Journal of Composite Materials,48*(11), 1297–1302. https://doi.org/10.1177/0021998313485263.
15. Rajkumar, M., Meenakshisundaram, N., & Rajendran, V. (2011). Development of nanocomposites based on hydroxyapatite/sodium alginate: Synthesis and characterisation. *Materials Characterization,62*(5), 469–479.
16. Usha, R., Jaimohan, S. M., Rajaram, A., & Mandal, A. B. (2010). Aggregation and self-assembly of non-enzymatic glycation of collagen in the presence of amino guanidine and aspirin: An in vitro study. *International Journal of Biological Macromolecules,47*, 402–409.
17. Chauhan, S., Bansal, M., Khan, G., Yadav, S. K., Singh, A. K., Prakash, P., et al. (2018). Development, optimization and evaluation of curcumin loaded biodegradable crosslinked gelatin film for the effective treatment of periodontitis. *Drug Development and Industrial Pharmacy,44*(7), 1212–1221.
18. Lin, S., Chen, L., Huang, L., Cao, S., Luo, X., & Liu, K. (2015). Novel antimicrobial chitosan–cellulose composite films bioconjugated with silver nanoparticles. *Industrial Crops and Products,70*, 395–403.
19. Jiang, B., Li, S., Wu, Y., Song, J., Chen, S., Li, X., et al. (2018). Preparation and characterization of natural corn starch-based composite films reinforced by eggshell powder. *CyTA-Journal of Food,16*(1), 1045–1054.
20. Günther, H. (2013). *NMR spectroscopy: Basic principles, concepts and applications in chemistry*. Wiley.
21. Yin, M., Lin, X., Ren, T., Li, Z., Ren, X., & Huang, T. S. (2018). Cytocompatible quaternized carboxymethyl chitosan/poly(vinyl alcohol) blend film loaded copper for antibacterial application. *International Journal of Biological Macromolecules,120*, 992–998.
22. Sun, Z., Li, M., Jin, Z., Gong, Y., An, Q., Tuo, X., et al. (2018). Starch-graft-polyacrylonitrile nanofibers by electrospinning. *International Journal of Biological Macromolecules,120*, 2552–2559.
23. Haroon, M., Yu, H., Wang, L., Ullah, R. S., Haq, F., & Teng, L. (2019). Synthesis and characterization of carboxymethyl starch-g-polyacrylic acids and their properties as adsorbents for ammonia and phenol. *International Journal of Biological Macromolecules,138*, 349–358.
24. Hao, Y., Chen, Y., Li, Q., & Gao, Q. (2019). Synthesis, characterization and hydrophobicity of esterified waxy potato starch nanocrystals. *Industrial Crops & Products,130*, 111–117.
25. Tan, W., Li, Q., Dong, F., Chen, Q., & Guo, Z. (2017). Preparation and characterization of novel cationic chitosan derivatives bearing quaternary ammonium and phosphonium salts and assessment of their antifungal properties. *Molecules,22*(9), 1438.
26. Burgess, R. (2012). *Understanding nanomedicine, an introductory textbook* (1st ed.). Jenny Stanford Publishing. https://doi.org/10.1201/b12299
27. Haugstad, G. (2012). *Atomic force microscopy: Understanding basic modes and advanced applications*. Wiley.
28. Eaton, P., & West, P. (2010). *Atomic force microscopy*. Oxford university press.
29. Yu, H., & Rahim, N. A. A. (Eds.). (2013). *Imaging in cellular and tissue engineering*. CRC Press.
30. Gaczynska, M., & Osmulski, P. A. (2008). AFM of biological complexes: What can we learn? *Current Opinion in Colloid & Interface Science,13*(5), 351–367. https://doi.org/10.1016/j.cocis.2008.01.004.
31. Bonardd, S., Roble, E., Barandiaran, I., Saldías, C., Leiva, Á., & Kortaberria, G. (2018). Biocomposites with increased dielectric constant based on chitosan and nitrile-modified cellulose nanocrystals. *Carbohydrate Polymers,199*, 20–30.
32. Tang, R., Yu, Z., Renneckar, S., & Zhang, Y. (2018). Coupling chitosan and TEMPO-oxidized nanofibrilliated cellulose by electrostatic attraction and chemical reaction. *Carbohydrate Polymers,202*, 84–90.
33. Ni, P., Ba, H., Zhao, Ga, Han, Y., Wickramaratne, M. N., Dai, H., et al. (2019). Electrospun preparation and biological properties in vitro of polyvinyl alcohol/sodium alginate/nano-hydroxyapatite composite fiber membrane. *Colloids and Surfaces B: Biointerfaces,173*, 171–177.

34. Rhim, J. W. (2004). Physical and mechanical properties of water-resistant sodium alginate films. *LWT-Food Science and Technology,37*(3), 323–330.
35. Yang, L., Guo, J., Wu, J., Yang, Y., Zhang, S., Song, J., et al. (2017). Preparation and properties of a thin membrane based on sodium alginate grafting acrylonitrile. *RSC Advances*, *7*(80), 50626–50633.
36. Li, J., Chen, C., Wang, X., Gu, Z., & Chen, B. (2011). Novel strategy to fabricate PLA/Au nanocomposites as an efficient drug carrier for human leukemia cells in vitro. *Nanoscale Research Letters,6*(1), 29.
37. Cernencu, A. I., Lungu, A., Dragusin, D., Serafim, A., Vasile, E., Ionescu, C., et al. (2017). Design of cellulose–alginate films using PEG/NaOH aqueous solution as co-solvent. *Cellulose,24*(10), 4419–4431.
38. Keyse, R. (1997). *Introduction to scanning transmission electron microscopy* (1st ed.). Taylor & Francis group. https://doi.org/10.1201/9780203749890
39. Kumar, S., Krishnakumar, B., Sobral, A. J. F. N., & Koh, J. (2019). Bio-based (chitosan/PVA/ZnO) nanocomposites film: Thermally stable and photoluminescence material for removal of organic dye. *Carbohydrate Polymers,205*, 559–564.
40. Upadhyaya, L., Singh, J., Agarwal, V., Pandey, A. C., Verma, S. P., Das, P., et al. (2014). In situ grafted nanostructured ZnO/carboxymethyl cellulose nanocomposites for efficient delivery of curcumin to cancer. *Journal of Polymer Research,21*, 550.
41. Perotti, G. F., Tronto, J., Bizeto, M. A., Izumi, C. M. S., Temperini, M. L. A., Lugão, A. B., et al. (2014). Biopolymer-clay nanocomposites: Cassava starch and synthetic clay cast films. *Journal of the Brazilian Chemical Society*, 25, 320–330.
42. Rath, D., Chahataray, R., & Nayak, P. L. (2013). Synthesis and characterization of conducting polymers multi walled carbon nanotube-Chitosan composites coupled with poly (metachloroaniline). *Middle-East Journal of Scientific Research,18*(5), 635–641.
43. Agel, M. R., Baghdan, E., Pinnapireddy, S. R., Lehmann, J., Schäfer, J., & Bakowsky, U. (2019). Curcumin loaded nanoparticles as efficient photoactive formulations against gram-positive and gram-negative bacteria. *Colloids and Surfaces B: Biointerfaces,178*, 460–468.
44. Tiwari, N., Nawale, L., Sarkar, D., & Badiger, M. (2017). Carboxymethyl cellulose grafted mesoporous silica hybrid nanogels for enhanced cellular uptake and release of curcumin. *Gels,3*(1), 8.
45. Ni, P., Bi, H., Zhao, G., Han, Y., Wickramaratne, M. N., Dai, H., et al. (2019). Electrospun preparation and biological properties in vitro of polyvinyl alcohol/sodium alginate/nanohydroxyapatite composite fiber membrane. *Colloids and Surfaces B: Biointerfaces,173*, 171–177.
46. Fujimoto, J. G., & Farkas, D. (2009). *Biomedical optical imaging*. Oxford University Press.
47. Herman, B., & Lemasters, J. J. (Eds.). (2012). *Optical microscopy: Emerging methods and applications*. Elsevier.
48. Di Gianfrancesco, A. (2017). Technologies for chemical analyses, microstructural and inspection investigations. In *Materials for ultra-supercritical and advanced ultra-supercritical power plants* (pp. 197–245). Woodhead Publishing.
49. Ali, A., Yu, L., Liu, H., Khalid, S., Meng, L., & Chen, L. (2017). Preparation and characterization of starch-based composite films reinforced by corn and wheat hulls. *Journal of Applied Polymer Science,134*(32), 45159.
50. Ashok, A., Reddy, K. O., Tian, F. H., & Rajulu, A. V. (2019). Preparation and properties of cellulose/Thespesia lampas microfiber composite films. *International Journal of Biological Macromolecules,127*, 153–158.
51. Venkatesana, J., Ryu, B., Sudha, P. N., & Kim, S. (2012). Preparation and characterization of chitosan–carbon nanotube scaffolds for bone tissue engineering. *International Journal of Biological Macromolecules,50*, 393–402.
52. Qiu, T. Y., Song, M., & Zhao, L. G. (2016). Testing, characterization and modelling of mechanical behaviour of poly (lactic-acid) and poly (butylene succinate) blends. *Mechanics of Advanced Materials and Modern Processes,2*(1), 7.

53. Xu, A., Xu, J., Xiao, L., Li, Z., Xiao, Y., Dargusch, M., et al. (2018). Double-layered microsphere based dual growth factor delivery system for guided bone regeneration. *RSC Advances,8*, 16503–16512.
54. Di Gianfrancesco, A. (2017). Technologies for chemical analyses, microstructural and inspection investigations. In *Materials for ultra-supercritical and advanced ultra-supercritical power plants* (pp. 197–245). Woodhead Publishing. 10.1016/b978-0-08-100552-1.00008-7.
55. Yang, Z., Yu, W., Xu, D., Guo, L., Wu, F., & Xu, X. (2019). Impact of frozen storage on whole wheat starch and its A-Type and B-Type granules isolated from frozen dough. *Carbohydrate polymers,223*, 115142.
56. Li, Z., Ramay, H. R., Hauch, K. D., Xiao, D., & Zhang, M. (2005). Chitosan–Alginate hybrid scaffolds for bone tissue engineering. *Biomaterials,26*(18), 3919–3928.
57. Sujka, M., & Jamroz, J. (2013). Ultrasound-treated starch: SEM and TEM imaging, and functional behaviour. *Food Hydrocolloids,31*(2), 413–419.
58. Moshaverinia, A., Chen, C., Akiyama, K., Ansari, S., Xu, X., Chee, W. W., Schricker, S. R., & Shi, S. (2012). Alginate hydrogel as a promising scaffold for dental-derived stem cells: An in vitro study. *Journal of Materials Science: Materials in Medicine,23*(12), 3041–3051.
59. Liu, Y., Liu, A., Ibrahim, S. A., Yang, H., & Huang, W. (2018). Isolation and characterization of microcrystalline cellulose from pomelo peel. *International Journal of Biological Macromolecules,111*, 717–721.
60. Wasserman, L. A., Papakhin, A. A., Borodina, Z. M., Krivandin, A. V., Sergeev, A. I., & Tarasov, V. F. (2019). Some physico-chemical and thermodynamic characteristics of maize starches hydrolyzed by glucoamylase. *Carbohydrate Polymers,212*, 260–269.
61. Griffiths, P. R., & De Haseth, J. A. (2007). *Fourier transform infrared spectrometry*. Wiley. 10.1002/047010631x
62. Huth, F., Govyadinov, A., Amarie, S., Nuansing, W., Keilmann, F., & Hillenbrand, R. (2012). Nano-FTIR absorption spectroscopy of molecular fingerprints at 20 nm spatial resolution. *Nano Letters,12*(8), 3973–3978. https://doi.org/10.1021/nl301159v.
63. Materazzi, S. (1997). Thermogravimetry–infrared spectroscopy (TG-FTIR) coupled analysis. *Applied Spectroscopy Reviews,32*(4), 385–404. https://doi.org/10.1080/05704929708003320.
64. Mendes, J. F., Paschoalin, R. T., Carmona, V. B., Sena Neto, A. R., Marques, A. C. P., Marconcini, J. M., et al. (2016). Biodegradable polymer blends based on corn starch and thermoplastic chitosan processed by extrusion. *Carbohydrate Polymers, 137*, 452–458.
65. Kulig, D., Zimoch-Korzycka, A., Jarmoluk, A., & Marycz, K. (2016). Study on Alginate-Chitosan complex formed with different polymers ratio. *Polymers,8*(5), 167. https://doi.org/10.3390/polym8050167.
66. Behera, S. S., Das, U., Kumar, A., Bissoyi, A., & Singh, A. K. (2017). Chitosan/TiO_2 composite membrane improves proliferation and survival of L929 fibroblast cells: Application in wound dressing and skin regeneration. *International Journal of Biological Macromolecules, 98*, 329–340.

Properties of Cellulose Based Bio-fibres Reinforced Polymer Composites

M. Ramesh and C. Deepa

Abstract In recent years, both industrial and academia are focusing their attention towards the development of sustainable composites, reinforced with cellulose fibres. To make use of these fibres, the properties of these fibres must be evaluated. In this chapter, the various mechanical, thermal and morphological properties and other characteristics of cellulose fibre reinforced composites (CFRCs) carried out by various researchers have been discussed. Different factors have been addressed to improve the adhesion of the fibre matrix resulting in the improvement of the properties of the bio-composites. The chapter concludes that the CFRCs are one of the new fields of material science for use in various applications ranging from the automotive to the construction industries.

Keywords Cellulose fibres · Bio-composites · Mechanical properties · Thermal properties · Morphological studies · FTIR analysis

1 Introduction

Increasing environmental concerns, together with the drastic decline of fossil fuel, have made people use more renewable and natural resources to offset the development of new materials [1]. The viability of natural problems urged the need to look for new options that could alternate synthetic fibre reinforced composites (SFRCs) with environmental impact [2–5]. It is realized that the reinforcing of fibres into matrix has a few points of interest, particularly the properties of this distinctive materials. Recycling of SFRCs is quite difficult, on the grounds that the detachment of the fibres is very troublesome. Along these lines, SFRCs are frequently arranged in unsuitable routes, for example, landfills or burning which causes serious natural effects [6, 7].

M. Ramesh (✉)
Department of Mechanical Engineering, KIT-Kalaignarkarunanidhi Institute of Technology, Coimbatore 641402, Tamil Nadu, India
e-mail: mramesh97@gmail.com

C. Deepa
Department of Computer Science and Engineering, KIT-Kalaignarkarunanidhi Institute of Technology, Coimbatore 641402, Tamil Nadu, India

© Springer Nature Switzerland AG 2020
A. Khan et al. (eds.), *Biofibers and Biopolymers for Biocomposites*,
https://doi.org/10.1007/978-3-030-40301-0_3

71

The characteristics of cellulose fibres are mainly depend on the physical structure, composition, growing conditions, place of origin, weather conditions, fibre separation methods, etc. [8–11]. Incentives for the use of cellulose fibres as reinforcement are the lightweight potential and properties of CFRCs similar to those of SFRCs [3, 12–14]. Similar to synthetic fibres, one of the benefits of using cellulose fibres is their low density, which makes them excellent basic mechanical properties, better handling and storage, recyclability and good thermal and acoustic insulation [15–20]. Apart from these advantages, CFRCs have several disavantages that limit their application, such as low strength, quality variability, high absorption of moisture, limited temperature of processing, and lower durability and incompatibility between fibres and matrices [17, 21]. However, researchers have come out with a number of methods and treatments through continuous studies to improve CFRC performance. This chapter provides an overview of CFRC's physical, electrical, thermal and other properties. Often discussed are the parameters that control the properties of CFRCs. After that, the microstructures of several most commonly used cellulose fibres and its composites were presented.

2 Properties of CFRCs

Synthetic fibres have a more precise strength compared to cellulose fibres, but they are not environmentally friendly or economical, resulting in the researchers showing a great deal of interest in cellulose fibres [22, 23]. Cellulose fibres are characterized by cellular structures made up of cells comprising crystalline and amorphous regions interconnected by fragments of lignin and hemicelluloses [24]. The final strengths of CFRCs depend on the manufacturing process, apart from the fibre and matrix components. The main objectives to be accomplished in order to develop composites with good mechanical properties are: (i) a homogeneous dispersion of the fibres in the matrix, (ii) a well-balanced interaction between the matrix and the fibres to facilitate fibre pull-out, (iii) a low matrix porosity and (iv) an optimized percentage of fibres: enough to reinforce the material while allowing continuity of the matrix [25]. The stems of the plant show high bending and torsional rigidity, considerable toughness, humid vibration and tolerance to defects [26].

2.1 Mechanical Properties

The mechanical properties of natural fibre-reinforced bio-composites are typically comparable with those of traditional fibre-reinforced composites. It is very difficult to predict mechanical properties of CFRCs. The modulus of elasticity of CFRCs has already been successfully modeled by some authors [27–30]. The mechanical properties of cellulose fibres are mostly relied upon their chemical composition, for example, fibre structure and cellulose quantity, planting region, origination, atmospheric condition and the techniques for fibre removal and storage [31]. Fibre-matrix bonding is a component of several variables, including mechanical interlocking, surface roughness, and polarity. The surface moisture can provide some information on the fibre-matrix bonding [32]. The strength of the materials is constantly shown as far as their mechanical properties, for example, elasticity, flexural strength, compression strength, impact strength and frictional characteristics. These are vital to decide load carrying capacity, particularly extraordinary and heavy loading conditions, directly associated with industrial applications. For last several years, various experiments have been conducted on CFRCs, to describe its attributes, which incorporate elasticity, breaking strain, flexural modulus, impact quality, and so on [33].

Identify structural characteristics and measure the mechanical properties of bio-composites designed for specific conditions of loading [34]. Petioles are given leaves of more than 1 m² in length to support Fig. 1 which is replicated from [35] and must therefore withstand large loads of bending, compression and torsion, especially in the case of wind or rain. Simple composites are then used as models to control and vary specific design features, such as reinforcing fibres' mechanical properties. A rhubarb petiole's cross-section is shown in Fig. 1b demonstrating the distribution in parenchymatic tissue of bark fibre bundles and vascular bundles. The cross-section

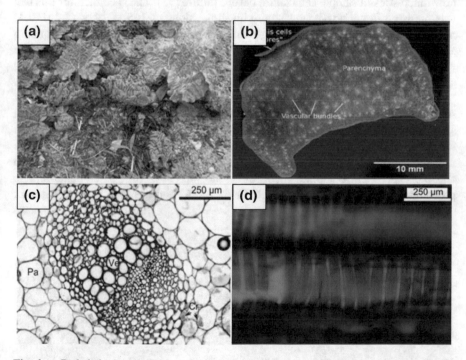

Fig. 1 a Red rhubarb petioles carry large leaves, subject to bending and compressive loads; **b** Rhubarb petiole cross-section; **c** thin cross-section stained with Safranin-Astra blue; and **d** longitudinal thin section along a red rhubarb petiole stained with Safranin-Astra blue

stained with the blue Safranin-Astra shown in Fig. 1c, displaying parenchyma, vessel components, and crystal-like structures in a vascular bundle. Examine the effect of incorporating specific design features of petioles into technical composites and improve the structure–function relationships within the petioles themselves. The thin section in the longitudinal direction a red rhubarb petiole stained with Safranin-Astra blue is depicted along a vascular bundle in Fig. 1d. From the picture, the helically arranged fibres around the xylem vessels are clearly visible in red. It illustrates how a simple spatial arrangement of fibre-like reinforcements with different properties can contribute to a combination of mechanical properties that is otherwise difficult to achieve.

The testing strategy includes settling the specimen in a setup and subjecting it to the load along the axial direction until the point that it breaks. The value is reported as a gauge length expansion function. The gauge segment lengthening is reported against the load applied to the sample during loading [36]. The tensile properties of polymer composites based on wood fibre were investigated and it was found that the tensile strength did not change with the fibre content [37]. Depending on tensile strength and modulus, the elastic properties of structural materials have been overcome. The composite's elasticity depends on how well the strain can be transferred between the damaged fibres and the remaining fibres by shearing the resin at the interface and how much stress a sample can endure before failure [38]. The specimen for this test was prepared in the dog-bone shape with dimensions of $165 \times 19 \times (3.2 \pm 0.4)$ mm^3. The composites' tensile strength and elastic modulus were carried out using a 100 kN universal testing machine (UTM) with a 1 mm/min crosshead speed. The value of the strain was measured using an extensometer attached to the specimen's gauge size. The sample will be shown in Fig. 2 before and after the tensile tests [39].

Flexural properties are primarily used to determine the material's suitability for structural applications by evaluating its flexural strength and modulus. The flexural strength defines CFRC's ability to withstand the bending before the full elastic limit is reached. The three-point flexural test is the commonly used method for composite material testing. The deviation in the sample is taken from the cross-head position and the experiment outcomes incorporate the final strength and elongation. The procedure includes setting the sample in the UTM and the load is applying until

Fig. 2 Tensile strength specimen **a** before and **b** after testing

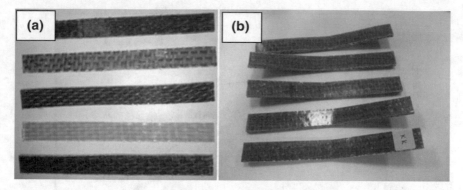

Fig. 3 Flexural strength test samples **a** before and **b** after the test

the point that it failures [36]. These experiments were conducted with 2.5 mm/min cross-head speed, as per the ASTM D790-10 standards. The specimen with the size of $125 \times 13 \times 3.2$ mm^3 were used for testing. The normal distance between the supports was set at 52 mm, with a ratio of 16:1. Mansour et al. tested the flexural properties of Alfa fibre reinforced composites [40] and found that flexural strength and modulus of NaOH treated fibre composites increased twice that of untreated composites. The long fibres gave better properties than the particle fibre composites and the flexural strength relies upon the fibre arrangement and the resin rich areas. The flexural strength tested specimens are depicted in Fig. 3 [39].

The CFRCs impact test is conducted to evaluate the sudden load carrying capacity. It is defined as the energy required per unit area under impact loading to break a specimen [34]. The specimen must be placed in the test set-up during the process and allows the pendulum to hit it until it splits or cracks. The impact rate on break and ductile nature can be investigated by utilizing this method [36]. The resistance due to impact loading of the composite materials can be measured by various strategies, such as: (i) pendulum type test; (ii) dropping weight test; (iii) projectile test; (v) uniform strain rate test; (iv) explosion test; (vi) Hopkinson bar test; and (vii) instrumental pendulum type test [41]. The impact force of composites is calculated using any of the following criteria; (i) the energy required to break the specimen; (ii) the amount of blows required to achieve a specified rate of distress; and (iii) the extent of the damage or the size and speed of the spall after the sample is subjected to stacking of the surface effect [42]. The impact test of piassava fibre reinforced polymer composites were conducted by Cristina et al. [43]. From the experiment, it is found that the impact value is increased with the amount of piassava fibres. The repulsive impact force of silica nano-particles incorporated sisal fibre composites were determined and found that the untreated silica nano-particles incorporated sisal fibres reinforced composites produced high impact resistance and the porosity of composites is minimized considerably [44]. Ray et al. studied the impact behavior of reinforced vinylester composites of untreated and alkali-treated jute fibres [45].

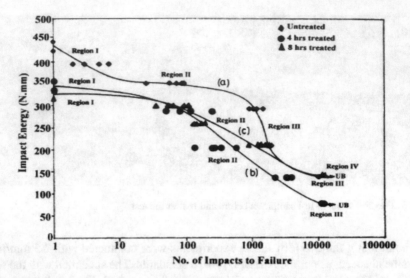

Fig. 4 Impact energy of jute fibre reinforced vinylester composites [45]

Studies have shown that the alkali treated fibre composites to improve the index of crystallinity and result in better dispersion of fibre into the matrix shown in Fig. 4.

2.2 Fatigue Properties

Dissimilar to the investigations made widely on the fatigue properties of engineered fibre composites, almost no consideration was given to the CFRCs [46]. These composites having discovered potential applications as a structural material, an enthusiasm for them has developed as of late. The multi-axial fatigue testing has pulled in the scientists because of a few reasons [47]. For instance, numerous parts in use are in reality subjected several types of loadings which are presented in this test. The weakening of a material's stiffness is defined by most methods used to assess the damage caused by cumulative fatigue. It is observed by following the change in the load of the specimen under a uniform stress, which indicates the calculation of the rigidity of the component in each loading period. The fatigue strength of hemp fibre with polypropylene matrix composites was assessed, and found that the strength is improved and more uniform stress in transient regions [48]. Research on multi-axial fatigue research has been done and a significant influence of shear pressure has been found on a part's fatigue life [49]. In addition, due to the reality of the non-uniform stress concentration on the specimen, the raw data of a specimen under multi-axial loading is also pointed out. Multi-axial tension–torsion fatigue testing details as the ratio between each loading category results in its applications, consolidating the different ratios at that stage to reflect the implications of multi-axial loading [50]. The fatigue life is found to be dependent on the loading ratio between the two types of loading used in multi-axial tests.

2.3 Interfacial Properties

The interfacial relationship at the interface of the fibre and the matrix can be enhanced by modifying the cellulose fibres with alkali solutions [51]. The research on the interfacial characteristics of various plant fibres with polymeric matrices conducted by Gassan [52]. From the test, it is found that the composites reinforced by flax and jute fibre had higher strength and modulus, good fibre-matrix adhesion, and resisted higher critical load during the material crack production. Mylsamy et al. performed interfacial experiments on raw and alkali-treated agave fibre-reinforced composites [53]. From the experiments it is found that the composites treated with alkali solution posses the good interfacial bonding when compared to untreated one and this leads to dissipate more energy during processing.

2.4 Thermal Properties

Thermal investigation analysis are one of the basic important properties, should be considered to recognize the general behavior of CFRCs. In order to determine the moisture content and volatile components current products, this research can be carried out. Therefore, the moisture content and volatile elements have a degrading effect on the properties of any material [54]. Consequently, three techniques were utilized to analyze the thermal properties, to be specific differential scanning calorimetry (DSC), thermogravimetric analysis (TGA), and dynamic mechanical analysis (DMA). From the TGA calculation, a fibre's weight shift can be calculated in a nitrogen atmosphere as a function of rising temperature [55]. Mass changes usually occur during the material's sublimation, evaporation, degradation and electrical transformation that is specifically identified with thermal stability. A few important parameters could then be evaluated from the DSC test, such as glass transition temperature (Tg), melting temperature (Tm), crystalline thickness, and oxidation [56]. Paul et al. [57] tested the thermal properties of banana fibre enhanced polypropylene composites. It is found from the tests that with fibre loading the thermal conductivity and thermal diffusivity are declined. The findings further showed that there were better thermal properties in the increase in NaOH concentration on banana fibre.

2.5   Sound Absorption Properties

The application of this technique is to decide the capacity of composites to absorb ordinary rate sound waves. The estimation of the sound absorption of CFRCs was carried out as per the ASTM E1050 guidelines. The Bruel and Kjaer instrument was utilized for the measurement of sound waves. This instrument incorporates one huge tube with 0.1 m diameter for the lower frequency range; and a little tube with 0.029 m cross section for higher frequency range within 500–6400 Hz. The coefficient of sound absorption versus frequency range were plotted for the average values of the repeated tests within an entire data transmission of the 1/3 band frequency for comparison. A value representing the proportion of the sound wave occurring on the front surface of an acoustically absorbent material to the sound wave transmitted from the back surface is defined as the transmission loss measured in decibels. It represents to the damping behavior of a material, which means the higher, very high sound is absorbed. The ASTM 5285 standards testing technique strategy has been utilized for sound transmission loss testing [58].

2.6   Fourier Transform Infra-Red (FTIR) Spectroscopy Analysis

FTIR provides quantitative and qualitative fibre analysis and detects chemical bonds in a fibre as well [55]. This research was done to verify composite chemical structures. By producing an infrared absorption spectrum, FTIR spectroscopy distinguishes chemical functional groups in a fibre. Using Shimadzu spectrometer (model: FTIR 8400S, Japan) [59, 60] the infrared absorption spectrum of cellulose fibres was observed. It is an important analytical technique to characterize their data on covalent bonding [61]. FTIR tests from the treated flax, hemp and wood fibres showed the presence of acetyl/propionyl and the crystallinity of fibres slowly decreased due to increased amorphous content during the esterification process [62]. The FTIR spectrum also indicates the presence of strong hydrogen bonds on treated fibres, resulting in good mechanical actions of borassus fruit fibres [63]. FTIR spectra indicated that the interfacial bonding of Acacia leucophloea fibres with the epoxy matrix improved by 5% treatment with NaOH, resulting in superior mechanical and thermal properties [64, 65]. The infrared spectrum reveals that alkali-treated Hildegardia lignocellulosic fibres decreased hemicellulose and increased water absorption [66].

2.7   X-Ray Diffraction (XRD) Analysis

XRD analysis was conducted with a X'Pert Pro diffractometer device (powder XRD) followed by the monochromatic frequency of CuKα radiation with a wavelength of

0.154 nm from 10 to 80 °C [59, 60]. The following formula was used to determine the crystallinity index [67, 68]:

$$CI = (H_{22.59} - H_{18})/(H_{22.59}) \tag{1}$$

$H_{22.59}$, H_{18} were heights of the peaks at the respective angles. This analysis confirms that closer cellulose packaging enhances the crystallinity index of cellulose fibres and also improves mechanical properties and break elongation [64, 65]. This states that cellulose microfibrils from fibres show better index compared to raw fibre and regenerated composite cellulose film mainly due to the absence of non-cellulose materials [69]. XRD shows that the surface of the fibres was rough and only after 8 h of alkali treatment showed an increase in crystallinity [66]. It is stated that, due to the absence of amorphous material in the fibres, the crystallite size and index of alkali-treated jute fibres gradually increased over time [70]. It is found that there would be a small decrease in hemp fibre crystallinity but a slightly improved 0.8–30% in sisal, jute and kapok fibres [71].

2.8 Water Absorption Characteristics

Water retention behavior is one of the primary concerns particularly for CFRCs in composite specific applications. A major concern is the water assimilation of CFRCs; in particular for their outdoor applications. The water retention attributes for a given composite material depend on the fibre quantity, fibre orientation, ambient temperature, exposed fibre layer, fibre permeability, void substance, and hydrophilicity nature of fibre and matrix [72–76]. The water assimilation tests were completed as illustrated in the ASTM D570 standards. To begin with, at room temperature, the samples are submerged in a holder filled with tap water. From that point the specimens were pulled back from the water to monitor the mass in the middle of the ageing process, wiped dry to remove any surface moisture, and then measured using a 4-digit weight balancing system of high precision.

As far as CFRCs are concerned, the presence of polar groups in fibres helps them to absorb large amounts of water and also makes them incompatible with the matrix. Together, these two factors result in a reduction in the interfacial binding strength between fibre/matrix resulting in poorer composite mechanical quality [77, 78]. The water assimilation is not fundamentally changed by any handling condition aside from by pressurization at low fibre quantity and increases with increase in fibre volume. Water retention causes the size of the composites to increase and also change the shape of the material. Nonetheless, the small increase in dimensions of the sample is known as swelling of the material [79]. Hemp fibre-reinforced polyester composites' water absorption behavior was analyzed and concluded that increased voids and cellulose content resulted in increased water intake characteristics of the composites [80]. By immersing the specimen in running tap water at room temperature, the water retention behavior of the cellulose fibres was obtained until their water content saturation point was reached. The dry and water-borne specimens were subjected to tensile and flexural loads [39]. The results show that water penetrating the fibre and matrix interface reduced the tensile and flexural strength of the composites by longer water immersion times.

2.9 Wear Behavior

A pin-on-disc setup is utilized to carry out the wear tests and the surfaces of the composite samples of measurements $10 \times 10 \times 20$ mm^3 was contemplated against a surface plate made of stainless steel. Prior to the test, rough sheet of emery paper was utilized to smooth the rough surface then the impurities on the surface were cleaned by a wet cloth. Then the surface was heated by utilizing air at 100 °C for over 15 min. Mahr Perthometer was used to measure the roughness of the exposed surface before and after the test. The specimen contact layer was washed with a different grade of rough paper for excellent contact between the sample and the stainless steel plate and then cleaned with a dry delicate brush. From the analysis, it is observed that the orientation varied in the roughness of the exposed composite surface. The weight balance was used to determine the sample weights before and after the test and then to determine the weight reduction.

The frictional force was determined by a load cell mounted in the middle of an arm and an infrared thermometer was used to measure the temperature of the initial interface and calibration was performed to determine the final temperature of the interface. The infrared thermometer was indicated at the midpoint of the interface between the specimen and the counterface during calibration. Using an external heating source, the counterface was heated up. While the counterface was heated, between the specimen and the counterface a thermo-couple was held. The temperatures from both the infrared and the thermo-couple were registered all the time until the temperature of the interface was about 80 °C. For several times this process has been replicated and the average values have been calculated. The temperatures measured with the aid of the thermocouple were plotted against the infrared ones and the appropriate line was calculated with the calibration equation [81].

2.10 Morphological Properties

Scanning electron microscopy (SEM) is a typical technique to analyze the strength at the attachment between fibre and matrix and the pictures were taken at the interface to see the internal structure. With a specific goal to decide the level of attachment between the fibres and the matrix, the surface morphology of broken samples was inspected [82]. The broken surfaces were coated in chemical solution for 24 h before scanning in the microscope. It is noticed that a microscope can check just the damaged surfaces of the broken samples [83]. The specimen with cracked surfaces was fixed on the table in the microscope during observation [84]. The morphology of the cracked surfaces was observed at the ambient room temperature. The scanned images were utilized to decide the deformities in the composites and to check the outcomes acquired from the optical magnifying lens in the microscope. The measurements for samples utilized for SEM observation were 0.5×1.5 cm^2, at various thicknesses of 0.1–0.5 cm. The observing surfaces of the samples were covered with gold or palladium or blend of both [85]. Then observation was conducted at an accelerating voltage of 2 kV [86]. The objective of this coating was to get some data with respect to filler scattering and bonding quality amongst filler and matrix [87]. The pictures of the broken surfaces of the CFRCs are presented in Fig. 5 [10].

Figure 6 compares the SEM images of a composite facial fracture packed with 5% and 30% of cellulose fibres. In the case of the composite filled with a low concentration of cellulose fibres (5%), a fairly homogeneous structure with uniform fibre dispersion is obtained as expected. However, the existence of certain clusters of fibres is revealed at high fibre content. Their abundance and size appear to increase with the fibre content, resulting in a slightly poor fibre/matrix interface adhesion. After cutting the sample for the facial fracture image, some holes are clearly seen [88].

Pictures of SEM are given in Fig. 7, replicated in Piltonen et al. [89], shows the morphological development of cross-sectional CFRC surfaces developed at different dissolution times. The layer of untreated cellulose is shown in Fig. 7a, the fibrous structure of the starting material is clearly seen. The fibre structure was still clearly

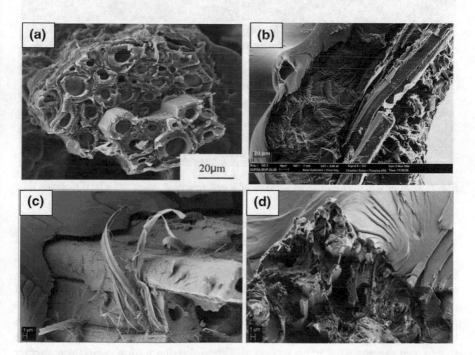

Fig. 5 SEM images of the surfaces of CFRCs after fracture

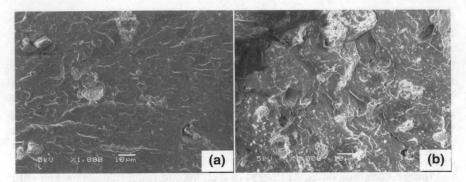

Fig. 6 SEM micrographs of CFRCs with **a** 5%; and **b** 30% cellulose fibre [88]

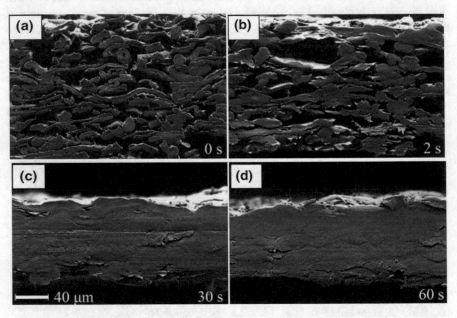

Fig. 7 SEM images of cross-section surfaces of **a** cellulose fibre sheet; and after dissolution time of **b** 2 s; **c** 30 s; and **d** 60 s

visible after a treatment period of 2 s, which suggests that there was no active development of the matrix process during the first 2 s of dissolution. From the SEM images, the thickness of the fibre network has been decreased, resulting in an increased density of CFRCs and hydrogen bonding between fibres. The visible fibre structure had completely disappeared after 30 s of dissolution and it was not possible to see distinctive matrix or fibre reinforcement phases. It indicates a very strong correlation between the fibres and the matrix because the fibres are completely integrated in the matrix and therefore not to be separated. When the dissolution period was further increased from 30 s, the morphology of the CFRCs remained unchanged [89].

Fig. 8 SEM micrographs of **a** elongated epidermis cells; **b** cross-section of a dried rhubarb petiole with several vascular bundles; and **c** the spiral texture of the fibres surrounding the vascular cells

SEM images of the woven composites are shown in Fig. 8 following their fracture under tensile and flexural loading [39, 90]. The fragmented surfaces of the highest composite tensile strength under dry conditions and vice versa for the wet condition were chosen as representative samples to better understand the effect of water absorption on the fibre/matrix bond. Near bonding between the fibre and the matrix, as shown in Fig. 8a, led to improved composite tensile quality under dry conditions in Fig. 8b. The dry test had a less hollow layer relative to the liquid submerged in the sample. This hollowed section revealed the pull-out phenomenon which occurred primarily in water-borne samples, as shown in Fig. 8c. Cracking and delamination matrix as shown in Fig. 8c among the physical damage done by the absorption of water, which further increases the fibre/matrix detachment and decreases the tensile properties of water samples [39].

During impact testing, SEM images of fractured specimens revealed that the helical fibres are stretched and pulled out. In Fig. 8 [35] increase the total pre-failure elongation and provide an additional hierarchical level for energy dissipation during fracture. Helical fibres were absent in bark fibres, which also appeared to be less hollow compared to the vessels, probably increasing Young's apparent fibres modulus, which was calculated using the fibre cross-sectional area [91]. Figure 9 Displays wetted fibre optical images. The increased level of wetting of treated fibre suggests increased interfacial interactions with polymers [92].

3 Conclusion

The primary motivation for the design of bio-composites was and continues to be the construction of a new generation of fibre-reinforced composites, but compliant with the environment in terms of production, use and removal. Much effort has been made to produce these new composites in this direction; however, there are still many technical and economic issues to be addressed before the CFRCs can be successfully commercialized. Different types of cellulose fibre and their properties are studied

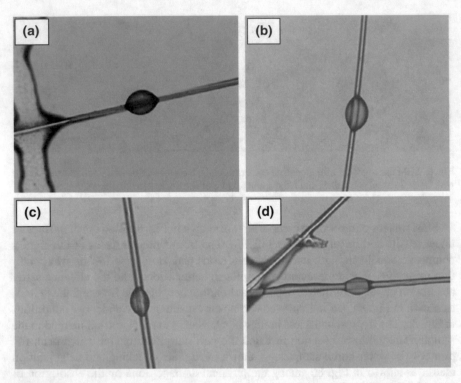

Fig. 9 Optical images of surface morphology of **a**, **b** pristine ($\theta = 51.6°$, $36.6°$) and **c**, **d** treated one wetted by polymer droplets with the scale bar of 20 μm ($\theta = 35.8°$, $21.3°$)

as a possible replacement for synthetic fibres such as glass and carbon fibres. The chapter's key findings are as follows:

(i)   The chemical compositions of cellulose fibres are cellulose, hemi-cellulose, lignin, wax and pectin as the key constituents and small quantities of carbohydrates, starch proteins in different quantities.

(ii) Cellulose fibres have a high variance in the mechanical properties of the conditions of the crop, the geographical and climatic conditions and the variety of plants, agricultural variables such as soil quality, weather conditions, plant maturity and the performance of the retardation process and the conditions of measurement which include or exclude moisture.

(iii) Polymer matrix cellulose fibres composites exhibited promising properties such as compressive properties, effects, tensile properties, flexural properties, thermal properties and resistance.

(iv) CFRCs may therefore be regarded as a valid alternative or even superior to SFRCs.

References

1. Madhu, P., Sanjay, M. R., Senthamaraikannan, P., Pradeep, S., Saravanakumar, S. S., & Yogesha, B. (2017). A review on synthesis and characterization of commercially available natural fibres: Part II. *Journal of Natural Fibers.* https://doi.org/10.1080/15440478.2017.1379045.
2. Bledzki, A. K., & Gassan, J. (1999). Composites reinforced with cellulose based fibres. *Progress in Polymer Science, 24*(2), 221–274.
3. Faruk, O., Bledzki, A. K., Fink, H. P., & Sain, M. (2012). Biocomposites reinforced with natural fibres: 2000–2010. *Progress in Polymer Science, 37*(11), 1552–1596.
4. John, M. J., & Thomas, S. (2008). Biofibres and biocomposites. *Carbohydrate Polymers, 71*(3), 343–364.
5. La Mantia, F. P., & Morreale, M. (2011). Green composites: A brief review. *Composites Part A, 42*(6), 579–588.
6. Conroy, A., Halliwell, S., & Reynolds, T. (2006). Composite recycling in the construction industry. *Composites Part A, 37*(8), 1216–1222.
7. Pickering, S. J. (2006). Recycling technologies for thermoset composite materials-current status. *Composites Part A, 37*(8), 1206–1215.
8. Meghdad, K. M., & Mortazavi, S. M. (2016). Physical and chemical properties of natural fibres extracted from Typha Australis leaves. *Journal of Natural Fibers, 13*, 353–361.
9. Palanikumar, K., Ramesh, M., & Reddy, K. H. (2016). Experimental investigation on the mechanical properties of green hybrid sisal and glass fibre reinforced polymer composites. *Journal of Natural Fibers, 13*(3), 321–331.
10. Ramesh, M. (2016). Kenaf (Hibiscus cannabinus L.) fibre based bio-materials: A review on processing and properties. *Progress in Materials Science, 78–79*, 1–92.
11. Ramesh, M., Deepa, C., Aswin, U. S., Eashwar, H., Mahadevan, B., & Murugan, D. (2017). Effect of alkalization on mechanical and moisture absorption properties of Azadirachta indica (Neem Tree) fibre reinforced green composites. *Transactions of the Indian Institute of Metals, 70*(1), 187–199.
12. Faruk, O., Bledzki, A. K., Fink, H. P., & Sain, M. (2014). Progress report on natural fibre reinforced composites. *Macromolecular Materials and Engineering, 299*(1), 9–26.
13. Furtado, S. C., Araujo, A. L., Silva, A., Alves, C., & Ribeiro, A. M. R. (2014). Natural fibre-reinforced composite parts for automotive applications. *International Journal of Automotive Composites, 1*(1), 18–38.
14. Pilla, S. (2011). Engineering applications of bioplastics and biocomposites—An overview. In *Handbook of bioplastics and biocomposites engineering applications* (pp. 1–15). Wiley.
15. Akin, D. E., Foulk, J. A., Dodd, R. B., & Epps, H. H. (2006). Enzyme-retted flax using different formulations and processed through the USDA flax fibre pilot plant. *Journal of Natural Fibers, 3*, 55–68.
16. Biagiotti, J., Puglia, D., & Kenny, J. M. (2004). A review on natural fibre based composites-Part I. *Journal of Natural Fibers, 1*(2), 37–68.
17. Dittenber, D. B., & GangaRao, H. V. S. (2012). Critical review of recent publications on use of natural composites in infrastructure. *Composites Part A, 43*, 1419–1429.
18. Yan, W., Liang, L., Yunfei, Z., Maoxing, X., & Qing, S. (2013). Techno-economic analysis of fibre reinforced polymer substation. *Architecture, Building Materials and Engineering Management, 1–4*(357–360), 1194–1199.
19. Yan, L., Chouw, N., & Jayaraman, K. (2014). Effect of triggering and polyurethane foam filler on axial crushing of natural flax/epoxy composite tubes. *Materials and Design, 56*, 528–541.
20. Yan, L., Chouw, N., & Jayaraman, K. (2014). Flax fibre and its composites: A review. *Composites Part B: Engineering, 56*, 296–317.
21. Azwa, Z. N., Yousif, B. F., Manalo, A. C., & Karunasena, W. (2013). A review on the degradability of polymeric composites based on natural fibres. *Materials and Design, 47*, 424–442.

22. Obi Reddy, C., Umamaheswari, E., Muzenda, M., Shukla, & Rajulu, A. V. (2016). Extraction and characterization of cellulose from pretreated ficus (peepal tree) leaf fibres. *Journal of Natural Fibers, 13*, 54–64.
23. Samson, R., & Tomkova, B. (2015). Morphological, thermal, and mechanical characterization of Sansevieria trifasciata fibres. *Journal of Natural Fibers, 12*, 201–210.
24. Pereira, P. H. F., Rosa, M. D. F., Cioffi, M. O. H., Benini, K. C. C. D. C., Milanese, A. C., Voorwald, H. J. C., et al. (2015). Vegetal fibres in polymeric composites: a review. *Polimeros, 25*(1), 9–22.
25. Ardanuy, M., Claramunt, J., & Filho, R. D. T. (2005). Cellulosic fibre reinforced cement based composites: A review of recent research. *Construction and Building Materials, 79*, 115–128.
26. Speck, T., & Burgert, I. (2011). Plant stems: functional design and mechanics. *Annual Review of Materials Research, 41*, 169–193.
27. Facca, A., Kortschot, M., & Yan, N. (2006). Predicting the elastic modulus of natural fibre reinforced thermoplastics. *Composites Part A, 37*, 1660–1671.
28. Madsen, B., & Lilholt, H. (2003). Physical and mechanical properties of unidirectional plant fibre composites: an evaluation of the influence of porosity. *Composites Science and Technology, 63*, 1265–1272.
29. Mussig, J., Rau, S., & Herrmann, A. S. (2006). Influence of fineness, stiffness and load displacement characteristic of natural fibres on the properties of natural fibre-reinforced polymers. *Journal of Natural Fibers, 3*, 59–80.
30. Sawpan, M., Pickering, K., & Fernyhough, A. (2007). Hemp fibre reinforced poly (lactic acid) composites. *Advances in Materials Research, 29–30*, 337–340.
31. Ramesh, M., Palanikumar, K., & Reddy, K. H. (2017). Plant fibre based bio-composites: Sustainable and renewable green materials. *Renewable and Sustainable Energy Reviews, 79*, 558–584.
32. Park, J. M., Son, T. Q., Jung, J. G., & Hwang, B. S. (2006). Interfacial evaluation of single ramie and kenaf fibre/epoxy resin composites using micromechanical test and nondestructive acoustic emission. *Composite Interfaces, 13*, 105–129.
33. Ochi, S. (2008). Mechanical properties of kenaf fibres and kenaf/PLA composites. *Mechanics of Materials, 40*, 446–452.
34. Graupner, N., Labonte, D., & Mussig, J. (2017). Rhubarb petioles inspire biodegradable cellulose fibre-reinforced PLA composites with increased impact strength. *Composites Part A, 98*, 218–226.
35. Huber, T., Graupner, N., & Mussig, J. (2009). As tough as it is delicious? A mechanical and structural analysis of red rhubarb (Rheum rhabarbarum). *Journal of Materials Science, 44*(15), 4195–4199.
36. Ramesh, M., Palanikumar, K., & Reddy, K. H. (2013). Mechanical property evaluation of sisal-jute-glass fibre reinforced polyester composites. *Composites Part B: Engineering, 48*, 1–9.
37. Jayaraman, K., & Bhattacharya, D. (2004). Mechanical performance of wood fibre-waste plastic composite materials. *Resources, Conservation and Recycling, 41*, 307–319.
38. Yahaya, R., Sapuan, S. M., Jawaid, M., Leman, Z., & Zainudin, E. S. (2015). Effect of layering sequence and chemical treatment on the mechanical properties of woven kenaf-aramid hybrid laminated composites. *Materials and Design, 67*, 173–179.
39. Maslinda, A. B., Majid, M. S. A., Ridzuan, M. J. M., Afendi, M., & Gibson, A. G. (2017). Effect of water absorption on the mechanical properties of hybrid interwoven cellulosic-cellulosic fibre reinforced epoxy composites. *Composite Structures, 167*, 227–237.
40. Mansour, R., Hocine, O., Abdellatif, I., & Noureddine, B. (2011). Effect of chemical treatment on flexure properties of natural fibre-reinforced polyester composite. *Procedia Engineering, 10*, 2092–2097.
41. Gopalaratnam, V. S., Shah, S. P., & John, R. (1984). A modified instrumented charpy test for cement based composites. *Experimental Mechanics, 24*, 102–111.
42. Balaguru, P. N., & Shah, S. P. (1992). *Fibre-reinforced cement composites*. UK: McGraw Hill Inc.

43. Cristina D. O.N., Ailton, S. F., Sergio, N. M., Regina Coeli, M. P., & Satyanarayana, G., (2012). Studies on the characterization of piassava fibres and their epoxy composites. *Composites Part A: Applied Science and Manufacturing, 43*(3), 353–362.
44. Vieira. L.M.G., J.C.d. Santos, T. H. Panzera, A.L. Christoforo., V. Mano, J.C.C. Rubio., F. Scarpa. (2016). Hybrid composites based on sisal fibres and silica nanoparticles. *Polym. Compos.* doi: 10.1002/pc.23915.
45. Ray, D., Sarkar, B. K., & Bose, N. R. (2002). Impact fatigue behaviour of vinyl ester resin matrix composite reinforced with alkali treated jute fibres. *Compos Part A-Appl. Sci. Manufact., 33*(2), 233–241.
46. Gamstedt, E. K., & Talreja, R. (1999). Fatigue damage mechanisms in unidirectional carbon-fibre-reinforced plastics. *Journal of Materials Science, 34*, 2535–2546.
47. Quaresimin, M. (2015). Multi-axial fatigue testing of composites: from the pioneers to future directions. *Strain, 51*, 16–29.
48. Ferreira, J. M., Silva, H., Costa, J. D., & Richardson, M. (2005). Stress analysis of lap joints involving natural fibre reinforced interface lay ers. *Composites Part B: Engineering, 36*, 1–7.
49. Quaresimin, M., Susmel, L., & Talreja, R. (2010). Fatigue behaviour and life assessment of composite laminates under multi-axial loadings. *International Journal of Fatigue, 32*, 2–16.
50. Amijima, S., Fujii, T., & Hamaguchi, M. (1995). Static and fatigue tests of a woven glass fabric composite under biaxial tension-torsion loading. *Composites, 22*, 281–289.
51. Ramesh, M. (2018). Flax (*Linum usitatissimum L.*) fibre reinforced polymer composite materials: A review on preparation, properties and prospects. *Progress in Materials Science.* https://doi.org/10.1016/j.pmatsci.2018.12.004.
52. Gassan, S. (2002). A study of fibre and interface parameters affecting the fatigue behaviour of natural fibre composites. *Composites Part A: Applied Science and Manufacturing, 33*(3), 369–374.
53. Mylsamy, K., & Rajendran, I. (2011). The mechanical properties, deformation and thermo mechanical properties of alkali treated and untreated Agave continuous fibre reinforced epoxy composites. *Materials and Design, 32*(5), 3076–3084.
54. Franc, P. H., & Vega, M. A. (1997). Effect of fibre treatment on the mechanical properties of LDPE henequen cellulosic fibre composites, *Journal of Applied Polymer Science, 10*, 197–207.
55. Manimaran, P., Senthamaraikannan, P., Murugananthan, K., & Sanjay, M. R. (2017). Physicochemical properties of new cellulosic fibres from azadirachta indica plant. *Journal of Natural Fibers.* https://doi.org/10.1080/15440478.2017.1302388.
56. Julkapli, N. M., & Akil, H. M. (2010). Thermal properties of kenaf-filled chitosan bio-composites. *Polymer-Plastics Technology and Engineering, 49*, 147–153.
57. Paul, S. A., Boudenne, A., Ibos, L., Candau, Y., Joseph, K., & Thomas, S. (2008). Effect of fibre loading and chemical treatments on thermo-physical properties of banana fibre/polypropylene commingled composite materials. *Composites Part A: Applied Science and Manufacturing, 39*(9), 1582–1588.
58. Hao, A., Zhao, H., & Chen, J. Y. (2013). Kenaf/polypropylene nonwoven composites: The influence of manufacturing conditions on mechanical, thermal and acoustical performance. *Composites Part B, 54*, 44–51.
59. Kumar, R., Hyness, N. R. J., Senthamaraikannan, P., Saravanakumar, S. S., & Sanjay, M. R. (2017). Physicochemical and thermal properties of ceiba pentandra bark fibre. *Journal of Natural Fibers.* https://doi.org/10.1080/15440478.2017.1369208.
60. Maheshwaran, M. V., Hyness, N. R. J., Senthamaraikannan, P., Saravanakumar, S. S., & Sanjay, M. R. (2017). Characterization of natural cellulosic fibre from Epipremnum aureum stem. *Journal of Natural Fibers.* https://doi.org/10.1080/15440478.2017.1364205.
61. Ishak, M. R., Leman, Z., Sapuan, S. M., Rahman, M. Z. A., & Anwar, U. M. K. (2013). Chemical composition and FTIR spectra of sugar palm (Arenga pinnata) fibres obtained from different heights. *Journal of Natural Fibers, 10*, 83–97.
62. Tserki, V., Zafeiropoulos, N. E., Simon, F., & Panayiotou, C. (2005). A study of the effect of acetylation and propionylation surface treatments on natural fibres. *Composites Part A, 36*, 1110–1118.

63. Bledzki, A. K., Mamun, A. A., Gabor, M. L., & Gutowski, V. S. (2008). The effects of acetylation on properties of flax fibre and its polypropylene composites. *Express Polymer Letters, 2*(6), 413–422.
64. Arthanarieswaran, V. P., Kumaravel, A., & Saravanakumar, S. S. (2015). Characterization of new natural cellulosic fibre from Acacia leucophloea bark. *International Journal of Polymer Analysis and Characterization, 20*, 367–376.
65. Arthanarieswaran, V. P., Kumaravel, A., Kathirselvam, M., & Saravanakumar, S. S. (2016). Mechanical and thermal properties of Acacia leucophloea fibre/epoxy composites: Influence of fibre loading and alkali treatment. *International Journal of Polymer Analysis and Characterization, 21*(7), 571–583.
66. Varada Rajulu, A., Devi, L. G., Rao, G. B., & Reddy, R. L. (2003). Chemical resistance and tensile properties of epoxy/unsaturated polyester blend coated bamboo fibres. *Journal of Reinforced Plastics and Composites, 22*(11), 1029–1034.
67. Saravanakumar, S. S., Kumaravel, A., Nagarajan, T., Sudhakar, P., & Baskaran, R. (2013). Characterization of a novel natural cellulosic fibre from prosopis juliflora bark. *Carbohydrate Polymers, 92*, 1928–1933.
68. Hyness, N. R. J., Vignesh, N. J., Senthamaraikannan, P., Saravanakumar, S. S., & Sanjay, M. R. (2017). Characterization of new natural cellulosic fibre from heteropogon contortus plant. *Journal of Natural Fibers.* https://doi.org/10.1080/15440478.2017.1321516.
69. Obi Reddy, K., Zhang, J., Zhang, J., & Varadarajulu, A. (2014). Preparation and properties of self-reinforced cellulose composite films from Agave microfibrils using an ionic liquid. *Carbohydrate Polymers, 114*, 537–545.
70. Aparna, R., Sumit, C., Prasad, K. S., Kumar, B. R., Basu, M. S., & Adhikari, B. (2012). Improvement in mechanical properties of jute fibres through mild alkali treatment as demonstrated by utilisation of the Weibull distribution model. *Bioresource Technology, 107*, 222–228.
71. Belouadaha, Z., Ati, A., & Rokbi, M. (2015). Characterization of new natural cellulosic fibre from Lygeum spartum L. *Carbohydrate Polymers, 134*, 429–437.
72. Beg, M. D. H., & Pickering, K. L. (2008). Mechanical performance of kraft fibre reinforced polypropylene composites: influence of fibre length, fibre beating and hygrothermal ageing. *Composites Part A, 39*, 1748–1755.
73. Chow, C. P. L., Xing, X. S., & Li, R. K. Y. (2007). Moisture absorption studies of sisal fibre reinforced polypropylene composites. *Composites Science and Technology, 67*, 306–313.
74. Dhakal, H. N., Zhang, Z. Y., & Richardson, M. O. W. (2007). Effect of water absorption on the mechanical properties of hemp fibre reinforced unsaturated polyester composites. *Composites Science and Technology, 67*, 1674–1683.
75. Doan, T. T. L., Brodowsky, H., & Mader, E. (2001). Jute fibre/polypropylene composites II: Thermal, hydrothermal and dynamic mechanical behavior. *Composites Science and Technology, 67*, 2707–2714.
76. Talavera, F. J. F., Guzman, J. A. S., Richter, H. G., Duenas, R. S., & Quirarte, J. R. (2007). Effect of production variables on bending properties, water absorption and thickness swelling of baggase/plastic composite boards. *Industrial Crops and Products, 26*, 1–7.
77. Haameem, M. J. A., Majid, M. S. A., Afendi, M., Marzuki, H. F. A., Hilmi, E. A., Fahmi, I., et al. (2016). Effects of water absorption on Napier grass fibre/polyester composites. *Composite Structures, 144*, 138–146.
78. Salleh, Z., Taib, Y. M., Hyie, K. M., Mihat, M., Berhan, M. N., & Ghani, M. A. A. (2012). Fracture toughness investigation on long kenaf/woven glass hybrid composite due to water absorption effect. *Procedia Engineering, 41*, 1667–1673.
79. Rassmann, S., Reid, R. G., & Paskaramoorthy, R. (2010). Effects of processing conditions on the mechanical and water absorption properties of resin transfer moulded kenaf fibre reinforced polyester composite laminates. *Composites Part A, 41*, 1612–1629.
80. Dhakal, H. N., Sarasini, F., Santulli, C., Tirillo, J., Zhang, Z., & Arumugam, V. (2015). Effect of basalt fibre hybridisation on post-impact mechanical behaviour of hemp fibre reinforced composites. *Composites Part A: Applied Science and Manufacturing, 75*, 54–67.

81. Chin, C. W., & Yousif, B. F. (2009). Potential of kenaf fibres as reinforcement for tribological applications. *Wear, 267*, 1550–1557.
82. Szopa, J., Kwiatkowska, M. W., Kulma, A., Zuk, M., Telichowska, K. S., Dyminska, L., et al. (2009). Chemical composition and molecular structure of fibres from transgenic flax producing polyhydroxybutyrate, and mechanical properties and platelet aggregation of composite materials containing these fibres. *Composites Science and Technology, 69*, 2438–2446.
83. Abdullah, A. H., Khalina, A., & Ali, A. (2011). Effects of fibre volume fraction on unidirectional kenaf/epoxy composites: The transition region. *Polymer-Plastics Technology and Engineering, 50*(13), 1362–1366.
84. Taib, R. M., Hassan, H. M., & Ishak, Z. A. M. (2014). Mechanical and morphological properties of polylactic acid/kenaf bast fibre composites toughened with an impact modifier. *Polymer-Plastics Technology and Engineering, 53*(2), 199–206.
85. Suriani, M. J., Ali, A., Khalina, A., Sapuan, S. M., & Abdullah, S. (2012). Detection of defects in kenaf/epoxy using infrared thermal imaging technique. *Procedia Chemistry, 4*, 172–178.
86. Sarikanat, M., Seki, Y., Sever, K., & Kahya, C. D. (2014). Determination of properties of Althaea officinalis L. (Marshmallow) fibres as a potential plant fibre in polymeric composite materials. *Composites Part B: Engineering, 57*, 180–186.
87. Cao, X. V., Ismail, H., Rashid, A. A., Takeichi, T., & Huu, T. V. (2012). Maleated natural rubber as a coupling agent for recycled high density polyethylene/natural rubber/kenaf powder bio-composites. *Polymer-Plastics Technology and Engineering, 51*(9), 904–910.
88. Hadjadj, A., Jbara, O., Tara, A., Gilliot, M., Malek, F., Maafi, E. M., et al. (2016). Effects of cellulose fibre content on physical properties of polyurethane based composites. *Composite Structures, 135*, 217–223.
89. Piltonen, P., Hildebrandt, N. C., Westerlind, B., Valkama, J. P., Tervahartiala, T., & Illikainen, M. (2016). Green and efficient method for preparing all-cellulose composites with NaOH/urea solvent. *Composites Science and Technology, 135*, 153–158.
90. Athijayamani, A., Thiruchitrambalam, M., Natarajan, U., & Pazhanivel, B. (2009). Effect of moisture absorption on the mechanical properties of randomly oriented natural fibres/polyester hybrid composite. *Materials Science and Engineering A, 517*, 344–353.
91. Memona, A., & Nakai, A. (2013). Mechanical properties of jute spun yarn/PLA tubular braided composite by pultrusion molding. *Energy Procedia, 34*, 818–829.
92. Ying, Z., Wu, D., Zhang, M., & Qiu, Y. (2017). Polylactide/basalt fibre composites with tailorable mechanical properties: Effect of surface treatment of fibres and annealing. *Composite Structures, 176*, 1020–1027.

Biocomposites from Biofibers and Biopolymers

K. Gopalakrishna, Narendra Reddy and Yi Zhao

Abstract Biobased composites are developed using either the reinforcement and/or matrix from renewable and biodegradable polymers. Although there are plenty of biobased resources available as reinforcements, there are limited numbers of bioresins. Also, the properties of composites developed using biobased resins are not suitable for commercial applications and biobased resins are expensive compared to common synthetic polymer based resins. Considerable efforts are being made to develop biobased resins or modify existing resins to reduce cost and improve performance. In this chapter, we report the latest developments in developing biobased composites classified based on the matrix used. In addition, the performance of the biobased composites under various environmental conditions has also been discussed. Due to the extensive literature available, we have considered studies that are distinct and have been reported in the recent years.

1 Polylactic Acid as Matrix

PLA is one of the most widely used synthetic biopolymer since it is derived from a renewable resource, easily biodegradable and melts between 160 and 170 °C which is convenient to use with a large variety of biomass as reinforcements. In addition to being a resin or matrix for composites, PLA has been used in the manufacture of commodity plastics and also as textile fibers. The structure and properties of PLA have been varied to obtain desired properties and achieve easy processability and hence, PLA is available in different forms and with distinct properties (Table 1). Several biomasses have been combined with PLA to develop biocomposites. Properties of

K. Gopalakrishna · N. Reddy (✉)
Center for Incubation, Innovation, Research and Consultancy, Jyothy Institute of Technology, Thataguni Post, Bengaluru 560082, India
e-mail: narendra.r@ciirc.jyothyit.ac.in

Y. Zhao
Department of Nonwoven Materials and Engineering, College of Textiles, Donghua University, Shanghai 201620, China

© Springer Nature Switzerland AG 2020
A. Khan et al. (eds.), *Biofibers and Biopolymers for Biocomposites*,
https://doi.org/10.1007/978-3-030-40301-0_4

Table 1 Properties of three different forms of PLA used as matrix for composites (Siakeng 2018). Reproduced with permission from Society of Plastics Engineers, through Open Access Publishing

Properties	PDLLA—Poly(D,L-lactic acid)	PDLA—Poly(D-lactic acid)	PLLA—Poly(L-lactic acid)
Structure	Amorphous	Crystalline	Hemicrystalline
Melting temperature T_M/°C	120–170	120–150	173–178
Glass transition (Tg)/ °C	43–53	40–60	55–80
Density (g/cm^3)	1.25	1.248	1.290
Decomposition temperature (°C)	185–200	200	200
Elongation at break (%)	Variable	20–30	20–30
Breaking strength (g/d)	Variable	4–5	5–6
Half-life in 37 °C saline	2–3 months	4–6 months	4–6 months
Solvents	Acetone, ethyl lactate, tetrahydrafuran, dimethylformamide, N,N xylene		Chloroform, furan, dioxaneanddioxolane

PLA containing composites reinforced with different biomasses are mainly dependent on the inherent properties of the biomass (Fig. 1 and Table 2). Relatively poor compatibility between the hydrophobic PLA and hydrophilic biomass also needs to be addressed to achieve desired tensile and flexural properties [1]. However, the availability of raw materials with unique and distinct properties should enable the development of composites with desired properties. As shown in Table 2, PLA has

Fig. 1 Comparison of the tensile stress and strain between composites (densities between 1.299 and 1.305) containing different biomasses used as reinforcement for PLA [1]

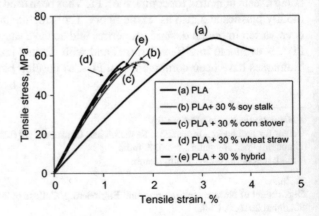

Table 2 Selected list of reinforcements, processing conditions and properties of biocomposites obtained using PLA as the matrix [2]. Reproduced with permission from Society of Plastics Engineers, through Open Access Publishing

Fibers	Fiber (%)	Processing	Tensile strength (MPa)	Young's modulus (GPa)	Flexural strength (MPa)	Flexural modulus (GPa)	Impact strength (kJ/m²)
Chopped recycled newspaper cellulose fiber	30	Injection molding	67.9 ± 0.5	5.3 ± 0.4	106.2 ± 1.8	5.4	23.5 ± 0.4 J/m
Coconut	0.5	Extrusion + compression molding	67.99 ± 3.75	2.37 ± 0.15	102.9 ± 1.3		81.37 ± 1.23 kJ/mm²
Cordenka	25	Injection molding	108	4.2			8.5
Cotton		Compression molding	4.12 ± 2	4.24 ± 0.64			28.7 ± 4.4
Man-made cellulose	30	Injection molding	92	8.032	152	7.89	7.9
Man-made cellulose fibers		Injection molding	92 ± 4.7	5.8 ± 0.15			11.25
Abaca	30	–	74	5.85	124	6.51	5.3
Recycled cellulose fibers	30	Injection molding			82.6 ± 3.8	6.2 ± 0.1	21 J/m
Flax (Random orientation)	30	Compression molding	53	8.3	–		–
Flax (Random orientation)	30	Injection molding	54.15	6.31	–		11.13
Flax (Random orientation)	25	Film-stacking	81.3	8.85	–		–

(continued)

Table 2 (continued)

Fibers	Fiber (%)	Processing	Tensile strength (MPa)	Young's modulus (GPa)	Flexural strength (MPa)	Flexural modulus (GPa)	Impact strength (kJ/m^2)
Cellulose Nanowhiskers freeze dried	5	Chloroform solution casting	37.23 ± 3.2	1.23 ± 0.2			
Hemp(Random orientation)	30	Injection molding	75	7.9	–		–
Hemp (multi-directional)	35	Compression molding	85	12.7	–		7.5
Hemp (multi-directional)	40	Compression molding	44.63	7.39	90		–
Corn stover + wheatstraw + soy stalk (10% + 10% + 10%)	30	Extrusion + injection molding	58	5.55	80	6.9	23 J/m
Jute (Uni-directional)	50	Compression molding	152	5.3	174		–
Jute (multi-directional)	40	Film-stacking	93.5	–	–		14.3
Jute (Random orientation)	30	Compression molding	48	–	101		8.3
Rice hull/CA (CA:PLA-grafted Maleic anhydride)			26.7 ± 1.4	2.76 ± 0.11	28.8 ± 3.14	1.63 ± 0.09	48.7 ± 4.16

(continued)

Table 2 (continued)

Fibers	Fiber (%)	Processing	Tensile strength (MPa)	Young's modulus (GPa)	Flexural strength (MPa)	Flexural modulus (GPa)	Impact strength (kJ/m²)
Kenaf (multi-directional)	40	Compression molding	82	7.6	126		14
Kenaf	70	Hot pressing	223	32	254	35.5	9.5 ± 1.3
Kenaf (Uni-directional)	35	Compression molding	131	15	160		–
Kenaf (Random orientation)	30	Compression molding	32	4.5			–
Kenaf	20	Injection molding			90 ± 1.99	4.5 ± 0.34	34 ± 2.98 J/m
Ramie (Random orientation)	30	Compression molding	52.5	–	170		10
Ramie (Random orientation)	30	Compression molding	52	–	105		9.3
Sisal (Random orientation)	30	Injection molding	–	–	97		3.3
Sisal	30	–	23.3 ± 0.3	3.5 ± 0.07			
Sugarbeet pulp + polymeric diphenyl methane diisocyanate (2%)	30	Extrusion + injection molding	61.1	5			3.25
Wood flour	30	Injection molding	58.28 ± 1.83	6.22 ± 0.54			35.96

Table 3 Properties of composites containing various levels of calcium hypophosphite (CaHP) as the flame retardant [3]

Sample	Composition (%)		LOI	UL-94 rating		
	PLA	CaHP		t1/t2	Dripping	Rating
PLA	100	0	19.5	–	Y	–
PLA/5% CaHP	95	5	24	12.9/2.0	Y	V-2
PLA/10% CaHP	90	10	25	11.7/2.2	Y	V-2
PLA/15% CaHP	85	15	25.5	13.4/2.2	Y	V-2
PLA/20% CaHP	80	20	25.5	6.9/2.6	Y	V-1
PLA/25% CaHP	75	25	26	5.3/3.0	Y	V-1
PLA/30% CaHP	70	30	26.5	0.9/4.8	N	V-0

been extensively used as matrix with a wide variety of biobased reinforcements including recycled paper [2]. However, PLA is considerably expensive than commodity synthetic polymers and also needs chemical and/or physical modifications to impart the desired properties to composites.

PLA is prone to combustion and tends to melt and drip making it unsuitable for some applications [3]. To overcome this limitation, PLA was combined with calcium hypophosphite to improve the flame retardant properties. After combining various amounts (5–30%) of the flame retardants, the mixture was pelletized and later hot pressed into sheets at 10 MPa for 10 min. Inclusion of the flame retardant led to increase in residual mass from 0.2 to 28% after burning at 800 °C. Flame resistant ratings of PLA composites changed from V-2 to V-0 when the hypophosphite content increased from 5 to 30% (Table 3). Similarly, the enthalpy decreased from 25.2 to 19.2 J/g and melting temperature increased from 167 to 170.8 °C. However, the mechanical properties of the composites decreased steadily with increasing content of CaHP.

2 Biobased Resins as Matrix

2.1 Bioepoxies

Synthetically derived epoxy is one to the most studied and commercially used binder for developing composites. However, epoxy is non-biodegradable, needs high concentrations to provide desired properties and is also relatively expensive. Hence, several attempts have been made to develop biobased resins as alternative to epoxy. A bioepoxy resin having viscosity between 200 and 500 mPas S, gel time of 40 min, flexural modulus of 3310 MPa, flexural stress of 94 MPa was used as a binder for hemp fabric. Composites were developed through the resin transfer molding approach and

Table 4 Comparison of the tensile properties of biobased resins reinforced with chicken feathers pyrolyzed to various extents [5]. Reproduced with permission from John Wiley and Sons

Fiber type	Fiber content (%)	Energy absorption (kJ/m^2)	Fracture stress (MPa)	Fracture strain
Pure matrix	0	0.600	0.50	0.051
Untreated feathers	5	2.202	1.86	0.055
Untreated feathers	32	6.966	5.30	0.048
Pyrolyzed feathers-2 h	5	1.377	1.27	0.049
Pyrolyzed feathers-2 h	8	2.733	2.18	0.057
Pyrolyzed feathers-2 h	19	4.260	3.73	0.053
Pyrolyzed feathers-2 h	32	7.566	6.22	0.052
Pyrolyzed feathers-10 h	5	1.243	1.46	0.036

were later subjected to various post-treatments to improve properties. Water absorption of the composites was at a maximum of 11% after immersion for about 800 h. The tensile strength of the composites was 63 MPa and modulus was 5870 MPa and flammability was between 14.28 and 14.85 mm/min. The water absorption and the flammability of the hemp-epoxy composites was considered to be non-satisfactory and needed further improvement [4]. Other than fibers and cellulosic materials, thermally treated chicken feathers have been used along with epoxydized soybean oil and methacrylated lauric acid as the matrix to develop completely biodegradable composites [5]. Pyrolysis of the feathers was done by treating at 215 °C for 2 h or 10 h under nitrogen atmosphere. Later, the reinforcing material was combined with a 50/50 ratio of the two bioresins and converted into composites having densities between 1.03 and 1.10 g/cm^3 using the resin transfer molding approach. Tensile properties of the composites were dependent on the extent of pre-treatment of the feathers (Table 4). A 15 times increase in tensile and storage modulii could be achieved using 32% pyrolyzed feathers. It was suggested that the composites could be suitable for light-weight and inexpensive applications.

2.2 Vegetable Oil Based Resins

Biobased thermoset resins were synthesized by the reaction between epoxidized soybean oil and methacrylic acid. The modified soybean oil based resin was obtained after heating the soybean oil for 12 h with excess methacrylic acid at 120 °C. This resin was further reacted with methacrylic anhyride or acetic anhydride for 4 h at 69 °C to obtain MMSO or AMSO, respectively [6]. These resins were used as matrix for flax fiber mats which were treated with 4% sodium hydroxide solution for 1 h. Composites were formed using the hand lay-up technique with a fiber to resin ratio of 60:40. In addition, polystyrene was included in some of the composites to increase the compatibility between the matrix and reinforcement. Flexural properties of the composites varied considerably depending on the type of matrix and reinforcement used (Fig. 2). Composites containing MSO and AMSO resins had higher impact strength but low thermal stability. Lignin obtained as a byproduct during processing of sugarcane bagasse was chemicallly modified using formaldehyde and made into a bioresin. Sisal fibers used as reinforcement were randomly added (30–70%) into the bioresin and mechanically stirred for 20 min. Later, the bioresin and fibers were cured at different temperature, pressure and time to form the composites [7]. Composites containing 40% sisal fibers had highest impact strength of 500 J/m and the strength decreased with further increase in fiber content. Morphological analysis did not show any fiber pullouts suggesting good compatibility between the matrix and fibers. Properties obtained in this study were considerably higher compared to previous reports which was suggested to be due to the presence of lignin in the matrix and fibers which would improve the adhesion between the two components. Lignin based resins would not only be biodegradable but will also be inexpensive compared to synthetic based resins. In a similar approach, tannins obtained from *Acacia mimosa* were modified using formaldehyde and used as matrix for sisal fibers [7]. Properties of the sisal fiber reinforced tannin-phenolic resin composites were betweeen 250 and 400 J/m and the composites also had good stiffness and lower loss modulus. Moisture sorption of the composites increased to about 20% when 70% resin was used. Good interaction between the resin and matrix was also observed and it was suggested that high level of biodegradability could be observed using this system. Since resins prepared from epoxidized soybean oil have inferior properties, a study was done to

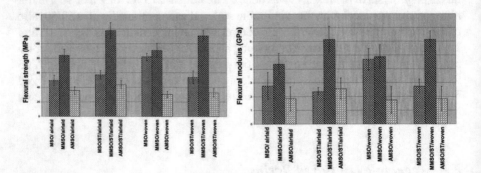

Fig. 2 Flexural strength and modulus of composites developed using various soybased bioresins as matrix and flax fibers and fabrics as reinforcement [6]. Reproduced with permission from John Wiley and Sons

Table 5 Comparison of the properties of epoxidized soybean oil resin modified using various levels of DGEBA [8]

DGEBA (%)	Tensile Strength (MPa)	Tensile modulus (MPa)	Elongation (%)	Flexural strength (MPa)	Flexural modulus (MPa)
0	22.6	648	7.3	38.3	775
10	33.1	735	6.4	61.6	1285
20	47.6	787	6.1	83.0	2004
30	41.7	841	4.9	90.1	2406

improve soy oil resin properties by the addition of diglycidylether of bisphenol-A (DGEBA). Extent of improvement in the properties of the bioresin was dependent on the amount of DGEBA. As seen from Table 5, high amounts of DGEBA imparts brittleness but increased glass transition temperature, thermal stability and crosslink density were possible [8].

In another study on developing biobased resins using lignins [9], various phenolics were selected and blended in different ratios and later subject to methacrylation to obtain methacrylated lignin-based bio-oil mimic (MBO). The MBO was used as low viscosity (cP of 30.3 at 25 °C) vinyl ester resin or as a blend with the regular epoxy resins. Average molecular weight for the MBO was 217 g/mol compared to 104 g/mol for styrene. MBO had a initial degradation temperature of 306 °C, maximum degradation temperature of 418 °C, glass transistion temperature of about 115 °C and storage modulus of about 2.5 GPa at 25 °C. The resin was able to form transparent thermosets and considered to be suitable as replacement for petroleum and vinyl ester based thermoset resins used for composite applications [9]. In another approach, lignin has been blended with poly(lactic acid) through graft polymerization. The chain length of PLA grafted onto lignin was controlled using different preacetylation treatments and by varying the lignin/lactide ratio [10]. Grafted copolymers had higher glass transistion temperature from 45 to 85 °C when lignin content was increased from 10 to 50%. Nearly all of UV C and UV B rays were blocked at a lignin content of 10%. Composites also had increased tensile strength and elongation without change in modulus. Tough and shape memory biobased resins were synthetisized by the reaction between castor oil and hyperbranched polyurethane in the presence of graphene oxide [11]. Addition of GO increased mechanical properties (Table 6) and also shape memory recovery of up to 99% (Fig. 3).

Table 6 Mechanical properties of the castor oil based hyperbranched polyurethane containing various levels of graphene oxide [11]. Reproduced with permission from Royal Society of Chemistry

Sample	HPU	HPU + GO0.05	HPU + GO1.0	HPU + GO2.0
σ_B (MPa)	7.06 ± 1.3	11.23 ± 2.2	14.05 ± 2.8	16.11 ± 3.1
E (MPa)	2.84 ± 0.2	4.2 ± 0.25	4.85 ± 0.37	6.55 ± 0.32
ε_B (%)	695 ± 43	735 ± 57	795 ± 36	810 ± 46
T (MJm^{-3})	2540 ± 128	4247 ± 164	5845 ± 108	6807 ± 211
Scratch hardness (kg)	5 ± 0.2	5.5 ± 0.1	5.5 ± 0.2	6.5 ± 0.2
Impact strength (cm)	>100	>100	>100	>100

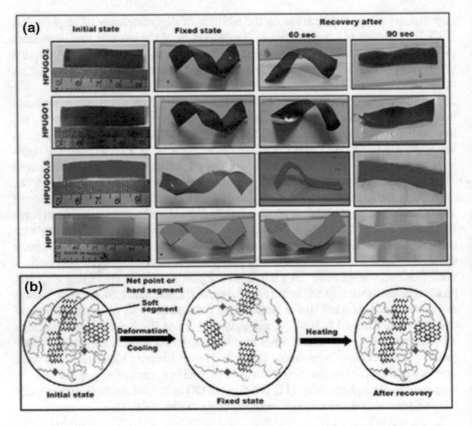

Fig. 3 Behavior and extent of recovery of nanocomposites prepared from biobased polyurethane and graphene oxide [11]. Reproduced with permission from Royal Society of Chemistry

2.3 Polysaccharides and Lignocelluloses as Resins

In a study that does not involve any thermoplastic biopolymer, sunflower stalks were powdered and combined with chitosan solution and made into a composite. The stalks were made into particles of about 1.6–6.3 mm and combined in 4.3–15.3 wt% with chitosan. The mixture was placed in a PVC mold and compacted at 20 °C for 1 min at pressure between 1×10^{-3} and 32×10^{-3} MPa and later dried in a oven at 50 °C for 50 h. Mechanical properties and thermal conductivity of the composites were studied and it was reported that the properties were similar to that of other biobased material (Table 7) in use. A acoustic coefficient of 0.2 considered low for insulation applications was reported for the 4.3% chitosan containing composites [12]. In another study, hemp shives having length between 0–5 and 0–20 mm were combined with wheat starch having a density of 453 kg/m³. To prepare the starch binder, the starch was heated in 60 °C water until gelatinization. The hemp shives were combined with starch and compressed under a pressure of 0.25 MPa to form the composites. Porosity and mechanical properties of the composites were dependent on the proportion of the hemp shives. With density of the composites being between 182 and 188 kg/m³, the tensile strength of the composites ranged between 0.08 and 0.11 MPa and compressive strength varied from 0.57 to 0.63 MPa. Comparitively, the modulus ranged from 2.04 to 2.47 MPa. The specific heat capacity for the composites was found to be 1264–1288 J/Kg K but was directly dependent on the temperature. Compared to lime as matrix, the wheat starch based composites had lower moisture buffer values 2.6–2.8 g/(m² %RH) suitable for indoor applications [13]. To enhance the properties of hemp shive composites, starch as matrix was modified using alkali and a silane coupling agent [14]. For the modification, hemp shives were treated with 1 and 6% NaOH for 72 h and later dried at 60 °C for 48 h. Similarly, the shives were immersed in a aqueous solution of (3-glycidyloxypropyl) trimethoxysilane as

Table 7 Properties of composites made from sunflower stalk particles as reinforcement and chitosan as matrix [12]. Reproduced with permission from Elsevier

Properties	Compaction (10^{-3} MPa)	Size of particles (mm)	Chitosan/sunflower ratio (%)
Tensile modulus (MP)a	574	3.1	15.38
Tensile stress (MPa)	574	6.3	15.38
Thermal conductivity (Wm−1 K−1)	574	6.3	–
Compressive modulus (MPa)	574	6.3	4.38
Compressive strength (MPa)	574	6.3	15.38
Thermal conductivity ($Wm^{-1} K^{-1}$)	1	6.3	4.38

the coupling agent at pH 2.9 for 15 min. The hemp shives were combined with the binder in 60/40 ratio and compacted under a pressure of 0.25 MPa to form the composites. Treating the composites with alkali and compatibilizer increased the flexural strength from 0.15 to 0.25 MPa and compressive strength from 0.4 to 0.8 MPa. It was suggested that the chemical treatments cause the formation of covalent bonds between the starch and hemp leading to better composite properties [14].

In another study, particles made from hemp shive having density between 100 and 110 kg/m^3 and particle size of 4 mm were used as reinforcement with wheat straw as matrix. To prepare the composites, the wheat straw was combined with hemp with dry material to water ratio of 1 and compressed at 0.25 MPa at 180 °C for 100 min [15]. The process of hydrothermal treatment of the straw results in the removal of water soluble components such as pectins which become the binders for the hives. Compressive strength of the composites varied from 260 to 330 kPa and elastic modulus between 2.6 and 3.4 MPa depending on the ratio of the binder and reinforcement. The thermal conductivites of the composites were between 66.8 and 69.3 mWm^{-1} K^{-1} in the dry condition and between 71.4 and 75.9 mWm^{-1} K^{-1} when humidity was 50% similar to the specification for building materials (65 mWm^{-1} K^{-1}). Hemp shive particles were combined with corn starch having a amylose content of 40 and 26% moisture. The particles were mixed with starch in 10–50 wt % and heated up to 100 °C for 2 h and later hardened by treating at 160 °C for 1 h to form the composites. Compressive and bending strength of the composites varied from 2.5 to 2.6 MPa and 5.2 to 6.0 MPa suitable for use in softboards and self bearing structural materials [16]. In a similar approach, corn stalk particles were used as reinforcement for epoxy resin and gypsum. Thermal conductivity of some of the samples was below 0.1 W m^{-1} K^{-1} suitable for building insulation applications. Compressive strength ranged between 0.1 and 0.29 MPa and water absorption and flexural strength between 0.04 and 0.13 MPa. It was suggested that corn stalk reinforced epoxy composites could be useful as alternatives to gypsum based insulation materials [17]. Two polyamides derived from biological sources and commercially sold as PA 6.10 (vestamid Terra HS) and PA 10.10 (Vestamid Terra DS) were used as matrix for cellulose fibers having diameter of 12 µm and strength, elongation and modulus of 833 MPa, 13% and 20 GPa, respectively. The two polyamides had biodegradable content of 62 and 100%, respectively compared to 0% for PA 6 (Ultramid B27E). A two-step compounding process was used to combine various ratios of the matrix and reinforcement and later injection molded at 190–240 °C depending on the type of matrix [18]. Scanning electron images (Fig. 4) of the impact test fractured samples showed that the glass fibers had fewer breakages and lesser fiber pull outs compared to the biobased polyamides due to the differences in polarity. Tensile strength and tensile modulus of the biopolyamide composites (Fig. 5) was similar but energy absorption was higher when compared to that of the synthetic polyamide.

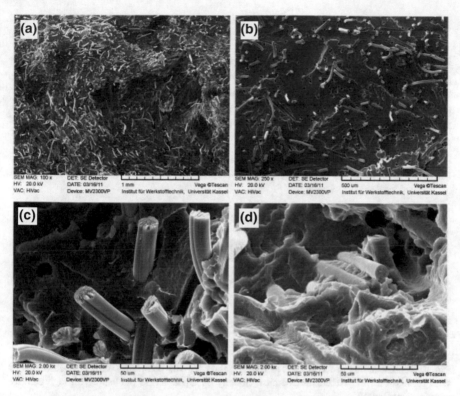

Fig. 4 SEM images of biopolyamide cellulose fiber composites containing 30 and 15% cellulose shows fiber breakage and pull outs (**a**, **b** and **c**) compared to the glass fiber composites (**d**) [18]. Reproduced with permission from Elsevier

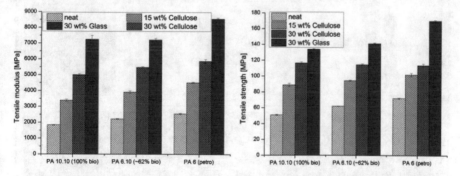

Fig. 5 Tensile modulus and strength of biopolyamide-cellulose fiber composites compared to synthetic polyamide composites [18]. Reproduced with permission from Elsevier

2.4 Proteins as Resins

In a unique approach, carbon nanofibrils prepared from norwegian spruce and scots pine softwood were obtained with length of 400–600 nm and height of 2–3 nm. These nanofibrils were combined with various ratios of silk fusion proteins (Z-silk and FN-silk) obtained from recombinant spider silk. Composite films and fibers of CNFs and silk were made by solution casting (Fig. 6) and by using a double flow focusing geometry device (Fig. 7), respectively. Carbon nanofibrils had storage modulus of about 350 Pa which increased to 450 Pa with the addition of silk. Although there was no significant increase in Young's modulus, the tensile stress increased from about

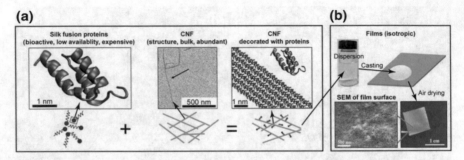

Fig. 6 Representation of the process of forming the CNF-silk composite films [19]. Reproduced with permission from American Chemical Soceity by Open access publishing through creative commons attribuiton

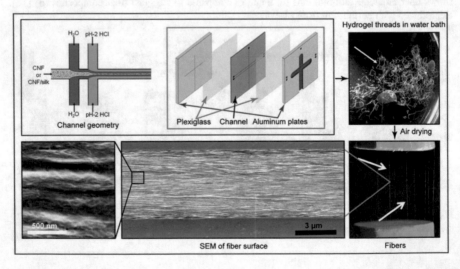

Fig. 7 Depiction of the formation of CNF-silk fiber composites and SEM image and highly oriented nature of the fibers in the composites [19]. Reproduced with permission from American Chemical Soceity by Open access publishing through creative commons attribuiton

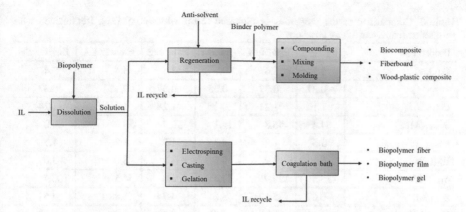

Fig. 8 Schematic representation of the possibilities of using ionic liquids to develop various types of biocomposites [20]

830–1100 MPa when Z-silk was added into the nanofibrils [19]. The mechanical properties of self assembled carbon nanofibrils obtained in this study were reported to be the highest compared to previous reports.

2.5 *Ionic Liquid Processed Biocomposites*

Ionic liquids have been extensively used to dissolve biopolymers to develop materials for a wide range of applications. Relatively low cost, short dissolution time and easy recovery of the solvent are some of the advantages of using ILs. Several studies have also been conducted on dissolving or modifying biopolymers using ionic liquids for developing biodegradable composites. A schematic representation of the various processes used to develop biocomposites after treating with the ionic liquids is shown in Fig. 8.

3 Performance of Biodegradable Resins and Composites

Ability to resist moisture at different temperatures and humidities is one of the most basic criteria for biopolymers and biocomposites. Since biobased materials are hydrophilic, the performance of biocomposites at high humidities and in aqueous conditions is not satisfactory. An extensive investigation was conducted to determine the water sorption and changes in properties of commercially available biobased resins and cellulose fibers [21]. Mechanical properties of the various biobased resins are given in Table 8 and the influence of moisture on the properties of these resins is listed in Table 9. Bioresins with a wide variety of properties and suitable for different

Table 8 Comparison of the properties of biobased resins with epoxy [21]. Reproduced with permission from John Wiley and Sons

Property	Epoxy	EpoBioX	Tribest	Palapreg	Envirez SA	Envirez SB
E (GPa)	3.2	3.6	0.7	3.5	3.4	3.4
v	0.37	0.37	0.35	0.37	0.37	0.37
G (GPa)	1.17	1.31	0.26	1.28	1.24	1.24
σ_{max} (MPa)	84.4	56.8	14.1	50.7	39.4	49.0
$\varepsilon\sigma_{max}$ (%)	5.3	1.8	4.3	1.7	1.2	1.7
E_B (GPa)	3.1	3.2	0.6	3.2	3.0	3.0
σ_B (MPa)	135.5	125.9	22.4	108.8	54.5	100.4
ε_B (%)	6.67	5.57	8.01	4.74	1.88	4.51
K_{IC} (MPa m$^{1/2}$)	0.31	0.87	0.37	0.17	0.32	0.43
a_{cU} (KJ/m^2)	42.7	30.2	23.2	16.2	10.1	15.5
a_{cU} (KJ/m^2)-90% RH	18.5	16.1	18.9	3.6	3.3	5.9

Table 9 Performance of biobased resins under different humidities compared to epoxy [21]. Reproduced with permission from John Wiley and Sons

Material	Humidity	Elastic modulus, GPa	Poisson's ratio	Max stress (MPa)	Max strain (%)
Epoxy	–	3.20	0.37	84.4	5.33
	41	3.06	0.32	78.6	4.23
	70	3.17	0.34	57.7	2.38
EpoBioX	–	3.64	0.37	56.8	1.82
	41	3.44	0.35	60.7	2.01
	70	3.58	0.35	55.3	1.83
Tribest	–	0.69	0.36	14.1	4.33
	41	0.50	0.29	10.2	2.60
	70	0.42	0.26	9.1	2.91

applications are available. Similarly, some of the resins showed excellent resistance of moisture, similar to epoxy and hence considered for development of composites with flax, regenerated cellulose and e-glass fibers as the reinforcement [22]. The fibers were also subject to chemical treatments to improve performance properties. Tensile properties and susceptibility to moisture of the composites was dependent on the type of reinforcement and the extent of fiber modifications. Although the raw resins had good resistance to humidity and water, the mechanical properties of the composites was considerably affected due to humidity with more than 100% reduction in stiffness. Fiber treatments also did not show any postive influence in improving the water resistance of the composites (Tables 10 and 11) [22].

Table 10 Changes in the tensile properties at two different humidities of composites containing glass fibers (GF), cellulose fibers (RCF) combined with various biodegradable resins [22]. Reproduced with permission from John Wiley and Sons

Material	RH (%)	E_L (GPa)	E_U (GPa)	v	σ_{max} (MPa)	$\varepsilon_{\sigma max}$ (%)
GF/Tribest	41	41.7 ± 1.1	41.9 ± 0.9	0.43 ± 0.09	831 ± 77	2.1 ± 1.2
RCF/Tribest		14.5 ± 1.2	14.8 ± 1.2	0.37 ± 0.00	356 ± 82	7.9 ± 0.7
GF/EpoBioX		36.4 ± 3.1	36.4 ± 3.1	0.31 ± 0.00	833 ± 51	2.1 ± 0.2
RCF/EpoBioX		14.3 ± 1.2	14.7 ± 1.1	0.34 ± 0.11	245 ± 23	4.3 ± 0.6
GF/Tribest	70	41.5 ± 1.5	42.0 ± 1.3	0.31 ± 0.03	722 ± 42	1.8 ± 0.2
RCF/Tribest		12.5 ± 0.5	12.9 ± 0.5	0.42 ± 0.03	320 ± 50	8.0 ± 1.8
GF/EpoBioX		36.6 ± 2.9	36.7 ± 3.0	0.27 ± 0.02	748 ± 84	2.6 ± 0.3
RCF/EpoBioX		13.2 ± 0.3	13.6 ± 0.3	0.50 ± 0.12	229 ± 12	3.6 ± 2.4

Table 11 Influence of moisture on the tensile properties of flax fibers (FF) and cellulose fiber (RCF) reinforced biobased composites after treating the fibers to different extents (2–5%) [23]. Reproduced with permission from John Wiley and Sons

Material	RH (%)	E_L (GPa)	E_U (GPa)	σ_{max} (MPa)	$\varepsilon_{\sigma max}$ (%)
FF/Envirez	–	7.5 ± 0.2	8.0 ± 0.3	57 ± 1	1.9 ± 0.1
FF/Envirez-2%		5.6 ± 0.3	5.9 ± 0.4	29 ± 2	1.4 ± 0.5
FF/Envirez-4%		5.3 ± 0.1	5.4 ± 0.2	28 ± 2	2.0 ± 0.2
FF/Envirez-5%		3.9 ± 0.6	4.1 ± 0.7	21 ± 2	1.9 ± 0.4
FF/Envirez	98	3.9 ± 0.1	4.5 ± 0.2	66 ± 4	3.7 ± 0.8
FF/Envirez-2%		3.1 ± 0.1	4.3 ± 1.2	31 ± 1	5.0 ± 1.4
FF/Envirez-4%		2.7 ± 0.2	3.1 ± 0.3	28 ± 3	3.4 ± 1.4
FF/Envirez-5%		2.1 ± 0.2	2.4 ± 0.2	21 ± 2	7.9 ± 0.5
RCF/Envirez	–	8.0	8.7	174	7.7
RCF/Envirez-2%		8.3	8.7	146	7.0
RCF/Envirez-5%		7.7 ± 0.7	8.0 ± 0.9	145 ± 15	7.0 ± 1.3
RCF/Envirez	98	3.5 ± 0.1	3.8 ± 0.2	101 ± 2	10.1 ± 1.6
RCF/Envirez-2%		3.4 ± 0.6	3.7 ± 0.6	101 ± 2	11.0 ± 0.4
RCF/Envirez-5%		3.3 ± 0.3	3.5 ± 0.2	94 ± 3	11.0 ± 0.3

4 Environmental Degradation

One of the major limitations of biobased materials is their relatively fast degradation when exposed to external environments. However, there is not much information on the performance of biobased composites when exposed to heat, UV, humidity and high temperatures [24]. The performance of pure PHB films and hemp fiber reinforced PHB composites were studied after exposure to UV light, water, heat and high humidity. Composites were made using 40% hemp fibers and compression molded at 207 kPa at 180 °C. Weathering tests were done using xenon arc source at an irradiance of 0.85 W/m^2 for 30–1440 min. Considerable changes were observed in the structure and properties of the composites. Mass loss and increased fading was seen (Fig. 9) in the composites and the molecular weight of the PHB also decreased

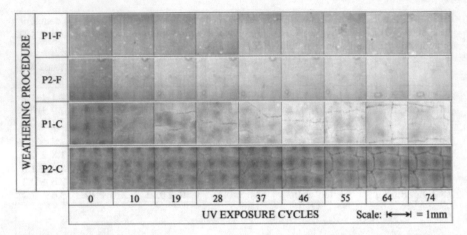

Fig. 9 Progressive fading and increased apprearnace of cracks in the PHB films and Hemp-PHB composites after exposure to increasing UV cycles [24]. Reproduced with permission from Elsevier

steadily with increased weathering. A 25–47% decrease in strength, 17–62% decrease in modulus but increase in elongation by about 30% was observed for the composites depending on the conditions during exposure. These changes in the composites were suggested to be due to the photooxidative and hydrolytic degradation of PHB and hygrothermal expansion and contraction of the hemp fibers.

5   Biodegradability

Despite the fact that biobased composites developed using natural and/or synthetic polymers are claimed to be environmentally friendly and biodegradable, there are very few reports on the biodegradability of biobased composites. Also, the biodegradabilty of the natural and synthetic polymers is highly dependent on the conditions prevalent during biodegradation. A study was conducted by Gomez and Michel to understand the biodegradability of conventional plastics and bioplastics under different environments [25]. The biodegradation profiles of various materials was investigated in comparison to cellulose paper and 100% polypropylene as positve and negative controls, respectively. Biodegradability was measured in soil media containing 43% certified organic top soil, 43% no-till farm soil and 14% sand with the addition of nutrients to maintain the C:N ratio of 20:1. Amount of CO_2 and/or CH_4 emitting from the samples was used to calculate the % biodegradation relative to the control. Extent of biodegradation varied considerable between aerobic, anaerobic and soil and compost conditions (Fig. 10). It was found that PHA showed lowest biodegradability next to corn based plastic containing co-polyester, PLA, paper pulp and natural fibers. Even under anaerobic conditions, the biobased materials showed

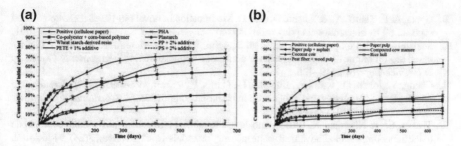

Fig. 10 Extent of biodegradation of various synthetic (**a**) and biobased materials (**b**) after 660 days in soil incubation [25]. Reproduced with permission from Elsevier

considerably higher biodegradability compared to conventional plastics with or without additives. However, some of the organic materials released methane, which is a major contributor to green house gas emissions. Hence, the biodegradability should be assessed considering the overall impact and not just the weight loss.

Acknowledgements Narendra Reddy thanks the Department of Biotechnology, Government of India for funding through the Ramalingaswami Reentry fellowship. Sincere thanks to Centre for Incubation, Innovation, Research and Consultancy and Jyothy Institute of Technology, Thataguni Bengaluru for facilitating the arrangments and encouraging us to complete the chapter.

References

1. Nyambo, C., Mohanty, A. K., & Misra, M. (2010). Polylactide-based renewable green composites from agricultural residues and their hybrids. *Biomacromolecules, 11*(6), 1654–1660.
2. Siakeng, R., Mohammad, J., Hidayah, A., Sapuan, S. M., Mohammad, A., & Naheed, S. (2019). Natural fiber reinforced polylactic acid composites: A review. *Polymer Composites, 40*(2), 446–463.
3. Tang, G., Xinjie, H., Houcheng, D., Xin, W., Shudong, J., Keqing, Z., et al. (2014). Combustion properties and thermal degradation behaviors of biobased polylactide composites filled with calcium hypophosphite. *RSC Advances, 4*(18), 8985–8993.
4. Di, L., & Gerardus, J. (2014). Composites with hemp reinforcement and bio-based epoxy matrix. *Composites Part B: Engineering, 67*, 220–226.
5. Senoz, E., Joseph, F. S., Kaleigh, H. R., Richard, P. W., & Melissa, E. N. M. (2012). Pyrolyzed chicken feather fibers for biobased composite reinforcement. *Journal of Applied Polymer Science, 128*(2), 983–989.
6. Adekunle, K., Åkesson, D., & Skrifvars, M. (2010). Biobased composites prepared by compression molding with a novel thermoset resin from soybean oil and a natural-fiber reinforcement. *Journal of Applied Polymer Science, 116*(3), 1759–1765.
7. Ramires, E. C., Jackson, D. M., Christian, G., Alain, C., & Frollini, E. (2010). Valorization of an industrial organosolv–sugarcane bagasse lignin: characterization and use as a matrix in biobased composites reinforced with sisal fibers. *Biotechnology and Bioengineering, 107*(4), 612–621.
8. Ramires, E. C., & Elisabete, F. (2012). Tannin–phenolic resins: synthesis, characterization, and application as matrix in biobased composites reinforced with sisal fibers. *Composites Part B: Engineering, 43*(7), 2851–2860.

9. Gupta, A. P., Sharif, A., & Anshu, D. (2011). Modification of novel bio-based resin-epoxidized soybean oil by conventional epoxy resin. *Polymer Engineering & Science, 51*(6), 1087–1091.
10. Stanzione, J. F., Philip, A. G., Joshua, M. S., John, J. L. S., & Richard, P. W. (2013). Lignin-based bio-oil mimic as biobased resin for composite applications. *ACS Sustainable Chemistry & Engineering, 1*(4), 419–426.
11. Chung, Y., Johan, O., Isson, R., Jingxian, L., Curtis, F., Robert, W., et al. (2013). A renewable lignin–lactide copolymer and application in biobased composites. *ACS Sustainable Chemistry & Engineering, 1*(10), 1231–1238.
12. Thakur, S., & Niranjan, K. (2013). Bio-based tough hyperbranched polyurethane–graphene oxide nanocomposites as advanced shape memory materials. *RSC Advances, 3*(24), 9476–9482.
13. Mati-Baouche, N., Hélène, D. B., André, L., Shengnan, S., Carlos, J. S. L., Philippe, L., et al. (2014). Mechanical, thermal and acoustical characterizations of an insulating bio-based composite made from sunflower stalks particles and chitosan. *Industrial Crops and Products, 58*, 244–250.
14. Bourdot, A., Tala, M., Alexandre, G., Chadi, M., Patricia, V., Céline, T., et al. (2017). Characterization of a hemp-based agro-material: Influence of starch ratio and hemp shive size on physical, mechanical, and hygrothermal properties. *Energy and Buildings, 153*, 501–512.
15. Sandrine, U., Benitha, V. I., Mai, T. H., & Maalouf, C. (2015). Influence of chemical modification on hemp–starch concrete. *Construction and Building Materials, 81*, 208–215.
16. Viel, M., Florence, C., Sylvie, P., & Christophe, L. (2019). Hemp-straw composites: gluing study and multi-physical characterizations. *Materials, 12*(8), 1199–1228.
17. Kremensas, A., Agnė, K., Saulius, V., Sigitas, V., & Giedrius, B. (2019). Mechanical performance of biodegradable thermoplastic polymer-based biocomposite boards from hemp shivs and corn starch for the building industry. *Materials, 12*(6), 845–857.
18. Hanifi, B., Orhan, A., & Ceyda, D. (2016). Mechanical, thermal and acoustical characterizations of an insulation composite made of bio-based materials. *Sustainable Cities and Society, 20*, 17–26.
19. Feldmann, M., & Andrzej, K. B. (2014). Bio-based polyamides reinforced with cellulosic fibres–processing and properties. *Composites Science and Technology, 100*, 113–120.
20. Mittal, N., Ronnie J., Mona W., Tobias B., Karl, M. O. H., Fredrik L., et al. (2017). Ultrastrong and bioactive nanostructured bio-based composites. *ACS Nano, 11*(5), 5148–5159.
21. Mahmood, H., Muhammad, M., Suzana, Y., & Tom, W. (2017). Ionic liquids assisted processing of renewable resources for the fabrication of biodegradable composite materials. *Green Chemistry, 19*(9), 2051–2075.
22. Pupure, L., Newsha D., & Roberts J. (2014). Moisture uptake and resulting mechanical response of biobased composites. I. constituents. *Polymer Composites, 35*(6), 1150–1159.
23. Doroudgarian, N., Liva P., & Roberts J. (2015). Moisture uptake and resulting mechanical response of bio-based composites. II. Composites. *Polymer Composites, 36*(8), 1510–1519.
24. Michel, A. T., & Billington, S. L. (2012). Characterization of poly-hydroxybutyrate films and hemp fiber reinforced composites exposed to accelerated weathering. *Polymer Degradation and Stability, 97*(6), 870–878.
25. Gómez, E., & Frederick, C. M. (2013). Biodegradability of conventional and bio-based plastics and natural fiber composites during composting, anaerobic digestion and long-term soil incubation. *Polymer Degradation and Stability, 98*(12), 2583–2591.

Influence of Fillers on the Thermal and Mechanical Properties of Biocomposites: An Overview

Thiagamani Senthil Muthu Kumar, Krishnasamy Senthilkumar, Muthukumar Chandrasekar, Saravanasankar Subramaniam, Sanjay Mavinkere Rangappa, Suchart Siengchin and Nagarajan Rajini

Abstract The mounting interests on the development of materials with superior performance has induced the expansion of filler reinforced composites market around the globe. The use of fillers in the polymeric materials helps the enhancement of the functional properties of the resulting composites. The primary concerns of the polymeric industry are poor material properties, degradability, and cost factors. Hence, embedding the polymer matrix with the fillers becomes inevitable. The polymeric materials with an appropriate filler, better filler/matrix interaction, along with advanced techniques, leads to the formation of superior performing composites for potential applications in various industries. Dedicated efforts have been made to understand the relationship between the filler particles in the polymers and their properties. Reports in the past conclude that the fillers play a vital role in the enhancement in the properties of the composites. This review article presents the influence of fillers on the thermal and mechanical properties of biocomposites.

Keywords Fillers · Biocomposites · Thermal properties · Mechanical properties

1 Introduction

The use of polymeric materials has become unavoidable in almost all possible applications due to their low cost, ease of processing lightweight, reproducibility, and excellent functional properties. However, on the other hand, most of the polymers

T. Senthil Muthu Kumar (✉) · K. Senthilkumar · N. Rajini
Department of Mechanical Engineering, Kalasalingam Academy of Research and Education, Krishnankoil 626126, Tamil Nadu, India
e-mail: tsmkumar@klu.ac.in

T. Senthil Muthu Kumar · K. Senthilkumar · S. Subramaniam · S. M. Rangappa · S. Siengchin
Department of Mechanical and Process Engineering, The Sirindhorn International Thai-German Graduate School of Engineering (TGGS), King Mongkut's University of Technology North Bangkok, 1518 Wongsawang Road, Bangsue, Bangkok 10800, Thailand

M. Chandrasekar
Department of Aerospace Engineering, Faculty of Engineering, University Putra Malaysia, 43400 UPM Serdang, Selangor, Malaysia

© Springer Nature Switzerland AG 2020
A. Khan et al. (eds.), *Biofibers and Biopolymers for Biocomposites*,
https://doi.org/10.1007/978-3-030-40301-0_5

111

used today are obtained from the depleting petroleum resources and pose serious threats to the environment [1]. The growing concern over the environmental protection paved the way for continuous research and development of polymer-based materials which can both serve the purpose of its intended use and also do not harm the environment [2, 3]. It is to be noted that the conventional composite structures were usually made of glass, carbon or aramid fibers being reinforced with epoxy, unsaturated polyester resins, polyurethanes, and phenolics, etc. The most important drawback of such composite materials is the problem of suitable elimination after the end of a lifetime, as the components are closely interconnected, relatively stable and therefore difficult to separate, reuse and recycle [2, 4]. Even though many successful attempts have been made on the development of novel biodegradable or environmentally friendly polymeric materials, their functional properties and cost factor remains unresolved [3, 5, 6]. Hence, in recent years, there is a growing interest in using fillers by the polymer-based industries. And it is due to the reinforcing effect of the fillers, which improves the material properties such as dimensional stability, tensile, compressive and impact strengths, resistance to abrasion and thermal stability. Apart from the above-said enhancements, fillers can also reduce the cost of the material due to the increase in the bulk volume [6, 7]. However, the real challenge lies in finding suitable applications which can use the developed material resourcefully to compete both in terms of performance and commercial aspects in the market.

Fillers have always played a vital role in the polymer industry. It can be seen in the past that the development of the industry would not have been possible without the improvement of properties of these fillers. However, in recent years, due to the scarcity of the petroleum resources and the increasing prices of the commodities have recognized the necessity for the extensive use of the fillers [8, 9]. Initially, the fillers used for polymer-based composites were mostly inorganic. The main purpose of their usage was to reduce the material cost and to improve some of the properties like rigidity significantly, thermal stability, etc. of the polymer matrix [10]. However, organic fillers have found their way into the composites recently and are also gaining more interest among the investigators. Moreover, it is because of their low cost, biodegradable and renewable in nature and ease of disposal [10–12]. Apart from the economic and environmental benefits, their lower density allows obtaining lighter materials, an attractive characteristic for the automotive, construction and packaging industries [13–16]. Many researchers are involved in the preparation, optimization, and testing of composite materials with the introduction of fibres, which also involves the extraction of fibres from various plant resources. However, with the poor mechanical properties of some thermoplastic polymers, the expensive extraction processes lead to economically non-competitive materials [16–18]. Hence, for increasing the suitability in specific industrial applications, it becomes inevitable to decrease the cost of the raw material and filler processing. With this objective, lignocellulosic agricultural residues were considered as potential sources of biomass or agro-based fillers [19–24]. Each type of filler possesses different characteristics which are influenced by their chemical composition, particle size, morphology, and surface area, etc. On the other hand, the filler dispersion, filler concentration or loading and their compatibility with the polymer matrix play a significant role in the properties of the composites [24, 25]. Hence with these insights, this review article focuses on the influence of the fillers on the thermal and mechanical properties of biocomposites. Furthermore, the changes in the properties are discussed concerning several factors like filler size, dispersion, morphology, chemical modification, compatibility with the polymer matrix, etc.

2 Overview of Fillers Used in Biocomposites

The filler is defined as a solid particulate material that may be irregular in shape and size. Fillers are not only used for cost reduction of the composites but also to enhance their performance. The fillers are used as reinforcement material along with the polymer matrix for specific applications, or they may be infused along with another reinforcement material and the polymer matrix to improve certain required properties which may not be achieved by the reinforcement and resin constituents alone. Fillers are often referred to as extending fillers and functional fillers. Extending fillers are merely used to increase the bulk volume and whereas the functional fillers are used for improving the properties of the material [26, 27]. However, some extending fillers when reduced to finer particles or surface treated can perform as a functional filler also. Fillers can be used to improve dimensional stability, mechanical properties, thermal stability, etc. However, the changes in properties of the composites are subjected to various factors such as the filler composition, surface, size, concentration, shape, and dispersion in the matrix.

3 Classification of Fillers

Fillers can be classified based on the chemical composition, shape, and size (Fig. 1).

It is to be noted that these fillers are typically rigid and immiscible with the polymer matrix and their effects on chemical composition, shapes and size may vary with of both the organic and inorganic components used as the filler [28].

4 Inorganic Fillers

Inorganic fillers were most commonly used in industrial applications. When used in composite laminates, inorganic fillers can account for 40 to 65% by weight. There are several numbers of inorganic filler materials that can be used with polymer matrices or composites. Reports have been made based on the toughening mechanism of semi-crystalline polymers with the addition of inorganic fillers [29–32]. Some of the commonly used inorganic fillers are discussed below:

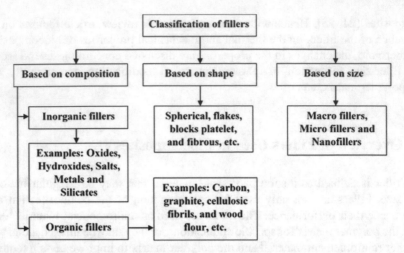

Fig. 1 Classification of fillers

- Calcium carbonate (CaCO₃) is the most widely used non-reactive mineral filler derived from limestone or marble and used in automotive parts. Previously, CaCO₃ was used as a cost reducing filler in composites however later the used of coupling agents along with CaCO₃ paved the way for its use as a toughening filler [33].
- Kaolin (Hydrous aluminosilicate) is also one of the commonly used industrial filler widely used in plastic and rubber industry having plate-like structure and high aspect ratio. Hence, the particle size and distribution play a vital role in achieving better properties. Its reinforcing effect can provide enhanced strength and stiffness of the materials [34].
- Silicon dioxide (SiO₂) is the most complex and abundantly available natural compound of silicon and oxygen, which is obtained from mining and purification of the mineral. It is widely used to reinforce rubber vulcanizates due to the enhanced thermal and tensile properties apart from tear and abrasion resistance [35, 36].
- Titanium dioxide (TiO₂) is a naturally occurring oxide of titanium which is non-toxic, chemically inert, corrosive resistant and inexpensive. It is used both in micro and level sizes with polymer matrices for enhanced mechanical and tribological properties due to its high hardness. Furthermore, it also provides photo stabilization and antibacterial activity [37, 38].
- Montmorillonite (MMT) is a mineral filler from the phyllosilicate group that is formed when it precipitates from water solution as microscopic crystals. They are soft and users are used in many applications such as absorbent for heavy metals, fillers for organic coatings, and as fillers for composites [39].

Even though there are many advantages of using inorganic fillers in composite materials for several applications, they cannot be completely degradable and some of the mineral fillers may be toxic to human health. Hence, there is a growing interest in using organic matters as fillers in composites for different applications.

5   Organic Fillers

Even though natural or organic fillers are not as popular as mineral or inorganic fillers, they are gaining more attention due to their advantages compared with the conventional fillers [5]. Their significant advantages include the least expensive, lower density, resistance to abrasion, low energy consumption, biodegradable, and renewable. It is to be noted that all organic fillers are not biodegradable. Material scientists have understood that the exploration natural fillers as reinforcing materials in polymer composites is inevitable and also an effective way to produce environmental friendly composites without disturbing their performance [39, 40]. The mounting research interests on organic fillers can be attributed to their renewability, ease of separation, carbon dioxide sequestration, and non-abrasive to equipment [41, 42]. Some of the commonly used organic fillers are discussed below:

- Carbon-based nanofillers such as single-walled carbon nanotube (SWCNT), multi-walled carbon nanotube (MWCNT), carbon nanofibres (CNFs), graphene and carbon black (CB) are extensively used in polymer matrix composites due to their excellent structural and functional properties and they are well suited for engineering applications [43–45].
- Wood flour (WF) is derived from various wood shavings, sawdust, chips and other residues of wood processing industries. The interest in WFs is because they provide dimensional stability and can enhance the elastic modulus. The main drawback of WF is that their poor adhesion with the polymer matrix and lower decomposition temperatures [46, 47].
- Lignocellulosic fillers also include wood, naturals fibres and trashes of variety of plants. These fillers are abundantly available, economical, and are renewable. Some examples of lignocellulosic fibres are cellulose, cotton, flax, sisal, kenaf, jute, hemp, bamboo, banana, pineapple, coir, ramie, and starch, etc. [48]. Further, many other agricultural wastes are also used as fillers such as the spent tea leaf powder, spent coffee bean powder, eggshell powder, banana peel powder, tamarind nut powder, and turmeric powder, etc.

The studies in the area of nanocomposites were dominated by the inorganic nanofillers such as nanoclay, MMT, silica, mica, etc. However, recently, there is a growing interest in using cellulose nanocrystals or nanofibrils as reinforcing fillers polymer composites [49, 50]. It is because of many attractive properties such as renewable and biodegradable, less expensive, available in abundance, high surface area, high aspect ratio, low density, high thermal stability, and high surface reactivity. These properties allow chemical modifications of the cellulose fibrils [51].

6   Factors Influencing the Properties of the Fillers

Many factors govern the properties of the fillers such as the source and chemical composition of the filler, size of the filler, aspect ratio, surface area, density, refractive index, hardness and the filler loading [8]. These factors can directly influence the performance of the composites. Hence it is imperative that these factors are handled with the utmost care in order to get the desired performance of the composites as an effect of the reinforcement of the fillers.

7   Influence of Fillers on the Thermal Properties of Biocomposites

Researchers reported that addition of a small number of fillers and nanomaterials within the matrix could enhance the (i) mechanical (ii) thermal and (iii) dynamic properties without changing the weight of the composites. Fillers are classified as particulate type fillers, fibers, and other fibers. Some examples of particulate type fillers are aluminum flakes and powders, aluminum borate whiskers, aluminum nitride, anthracite, antimonate of sodium, etc. In fibers such as aramid fibers, carbon fibers, carbon nanotubes, cellulose fibers, cellulose nanofibrils, etc. and other fibers such as vegetable, animal and mineral fibers.

This particular section focusses the thermal properties of biocomposites reinforced with fillers. The thermal properties of composites can be explored by techniques such as dynamic mechanical analysis (DMA), thermomechanical analysis (TMA), thermogravimetric analysis (TGA) and differential scanning calorimetry (DSC). Table 1 lists the essential outcomes/effects of materials by thermal analysis techniques, and the applications of similar techniques are also listed. The thermal properties of biocomposites much depend on many factors, including (i) fiber loading (ii) fiber orientation (iii) fiber layering sequence (iv) fiber surface treatments (v) the addition of fillers (vi) filler size (vii) filler loading, etc.

8   Effect of Filler Loading

Researchers studied the thermal properties of biocomposites using fillers/fibers as reinforcements. Some of the reported studies on TGA, DMA, DSC, and TMA of reinforced biocomposites tabulated in Table 2.

Saba et al. [52] fabricated hybrid composites using MH/kenaf fiber/epoxy matrix by hand lay-up technique with overall fiber loading were maintained as 40% by weight. The significant weight losses occurred between 300 and 400 °C due to the decomposition of MH to magnesium oxide (MgO) along with water loss by endothermic thermal dihydroxylation. The formation of MgO assisted in resisting

Table 1 Lists of outcome and applications of thermal analysis techniques

Technique	Outcome	Applications
DMA	• Energy dissipation behavior • Damping characteristic • Frequency • Force, and amplitude-dependent mechanical behavior of materials	Automobile and aerospace (composites), chemical, fats and oils, paints and lacquers, rubber (elastomers), ceramic materials, food industry, pharmaceuticals and plastics (thermosets, prepregs, coatings, films, adhesives, thermoplastics, packaging, and cables)
TMA	• Expansion or shrinkage of materials • Dimensional stability for the temperature	Plastics (elastomers, thermosets, thermoplastics), electronics industry, textile fibers, packaging (films) and chemical (organic and inorganic materials, metals, pharmaceutical products)
TGA	• Content analysis and thermal stability	Automotive, chemical, fats and oils, rubber, plastics, food industry, pharmaceutical, paints, and lacquers
DSC	• Glass transition • Melting • Crystallization behavior reaction enthalpies, and kinetics	Automotive, chemical, rubber, plastics, food industry and pharmaceutical

the temperature (even at elevated temperature of ~ 2800 °C) without undergoing any decomposition.

In another study, the researchers investigated the thermal stability of bayada clay/oil palm fiber/high-density polyethylene composites by varying the clay loading of 6.25–25 wt% [53]. It was reported that the clay formed a barrier to oil palm fiber composites against the polymer's exhausted gas, resulting in higher thermal stability of clay infused oil palm fiber composites.

DMA investigation revealed that 20% of MH filled kenaf/epoxy composites compensated the modulus incompatibility between fiber and matrix and improved the interactions between fiber/matrix bonding. Further, the authors reported that the occurrence of improved hydrogen bonding between the kenaf fiber and the epoxy matrix facilitated to decrease the segmental chain movement in the composites, caused higher E′ in MH filler filled composites [52]. The better interfacial bonding between the kenaf fiber and the matrix was also pronounced by having a broader tan delta curve.

In another study, Saba et al. [59] investigated the viscoelastic behavior of kenaf fiber reinforced hybrid composites using prior dried nanofillers such as (i) OPEFB (ii) MMT and (iii) OMMT. The fillers were uniformly dispersed at 3% (by weight) in the epoxy matrix. The researchers reported that the OMMT nanoclay restricted the movements of the epoxy matrix chains as they introduced between the phyllosilicate layers of OMMT nanoclay. Hence the highly stiffed OMMT/kenaf hybrid composites exhibited a remarkable stress transfer between the fiber and matrix and resulting in

Table 2 Reported work on thermal properties of bio-based composites

Technique	Type of composite	Range of filler loading	Significant effect	Reference
TGA	Kenaf fiber/magnesium hydroxide (MH)/epoxy	10, 15, 20 and 25 wt%	Maximum derivative weights: 18.1%/°C for kenaf/epoxy and 7.2%/°C for 20% MH/kenaf/epoxy	[52]
TGA	Oil palm fiber/bayada (clay)/high density polyethylene (HDPE)	6.25, 12.5, 18.75 and 25 wt%	Weight loss (%) increased in the order: 0:25 < 12.5:12.5 < 25:0	[53]
TGA	Palm fiber (PF)/acrylonitrile butadiene styrene (ABS)	5, 10 and 20%	10% of PF/ABS composites were thermally stable than the rest of the fiber/filler loaded composites	[54]
TGA	(a) Hemp/banana/flax/bagasse ash (b) Hemp/banana/kenaf/bagasse ash (c) Hemp/sisal/flax/bagasse ash (d) Hemp/sisal/kenaf/bagasse ash	1, 3, and 5 wt%	(a) With the addition of bagasse ash (1 wt% and 3 wt%) in hemp/banana/kenaf composites exhibited higher thermal resistance (b) With the addition of 5 wt% of bagasse ash in hemp/sisal/kenaf hybrid composites revealed higher thermal resistance	[55]
TGA	Cellulose/banana peel powder/silver nanoparticles	5–25% (order of 5%)	(a) Banana peel powder fillers exhibited higher thermal stability than the matrix and composites (b) Cellulose/banana peel powder/silver nanoparticles composites shown higher thermal stability than the matrix	[56]
DMA	Kenaf fiber/MH/epoxy	10, 15, 20 and 25 wt%	(a) 60–100 °C reported being the glassy region in storage modulus (E′) (b) 20% of MH/kenaf/epoxy produced higher E′ in the rubbery region (c) Loss modulus (E″) increased in the order of MH filled composites, 0% < 10% < 15% < 25% < 20% Damping increased in the order of MH filled composites, 20% < 25% < 15% < 10% < 0%	[51]

(continued)

Table 2 (continued)

Technique	Type of composite	Range of filler loading	Significant effect	Reference
DMA	Kenaf/epoxy, organically modified montmoril-lonite (OMMT)/kenaf/epoxy, montmorillonite (MMT)/kenaf/epoxy, nano-oil palm empty fruit bunch (OPEFB) kenaf/epoxy	0 and 3%	(a) E′ dropped between 75–100 °C (b) OMMT/kenaf/epoxy composites exhibited the highest E′ in rubbery region (c) Loss modulus increased in the order: kenaf/epoxy < nano-OPEFB/kenaf < MMT/kenaf < OMMT/kenaf (c) Tan delta increased in the order: OMMT/kenaf < MMT/kenaf < nano-OPEFB/kenaf < kenaf/epoxy	[57]
TMA	Kenaf/epoxy, OMMT/kenaf/epoxy, MMT/kenaf/epoxy, nano-oil palm empty fruit bunch (OPEFB) kenaf/epoxy	0 and 3%	(a) Dimension change (μm) decreased in the order, kenaf/epoxy < MMT/kenaf < nano-OPEFB/kenaf < OMMT/kenaf	[58]
TMA	OPEFB/epoxy	1, 3 and 5 wt%	Dimension change decreased in the order, 5 wt% < 1 wt% < 3 wt%	[59]
TMA	Palm fiber (PF)/acrylonitrile butadiene styrene (ABS)	5, 10 and 20%	No significant differences observed in Tg between the pure ABS and PF/ABS filled composites	[54]
DSC	Oil palm fiber/bayada (clay)/high density polyethylene (HDPE)	6.25, 12.5, 18.75 and 25 wt%	Values of specific enthalpy of melting [ΔHm (J/g)]: (a) HDPE = 183.5 (b) 25:0 = 143.0 (c) 12.5:12.5 = 148.8 (d) 0:25 = 137.1	[53]

increased E'. Regarding the damping factor, the stiffened and hardened nano-OPEFB filler restricted the movement of epoxy matrix molecules, resulting in improved E' than the E'' in OPEFB/kenaf hybrid composites.

In other findings, the researchers reported that the OMMT/kenaf/epoxy composites showed the highest dimensional stability [57]. Next, to the OMMT composites, the nano-OPEFB composites acted efficiently to control the thermal expansion behavior of kenaf/epoxy composites. It was attributed to (i) larger aspect ratio of nano-OPEFB filler and (ii) improved interfacial bonding between the kenaf fiber and the epoxy matrix. Researchers in other study investigated the coefficient of thermal expansion (CTE) by varying the wt% (1, 3 and 5 wt%) of oil palm nanofiber/epoxy composites [58]. It was revealed that until the 3 wt% of nanofiller filled composites, the CTE was decreased. Further increasing the filler loading, the CTE was found to be increased. It was corroborated to the (i) poor dispersion and (ii) accumulation of the 5 wt% of added filler particles, resulted in creating a free space for polymer segmental movement. Nevertheless, the oil palm nanocomposites possessed higher thermal stability than the epoxy matrix composites.

Hamid et al. [53] investigated the DSC by varying the bayada clay sizes from 0 to 25 wt% (order of 6.25 wt%) in oil palm fiber/high-density polyethylene (HDPE) composites. The authors reported that the melting temperature (Tm) of clay filled and fiber filled composites showed less significant compared to the HDPE matrix. The range of 'Tm' of HDPE, oil palm fiber, clay-fiber composites, and pure clay lied between 136 and 137 °C. Further, the ΔH_m of pure HDPE decreased while adding the oil palm fiber and clay at 25 wt% of loading (shown in Table 2). These observations indicated that the enrichment of thermal stability by incorporating the filler (fiber/clay) within the matrix and the clay composites.

9 Effect of Chemical Modification

Senthil et al. [60] investigated the TGA of cellulose/spent coffee bean powder (both untreated and alkali treated) composites. As expected, the alkali treated composites shown the highest thermal stability after a temperature of 350 °C. It was attributed to the existence of polyphenols presented in the spent coffee bean powder. Moreover, the authors reported that the alkali treated composites produced a higher residue. It could be due to (i) the presence of polyphenols and (ii) the removal of hemicelluloses. Many authors studied the thermal properties of biocomposites by using different chemical treatments. Some of the reported works tabulated in Table 3.

In another study, the thermal stability of *hildegardia populifolia* fiber/PPC composites (both treated and untreated) was explored [61]. The untreated fiber composites showed the highest thermal stability than the treated fiber composites. It was revealed that the existence of hemicellulose from the untreated fibers helped to improve the interfacial bonding between the fiber/matrix bonding, resulting in improved higher thermal resistance behavior. Continuation to the previous work, the authors [61] studied the visco-elastic behavior of *hildegardia populifolia* fiber/PPC composites and

Table 3 Reported work on thermal properties of bio-based composites

Technique	Details of composite	Details of chemical treatment	Significant effect	Reference
TGA	(a) Cellulose/spent coffee bean powder [5–25 wt% (order of 5 wt%)] (b) spent coffee bean powder, particle size = 35–40 μm	5% NaOH	Weight loss (%) increased in the order, untreated 25%/spent coffee bean powder composite < treated 25% spent coffee bean powder composite	[60]
TGA	(a) Poly (propylene carbonate) /Hildegardia populifolia; (b) Fiber length = 1–2 mm	2% NaOH treatment, 4 h	Thermal stability of untreated fiber composites higher than the NaOH treated fiber composites	[61]
DMA	(a) Poly (propylene carbonate) /Hildegardia populifolia; (b) Fiber length = 1–2 mm	2% NaOH treatment, 4 h	From the tan delta plot, no significant differences observed between the treated and untreated fiber composites	
DMA	(a) Natural rubber (NR)/gum, (b) NR/nanoclay (c) NR/jute (d) NR/nanoclay/jute	1% alkali treatment, 24 h at room temperature	(a) Storage modulus decreased in the order, NR/nanoclay/jute > NR/jute > NR/nanoclay > NR/gum (b) Loss modulus decreased in the order, NR/gum > NR/nanoclay > NR/jute > NR/nanoclay/jute	[62]

(continued)

Table 3 (continued)

Technique	Details of composite	Details of chemical treatment	Significant effect	Reference
DMA	(a) Polypropylene (PP)/pineapple leaf fiber (PALF)/nanoclay (b) Fiber length ~6 mm Nanoclay varied from 1 to 3 wt%	Maleic anhydride-grafted PP (Ma-g-PP)	Storage modulus increased in the order, PP/virgin < PP/PALF/Ma-g-PP < PP/PALF/Ma-g-PP/nanoclay	[63]
DSC	(a) PP/PALF/nanoclay (b) Fiber length ~6 mm (c) Nanoclay varied from 1 to 3 wt%	Maleic anhydride-grafted PP (Ma-g-PP)	(a) Melting temperature (°C) of composites decreased in the order, 178.68 (PP) > 175.62 (PP/PALF/Ma-g-PP) > 166.60(PP/PALF/Ma-g-PP/nanoclay) (b) Crystallization temperature (°C) increased in the order, 104.43 (PP) < 110.67 (PP/PALF/Ma-g-PP/nanoclay) <110.74 (PP/PALF/Ma-g-PP/nanoclay)	

reported that 20% of untreated and 20% of NaOH treated composites shown insignificant variations in tan delta values. It was attributed to the removal of lignin content from the fiber, resulting in no significant change in the hydrogen bond content.

In another study, the visco-elastic behavior of composites analyzed by adding the nanoclay with natural rubber and short jute fiber/natural rubber composites [62]. They observed that the natural rubber/nanoclay/jute fiber hybrid composites exhibited higher storage modulus from the glassy region to the rubbery region. It ascribed to the proper distribution of short jute fiber within the natural rubber in the presence of nanoclay. However, the natural rubber/clay shown higher storage modulus than the natural rubber/gum.

In another study, the effect of nanoclay addition was studied by Biswal et al. [63] in polypropylene/pineapple leaf fiber reinforced composites. They reported that good interfacial bonding of nanoclay with fiber was ensured by showing a lesser melting temperature (166.60 °C) among the fabricated samples. Moreover, the strong interaction of nanoclay, fiber, and polypropylene exhibited by increased order of crystallization temperature in composites (shown in Table 3).

10 Effect of Filler Size

Huda et al. [64] analyzed the DMA behavior by varying the particle sizes such as 12.5, 4.8, and 2.2 μm in poly(lactic) acid (PLA)/newspaper fiber reinforced composites. The product names of talc (i.e., filler) were Silverline 002, Nicron 403 and Mistron CB whereas the Mistron CB (2.2 μm) filler was chemically pre-treated (silane). Results revealed that incorporation of 2.2 μm filler in poly(lactic) acid/newspaper fiber composites exhibited higher storage modulus than the rest of the samples due to the existence of higher stiffer interface in the PLA matrix. In loss modulus, the filler infused composites shifted the glass transition temperature (T_g) to higher temperatures. It was attributed to the reduced movement of the polymer chains by the newspaper fibers. Also, the stresses surrounded by the fillers induced to shift the T_g to higher temperatures (between 65 and 75 °C). The storage modulus was increased in the order, virgin PLA < PLA/newspaper < PLA/newspaper/12.5 μm (filler 1) < PLA/newspaper/4.8 μm (filler 2) < PLA/newspaper/2.2 μm (filler 3).

Continuation of the previous work [64], the authors analyzed the DSC properties by varying the particle size. It was reported that the addition of fillers in poly(lactic) acid (PLA)/newspaper fiber hybrid composites, the T_g was found to be increased. When increasing the particle size, the T_g was further improved (i.e., T_g (2.2 μm) > T_g (4.8 μm)). Hence this improved of T_g supported to vary the soft and flexible properties to hard and tough. Regarding the T_c, the PLA/filler composites found to be decreased compared with the virgin PLA, because the fillers limiting the movement of PLA chains in the surfaces. Then reducing the filler size in PLA, the T_c was further decreased (i.e., T_c (2.2 μm) < T_c (4.8 μm)) in composites. Besides, the crystallinity of the filler filled composites was also found to be decreased, resulting in improved impact strength due to the reduction of crystallinity could reduce the flexibility of molecular chains.

In another study, Kim et al. [65] used three different sizes of silver nanoparticles (9, 65, and 300 nm) in paraffin and compared the thermal stability of TGA of phase change material (PCM). Also, the silver nanoparticles varied from 0.5 to 2 wt% (order of 0.5 wt%) in paraffin. The TGA results revealed that the decomposition temperature of silver nanoparticles/paraffin composites were improved (ranged between 5 and 50 °C) by varying the size of silver nanoparticles. This increase in temperature was due to the increased thermal stability of silver nanoparticles/paraffin composites. Among the different sizes of nanoparticles, the 65 nm exhibited higher thermal stability. It could be due to the effective interaction between the silver nanoparticles and paraffin. Researchers in another study [66], compared the decomposition temperature (T_d) of Polytetrafluoroethylene (PTFE)/SiO_2 composites by varying the particle size (5 and 25 µm) and the SiO_2 content (0–60 wt%). They observed that the T_d was not influenced by varying the particle size and content of SiO_2 due to the negligible interaction between the PTFE and SiO_2.

11   Influence of Fillers on the Mechanical Properties of Biocomposites

It is well known that the mechanical properties of the biocomposites are influenced by various factors about the fillers. This section discusses how different parameters influence the properties and the changes in the functional behavior of the bio-composite based on these factors.

12   Effect of Filler Size and Filler Loading

In an interesting study, it was found that flexural properties and impact strength of the PLA based composite reinforced with kenaf and rice husk fibres (RH) were dependent on the fibre aspect ratio (length: diameter). Kenaf/PLA had superior flexural strength, flexural modulus, and impact strength than the RH/PLA composite. Higher cellulose content and larger aspect ratio of the kenaf fibre and their fibrillation helped to withstand more load than the wider and shorter RH fibre [67]. In another study, the effects of filler particle size on the tensile and impact properties of the poplar sawdust/PP composites by varying the polar sawdust size to 40, 50 and 60 meshes examined. The fibre aspect ratio increases with the mesh size. It can be observed from their results that tensile strength and modulus increased with the use of bigger particle size and vice-versa in case of the impact strength [68]. Polytetrafluoroethylene (PTFE) was used as a filler with polyetheretherketone (PEEK), and the effect of varying concentrations of PTFE (0, 7.5, 15, 22.5 and 30 wt%) on the composites was investigated. It was found that the impact strength of the composites improved with increased filler addition while the tensile properties decreased. The higher impact strength of the composites was due to the higher molecular weight of the PFTE fillers added [69]. The effect of filler loading (5–25 wt%) on the PPC based composites with biofillers obtained from agro-wastes such as spent tea leaf powder (STLP) [24], tamarind nut powder (TNP) [70], spent coffee bean powder (SCBP) [1, 3, 5], banana peel powder (BPP) [71] and eggshell powder (ESP) [23] etc. were studied. According to their findings, there was a significant improvement in the tensile strength and modulus of the composites with increased filler concentration. It could be attributed to several reasons such as the uniform distribution of the fillers, better matrix filler bonding, and the presence of rigid phenolic components in the filler. In a similar manner, the effect of filler loading (5–25 wt%) on cellulose-based composites with the agro waste fillers such as STLP [72], SCBP [5], TNP [73], and BPP [19] was also studied. A same kind of trend was also reported that the increase in filler loading enhanced the tensile properties of the biocomposites.

Hybrid fillers were also used in composites to form hybrid filler composites. In a study, tamarind polysaccharide was used as a reducing agent for silver nitrate and copper sulfate to form silver nanoparticles and copper nanoparticles in PPC based composites through the in-situ generation process. It was reported that the increase in filler content improved the tensile properties which were attributed to the incorporation of the metal nanoparticles along with the cellulosic filler [74, 75]. Cellulosic filler banana peel powder was used as a reducing agent to in-situ generate silver nanoparticles in cellulose-based hybrid nanocomposites. It was found that the rigid phenolic compounds and the metal nanoparticles contributed to the enhanced tensile properties of the hybrid nanocomposites [56]. In another study, silver nanoparticles were in-situ generated with Napier grass microfibrils, and their effect on the tensile properties was investigated. It was found that the tensile strength and the modulus increased only for the lower filler loadings (up to 2 wt%). The tensile properties reduced with higher loading (above 2 wt%) and which was attributed to the random orientation of the fillers and possible agglomeration due to high filler content [76]. It was also reported that there was a significant effect of filler size and filler loading on multi-walled carbon nanotubes and graphene nanopowder epoxy composites. The incorporation of these fillers reduced the tensile and modulus to about 44% and 20% respectively—the main reason behind such a behavior entanglement of the nanotubes and agglomeration of the graphene particles. Hence, higher filler loading reduced the properties of the composites [77].

According to some researchers, increasing the rice husk (RH) content in the polyester matrix resulted in a drop in the tensile strength up to 20 wt% while the maximum Young's modulus was obtained at 15 wt% followed by a decline in values with a further increase of RH. The reasons for inferior strength and modulus at higher filler content are as follows: (a) poor interfacial adhesion between the hydrophilic natural filler and hydrophobic matrix and (b) agglomeration of fillers causes improper wetting of the resin and produces regions lacking in resin. These factors decrease the load bearing capability of the composite and affect the stress distribution within the matrix [78]. In another study, the tensile strength, modulus, and elongation at

break of cellulose/Thespesia short fibre composites witnessed a drop in the values
with increased filler loadings. The random orientation of the fibre and the presence
of amorphous lignin and hemicelluloses as impurities in the fibre eventually led to
the weakening of the composites indicating their negative effect [22].

13 Effect of Chemical Modification

Introduction of bamboo cellulose fibre (BCF) at 2 wt% in the PLA matrix resulted
in lower tensile strength and % elongation than the pure PLA. However, the Young's
modulus was slightly higher for the BCF/PLA composite, which was attributed to the
inherent stiffness of the bamboo fibre. They further demonstrated that mechanical
properties could be enhanced by fibre treatments. Fibre treatment with the potassium hydroxide (KOH) and silane solution improved both the strength and modulus
significantly by enhancing the interfacial adhesion between the filler and matrix.
KOH treatment removes fibre constituents such as hemicellulose, lignin, pectin,
waxes, and other surface impurities, allowing for better mechanical interlocking of
the treated fibres with the PLA matrix. In the case of silane treatment, silane-grafted
onto the fibre reacts and forms a chemical bond with the PLA matrix [79]. It was also
reported that the improvement in the flexural strength, modulus, and hardness for
the coir/PP composite reinforced with hydroxybenzene diazonium chloride treated
fibres. In addition to the better mechanical interlocking between the fibre and matrix,
chemical treatment of the fibre was also helpful in minimizing the micro-void at the
filler-matrix interface [80]. Another study showed that tensile strength of the wood
flour/PP composite could be improved by treating wood flour with 2 wt% NaOH
for 1 h and also by adding talc into the matrix [81]. PP-based composite with hemp
fibres treated by 0.05% potassium permanganate solution (KP) in acetone for 30 min
had a superior modulus of elasticity and modulus of rigidity than the composite
with the untreated fibres. The fibres undergo oxidation and forms into elementary
fibre bundles due to the fibre treatment in KP solution [82]. In another interesting
study, the effect of alkali treatment on *Sterculia urens* short fibres was studied on the
mechanical properties of the cellulose-based biocomposites. It was concluded that
the composites with treated *Sterculia urens* short fibres possessed improved tensile
properties than their untreated counterpart. However, the tensile properties were still
lower than the cellulose matrix. It was ascribed to the random orientation of the fillers
used [21].

14 Aging Effects

RH/HDPE composites subjected to the accelerated weathering possessed lower
impact strength than the pristine or dry specimens. The swelling and shrinking of RH
fibre due to the moisture absorption induce interfacial cracks such that the premature

failure occurs at lower loads. The samples subjected to weathering were found to have crazing and flaking of the RH fillers on the matrix, which is also believed to have caused a loss in the strength [83]. The decrease in modulus of rigidity, and Young's modulus of the rubber wood fibre/recycled PP composites exposed to the natural weathering was reported earlier. They also demonstrated that this resistance to moisture and temperature could be improved by adding 1 wt% UV stabilizer into the wood fibre/matrix during the fabrication [84]. Similarly, other researchers indicated that grafting maleic anhydride (MA) into the polyvinyl chloride reinforced with alfa fibres provided better resistance to the hydrothermal aging conditions by exhibiting higher strength and modulus than the composite without MA. MA improves the fibre compatibility between the fiber-matrix and acts a barrier to restrict the moisture diffusion into the composite; hence, lower loss in the mechanical properties [85]. Based on the reported studies in the literature, it is concluded that mechanical properties of the bio-composites are influenced by various factors such as filler size, filler loading, the addition of a compatibilizer, fibre treatments and exposure to the aging conditions.

15 Challenges, Opportunities, Current Developments, and Applications

There are some critical disadvantages of using organic fillers that limit their application. The major challenge of the application of organic fillers in polymer matrix composites lies in the compatibility of the filler and the matrix. Generally, the polymer matrix is hydrophobic, and the fillers are hydrophilic. Hence, the issue of compatibility between the polymer matrix and the filler arises [86]. The hydrophilic character of the filler results in high moisture absorption, poor matrix-filler interfacial adhesion, and poor filler dispersion. The mechanical properties of the resulting biocomposites largely depend on the interfacial adhesion between filler and matrix. Hence in order to overcome these limitation coupling agents can be used to modify the polymer matrix, or chemical modification of the filler can be done through treatments [87]. Another major disadvantage of organic fillers is high moisture absorption due to the presence of high hydroxyl groups. High moisture absorption leads to high swelling thereby affecting the performance of the resulting biocomposites. It could be reduced by chemical treatments of the hydroxyl groups present in the fillers. Further, the natural filler reinforced composites have other issues such as the poor resistance to adverse weather conditions, lower mechanical properties [88]. One more important issue of selection of suitable fillers for the application in composites is that the nonexistence of standards regarding the methods of collection, treatment, pre, and post-processing. Over recent years, interest to incorporate two or more fillers into a polymer matrix is increasing. It is also called as hybrid filler reinforcement. The main objective was to overcome the constraint of the single filler reinforced matrix with the other filler that has better properties when compared to the former filler [89]. Fillers having micro and nano-scaled sizes are also reinforced in a single polymer matrix to form hybrid nanocomposites. Most of the reports on the hybridized fillers used in polymer matrix composites proved to have improved the functional properties of the composites [56, 74, 75]. Natural or organic filler biocomposites are used almost in all applications. However, the prominent user is the automotive industry. These biocomposites also find their application on a variety of fields such as textile industry, construction, furniture, packaging, paper, healthcare, and energy sectors [89–91]. The main reasons for the substantial usage of natural fillers in these applications are the lower weight, renewable, and relatively lower cost. Majority of the components in an automotive vehicle such as the door panels, dashboard, armrests, body panels, seat bottoms, backrests etc. are made from the natural filler composites. Further, the natural fillers are used also used for the manufacture of musical instruments, furnishing for workplaces and homes, packing materials such as containers, cases, etc.

16   Concluding Remarks

Even though the inorganic fillers provide continuous performance improvements in the composite materials, the polymer composites filled with natural-organic fillers, are gaining more and more attention due to the mounting concerns over the environmental protection and also the cost aspects. The organic fillers can also provide performance characteristics at par with their inorganic counterpart; however, the selection of fillers for the particular application with suitable properties is critical. Although there are some limitations of using organic fillers in polymer matrices such as the elasticity, processability and dimensional stability, these limitations can be overcome through chemical modification of the filler, use of adhesion inhibitors and some additives. Furthermore, the performance of the composites lies in the selection of the most suitable matrix, filler, and optimization of all fabricating and processing parameters. Furthermore, the potential of application of natural fillers in polymer composites depends mostly on the increase of regulation and its commercial features. The development of information on natural fillers will lead to the standardization of the natural fillers and in turn, provides a higher level of confidence on the physio-mechanical properties of the resulting biocomposites. Still, many natural organic fillers are unexplored for commercial usage in large scale applications. Hence, there is a broad scope for the scientific community in the research and development of organic natural filler reinforced composites for large scale applications such as automotive, construction, and packaging industries.

Acknowledgement This research was supported by the King Mongkut's University of Technology North Bangkok (KMUTNB), Thailand through the Post Doc Program (Grant No. KMUTNB-61-Post-01 and KMUTNB-63-KNOW-001).

References

1. Muthu Kumar, S.T., Yorseng, K., Siengchin, S., Ayrilmis, N., & Rajulu, V. A. (2019). Mechanical and thermal properties of spent coffee bean filler/poly(3-hydroxybutyrate-co-3-hydroxyvalerate) biocomposites: Effect of recycling. *Process Safety and Environmental Protection, 124*, 87–195.
2. Mohanty, A. K., Misra, M., & Hinrichsen, G. I. (2000). Biofibres, biodegradable polymers and biocomposites: An overview. *Macromolecular Materials and Engineering, 276*, 11–24.
3. Thiagamani, S. M. K., Krishnasamy, S., & Siengchin, S. (2019). Challenges of biodegradable polymers: An environmental perspective. *Applied Science and Engineering Progress, 12*(3), 149.
4. Netravali, A. N., & Chabba, S. (2003). Composites get greener. *Materials Today, 4*(6), 22–29.
5. Kumar, T.S.M., Rajini, N., Huafeng, T., Rajulu, A.V., Ayrilmis, N., & Siengchin, S. (2018). Improved mechanical and thermal properties of spent coffee bean particulate reinforced poly(propylene carbonate) composites. *Particulate Science and Technology*.
6. Thiagamani, S. M. K., Nagarajan, R., Jawaid, M., Anumakonda, V., & Siengchin, S. (2017). Utilization of chemically treated municipal solid waste (spent coffee bean powder) as reinforcement in cellulose matrix for packaging applications. *Waste Management, 69*, 445–454.
7. Bledzik, A., & Gassan, J. (1993). Composites Reinforced with Cellulose Based Fibers. *Progress in Polymer Science, 24*(2), 221–274.
8. Onuegbu, G. C., & Igwe, I. O. (2011). The Effects of Filler Contents and Particle Sizes on the mechanical and end-use properties of snail shell powder filled polypropylene. *Materials Sciences and Applications, 2*(7), 810–816.
9. Katz, H. S., & Mileski, J. V. (1987). *Handbook of fillers for plastics*. Springer Science & Business Media.
10. La Mantia, F.P., Morreale, M., & Mohd Ishak, Z.A. (2005). Processing and mechanical properties of organic filler-polypropylene composites. *Journal of Applied Polymer Science, 96*(5), 1906–1913.
11. Rozman, H. D., Lai, C. Y., Ismail, H., & Ishak, Z. A. M. (2000). The effect of coupling agents on the mechanical and physical properties of oil palm empty fruit bunch polypropylene composites. *Polymer International, 49*(11), 1273–1278.
12. Canche-Escamilla, G., Rodriguez-Laviada, J., Cauich-Cupul, J. I., Mendizabal, E., Puig, J. E., & Herrera-Franco, P. J. (2002). Flexural, impact and compressive properties of a rigid-thermoplastic matrix/cellulose fiber reinforced composites. *Composites Part A: Applied Science and Manufacturing, 33*(4), 539–549.
13. Nair, K. C. M., Kumar, R. P., Thomas, S., Schit, S. C., & Ramamurthy, K. (2000). Rheological behavior of short sisal fiber-reinforced polystyrene composites. *Composites Part A: Applied Science and Manufacturing, 31*(11), 1231–1240.
14. Koronis, G., Silva, A., & Fontul, M. (2013). Green composites: A review of adequate materials for automotive applications. *Composites Part B: Engineering, 44*(1), 120–127.
15. Ashori, A. (2008). Wood–plastic composites as promising green-composites for automotive industries! *Bioresource Technology, 99*(11), 4661–4667.
16. Zah, R., Hischier, R., Leão, A. L., & Braun, I. (2007). Curauá fibers in the automobile industry—A sustainability assessment. *Journal of Cleaner Production, 15*(11–12), 1032–1040.
17. Jacob, A. (2006). WPC industry focuses on performance and cost. *Reinforced Plastics, 50*(5), 32–33.
18. Hosseinaei, O., Wang, S., Enayati, A. A., & Rials, T. G. (2012). Effects of hemicellulose extraction on properties of wood flour and wood–plastic composites. *omposites Part A: Applied Science and Manufacturing, 43*(4), 686–694.
19. Senthil Muthu Kumar, T., Rajini, N., Alavudeen, A., Siengchin, S., Rajulu, V., & Ayrilmis, N. (2019). Development and analysis of completely biodegradable cellulose/banana peel powder composite films. *Journal of Natural Fibers*, 1–10.

20. Senthil Muthu Kumar, T., Rajini, N., Obi Reddy, K., Varada Rajulu, A., Siengchin, S., & Ayrilmis, N. (2018). All-cellulose composite films with cellulose matrix and Napier grass cellulose fibril fillers. *International Journal of Biological Macromolecules.*

21. Jayaramudu, J., Reddy, G. S. M., Varaprasad, K., Sadiku, E. R., Sinha Ray, S., & Varada Rajulu, A. (2013). Preparation and properties of biodegradable films from *Sterculia urens* short fiber/cellulose green composites. *Carbohydrate Polymers, 93*(2), 622–627.

22. Ashok, B., Reddy, K. O., Madhukar, K., Cai, J., Zhang, L., & Rajulu, A. V. (2015). Properties of cellulose/Thespesia lampas short fibers bio-composite films. *Carbohydrate Polymers, 127,* 110–115.

23. Feng, Y., et al. (2014). Preparation and characterization of polypropylene carbonate bio-filler (eggshell powder) composite films. *International Journal of Polymer Analysis and Characterization, 19*(7), 637–647.

24. Xia, G., et al. (2015). Preparation and properties of biodegradable spent tea leaf powder/poly(propylene carbonate) composite films. *International Journal of Polymer Analysis and Characterization, 20*(4), 377–387.

25. Tjong, S. C., & Mai, Y.-W. (2008). Processing-structure-property aspects of particulate-and whisker-reinforced titanium matrix composites. *Composites Science and Technology, 68*(3–4), 583–601.

26. Park, S.-J., & Seo, M.-K. (2011). *Interface science and composites* (Vol. 18). Academic Press.

27. Plackett, D., Andersen, T. L., Pedersen, W. B., & Nielsen, L. (2003). Biodegradable composites based on L-polylactide and jute fibres. *Composites Science and Technology, 63*(9), 1287–1296.

28. Oksman, K., & Selin, J.-F. (2004). Plastics and composites from polylactic acid. In *Natural fibers, plastics and composites* (pp. 149–165). Springe.

29. Xanthos, M. (2010). *Functional fillers for plastics*. Wiley.

30. Zuiderduin, W. C. J., Westzaan, C., Huetink, J., & Gaymans, R. J. (2003). Toughening of polypropylene with calcium carbonate particles. *Polymer (Guildf), 44*(1), 261–275.

31. Tjong, S. C. (2006). Structural and mechanical properties of polymer nanocomposites. *Materials Science and Engineering R: Reports 53*(3–4), 73–197.

32. Eiras, D., & Pessan, L. A. (2009). Mechanical properties of polypropylene/calcium carbonate nanocomposites. *Materials Research 12*(4), 517–522.

33. Demjén, Z., Pukánszky, B., & Nagy, J. (1998). Evaluation of interfacial interaction in polypropylene/surface treated CaCO₃ composites. *Composites Part A: Applied Science and Manufacturing 29*(3), 323–329.

34. Fellahi, S., Chikhi, N., & Bakar, M. (2001). Modification of epoxy resin with kaolin as a toughening agent. *Journal of Applied Polymer Science, 82*(4), 861–878.

35. Shivamurthy, B., & Prabhuswamyc, M. S. (2009). Influence of SiO₂ fillers on sliding wear resistance and mechanical properties of compression moulded glass epoxy composites. *Journal of Minerals and Materials Characterization and Engineering, 8*(07), 513.

36. Peng, H., Liu, L., Luo, Y., Hong, H., & Jia, D. (2009). Synthesis and characterization of 3-benzothiazolthio-1-propyltriethoxylsilane and its reinforcement for styrene–butadiene rubber/silica composites. *Journal of Applied Polymer Science, 112*(4), 1967–1973.

37. Altan, M., & Yildirim, H. (2012). Mechanical and antibacterial properties of injection molded polypropylene/TiO₂ nano-composites: Effects of surface modification. *Journal of Materials Science and Technology, 28*(8), 686–692.

38. Friedrich, K., Fakirov, S., & Zhang, Z. (2005). *Polymer composites: From nano-to macro-scale*. Springer Science & Business Media.

39. Atta, A., El-Saeed, A., Al-Lohedan, H., & Wahby, M. (2017). Effect of montmorillonite nanogel composite fillers on the protection performance of epoxy coatings on steel pipelines. *Molecules, 22*(6), 905.

40. Majeed, K., et al. (2013). Potential materials for food packaging from nanoclay/natural fibres filled hybrid composites. *Materials and Design, 46,* 391–410.

41. Ku, H., Wang, H., Pattarachaiyakoop, N., & Trada, M. (2011). A review on the tensile properties of natural fiber reinforced polymer composites. *Composites Part B: Engineering, 42*(4), 856–873.

42. Kalia, S., Kaith, B. S., & Kaur, I. (2009). Pretreatments of natural fibers and their application as reinforcing material in polymer composites—A review. *Polymer Engineering & Science, 49*(7), 1253–1272.
43. S. Ojha, S. K. Acharya, & G. (2015). Raghavendra, Mechanical properties of natural carbon black reinforced polymer composites. *Journal of Applied Polymer Science, 132*(1).
44. Mittal, G., Dhand, V., Rhee, K. Y., Park, S.-J., & Lee, W. R. (2015). A review on carbon nanotubes and graphene as fillers in reinforced polymer nanocomposites. *Journal of Industrial and Engineering Chemistry, 21*, 11–25.
45. Hu, K., Kulkarni, D. D., Choi, I., & Tsukruk, V. V. (2014). Graphene-polymer nanocomposites for structural and functional applications. *Progress in Polymer Science, 39*(11), 1934–1972.
46. Carroll, D. R., Stone, R. B., Sirignano, A. M., Saindon, R. M., Gose, S. C., & Friedman, M. A. (2001). Structural properties of recycled plastic/sawdust lumber decking planks. *Resources, Conservation and Recycling, 31*(3), 241–251.
47. Matuana, L. M., & Stark, N. M. (2015). The use of wood fibers as reinforcements in composites. In *Biofiber reinforcements in composite materials* (pp. 648–688). Elsevier.
48. Sanjay, M. R., Madhu, P., Jawaid, M., Senthamaraikannan, P., Senthil, S., & Pradeep, S. (2018). Characterization and properties of natural fiber polymer composites: A comprehensive review. *Journal of Cleaner Production 172*, 566–581.
49. Siró, I., & Plackett, D. (2010). Microfibrillated cellulose and new nanocomposite materials: A review. *Cellulose, 17*(3), 459–494.
50. Lu, J., Wang, T., & Drzal, L. T. (2008). Preparation and properties of microfibrillated cellulose polyvinyl alcohol composite materials. *Composites Part A: Applied Science and Manufacturing, 39*(5), 738–746.
51. Wu, J., et al. (2010). Structure and properties of cellulose/chitin blended hydrogel membranes fabricated via a solution pre-gelation technique. *Carbohydrate Polymers, 79*(3), 677–684.
52. Saba, N., Alothman, O. Y., Almutairi, Z., & Jawaid, M. (2019). Magnesium hydroxide reinforced kenaf fibers/epoxy hybrid composites: Mechanical and thermomechanical properties. *Construction and Building Materials, 201*, 138–148.
53. Essabir, H., Boujmal, R., Bensalah, M. O., Rodrigue, D., Bouhfid, R., & el kacem Qaiss, A. (2016). Mechanical and thermal properties of hybrid composites: Oil-palm fiber/clay reinforced high density polyethylene. *Mechanics of Materials, 98*, 36–43.
54. Neher, B., Bhuiyan, M. M. R., Kabir, H., Gafur, M. A., Qadir, M. R., & Ahmed, F. (2016). Thermal properties of palm fiber and palm fiber reinforced ABS composite. *Journal of Thermal Analysis and Calorimetry, 124*(3), 1281–1289.
55. Vivek, S., & Kanthavel, K. (2019). Effect of bagasse ash filled epoxy composites reinforced with hybrid plant fibres for mechanical and thermal properties. *Composites Part B: Engineering, 160*, 170–176.
56. Thiagamani, S. M. K., Rajini, N., Siengchin, S., Varada Rajulu, A., Hariram, N., & Ayrilmis, N. (2019). Influence of silver nanoparticles on the mechanical, thermal and antimicrobial properties of cellulose-based hybrid nanocomposites. *Composites Part B: Engineering.*
57. Saba, N., Paridah, M. T., Abdan, K., & Ibrahim, N. A. (2016). Dynamic mechanical properties of oil palm nano filler/kenaf/epoxy hybrid nanocomposites. *Construction and Building Materials, 124*, 133–138.
58. Saba, N., Paridah, M. T., Abdan, K., & Ibrahim, N. A. (2016). Physical, structural and thermomechanical properties of oil palm nano filler/kenaf/epoxy hybrid nanocomposites. *Materials Chemistry and Physics, 184*, 64–71.
59. Saba, N., Jawaid, M., Paridah, M. T., & Alothman, O. (2017). Physical, structural and thermomechanical properties of nano oil palm empty fruit bunch filler based epoxy nanocomposites. *Industrial Crops and Products, 108*, 840–843.
60. Thiagamani, S. M. K., Nagarajan, R., Jawaid, M., Anumakonda, V., & Siengchin, S. (2017). Utilization of chemically treated municipal solid waste (spent coffee bean powder) as reinforcement in cellulose matrix for packaging applications. *Waste Management, 69*.
61. Li, X. H., Meng, Y. Z., Wang, S. J., Rajulu, A. V., & Tjong, S. C. (2004). Completely biodegradable composites of poly (propylene carbonate) and short, lignocellulose fiber Hildegardia populifolia. *Journal of Polymer Science Part B: Polymer Physics, 42*(4), 666–675.

62. Roy, K., Debnath, S. C., Das, A., Heinrich, G., & Potiyaraj, P. (2018). Exploring the synergistic effect of short jute fiber and nanoclay on the mechanical, dynamic mechanical and thermal properties of natural rubber composites. *Polymer Testing, 67*, 487–493.

63. Biswal, M., Mohanty, S., & Nayak, S. K. (2009). Influence of organically modified nanoclay on the performance of pineapple leaf fiber-reinforced polypropylene nanocomposites. *Journal of Applied Polymer Science, 114*(6), 4091–4103.

64. Huda, M. S., Drzal, L. T., Mohanty, A. K., & Misra, M. (2007). The effect of silane treated-and untreated-talc on the mechanical and physico-mechanical properties of poly(lactic acid)/newspaper fibers/talc hybrid composites. *Composites Part B: Engineering, 38*(3), 367–379.

65. Kim, I.-H., Sim, H.-W., Hong, H.-H., Kim, D.-W., Lee, W., & Lee, D.-K. (2019). Effect of filler size on thermal properties of paraffin/silver nanoparticle composites. *Korean Journal of Chemical Engineering, 36*(6), 1004–1012.

66. Chen, Y.-C., Lin, H.-C., & Lee, Y.-D. (2003). The effects of filler content and size on the properties of PTFE/SiO$_2$ composites. *Journal of Polymer Research, 10*(4), 247–258.

67. Yussuf, A. A., Massoumi, I., & Hassan, A. (2010). Comparison of polylactic acid/kenaf and polylactic acid/rise husk composites: The influence of the natural fibers on the mechanical, thermal and biodegradability properties. *Journal of Polymers and the Environment, 18*(3), 422–429.

68. Nourbakhsh, A., Karegarfard, A., Ashori, A., & Nourbakhsh, A. (2010). Effects of particle size and coupling agent concentration on mechanical properties of particulate-filled polymer composites. *Journal of Thermoplastic Composite Materials, 23*(2), 169–174.

69. Bijwe, J., Sen, S., & Ghosh, A. (2005). Influence of PTFE content in PEEK–PTFE blends on mechanical properties and tribo-performance in various wear modes. *Wear, 258*(10), 1536–1542.

70. Kumar, T. S. M., Rajini, N., Tian, H., Rajulu, A. V., Winowlin Jappes, J. T., & Siengchin, S. (2017). Development and analysis of biodegradable poly(propylene carbonate)/tamarind nut powder composite films. *International Journal of Polymer Analysis and Characterization*.

71. Senthil Muthu Kumar, T., Rajini, N., Siengchin, S., Varada Rajulu, A. & Ayrilmis, N. (2019). Influence of Musa acuminate bio-filler on the thermal, mechanical and visco-elastic behavior of poly (propylene) carbonate biocomposites. *International Journal of Polymer Analysis and Characterization*, 1–8.

72. Duan, J., Reddy, K. O., Ashok, B., Cai, J., Zhang, L., & Rajulu, A. V. (2016). Effects of spent tea leaf powder on the properties and functions of cellulose green composite films. *Journal of Environmental Chemical Engineering, 4*(1), 440–448.

73. Senthil Muthu Kumar, T., Rajini, N., Jawaid, M., Varada Rajulu, A., & Winowlin Jappes, J. T. (2018). Preparation and properties of cellulose/tamarind nut powder green composites: (Green composite using agricultural waste reinforcement). *Journal of Natural Fibers*.

74. Nallamuthu, I. D. M. P., et al. (2019). Antimicrobial properties of poly(propylene) carbonate/Ag nanoparticle-modified tamarind seed polysaccharide with composite films. *Ionics (Kiel)*.

75. Indira Devi, M. P., et al. (2019). Biodegradable poly(propylene) carbonate using in-situ generated CuNPs coated Tamarindus indica filler for biomedical applications. *Materials Today Communications*.

76. Indira Devi, M. P., Nallamuthu, N., Rajini, N., Varada Rajulu, A., Hari Ram, N., & Siengchin, S. (2018). Cellulose hybrid nanocomposites using Napier grass fibers with in situ generated silver nanoparticles as fillers for antibacterial applications. *International Journal of Biological Macromolecules, 118*, 99–106.

77. Ervina, J., Mariatti, M., & Hamdan, S. (2016). Effect of filler loading on the tensile properties of multi-walled carbon nanotube and graphene nanopowder filled epoxy composites. *Procedia Chemistry, 19*, 897–905.

78. Hardinnawirda K., & SitiRabiatull Aisha, I. (2012). Effect of rice husks as filler in polymer matrix composites. *Journal of Mechanical Engineering Science, 2*, 181–186.

79. Lu, T., et al. (2014). Effects of modifications of bamboo cellulose fibers on the improved mechanical properties of cellulose reinforced poly(lactic acid) composites. *Composites Part B: Engineering, 62*, 191–197.

80. Islam, M. N., Rahman, M. R., Haque, M. M., & Huque, M. M. (2010). Physico-mechanical properties of chemically treated coir reinforced polypropylene composites. *Composites Part A: Applied Science and Manufacturing, 41*(2), 192–198.
81. Gwon, J. G., Lee, S. Y., Chun, S. J., Doh, G. H., & Kim, J. H. (2010). Effects of chemical treatments of hybrid fillers on the physical and thermal properties of wood plastic composites. *Composites Part A: Applied Science and Manufacturing 41*(10), 1491–1497.
82. Panaitescu, D. M., et al. (2016). Influence of hemp fibers with modified surface on polypropylene composites. *Journal of Industrial and Engineering Chemistry, 37*, 137–146.
83. Rahman, W., Sin, L. T., Rahmat, A. R., Isa, N. M., Salleh, M. S. N., & Mokhtar, M. (2011). Comparison of rice husk-filled polyethylene composite and natural wood under weathering effects. *Journal of Composite Materials, 45*(13), 1403–1410.
84. Homkhiew, C., Ratanawilai, T., & Thongruang, W. (2014). Effects of natural weathering on the properties of recycled polypropylene composites reinforced with rubberwood flour. *Industrial Crops and Products, 56*, 52–59.
85. Hammiche, D., Boukerrou, A., Djidjelli, H., Corre, Y.-M., Grohens, Y., & Pillin, I. (2013). Hydrothermal ageing of alfa fiber reinforced polyvinylchloride composites. *Construction and Building Materials, 47*, 293–300.
86. Arjmandi, R., Hassan, A., Majeed, K., & Zakaria, Z. (2015). Rice husk filled polymer composites. *International Journal of Polymer Science, 2015*.
87. Hamim, F. A. R., Ghani, S. A., & Zainudin, F. (2016). Properties of recycled high density polyethylene (RHDPE)/ethylene vinyl acetate (EVA) blends: The effect of blends composition and compatibilisers. *Journal of Physical Science, 27*(2), 23.
88. Zaaba, N. F., & Ismail, H. (2019). Thermoplastic/natural filler composites: A short review. *Journal of Physical Science, 30*, 81–99.
89. Mochane, M. J., Mokhena, T. C., Mokhothu, T. H., Mtibe, A., Sadiku, E. R., Ray, S. S., et al. (2019). Recent progress on natural fiber hybrid composites for advanced applications: A review. *eXPRESS Polymer Letters, 13*(2), 159–198.
90. Peças, P., Carvalho, H., Salman, H., & Leite, M. (2018). Natural fibre composites and their applications: A review. *Journal of Composites Science, 2*(4), 66.
91. Holbery, J., & Houston, D. (2006). Natural-fiber-reinforced polymer composites in automotive applications. *JOM Journal of the Minerals Metals and Materials Society, 58*(11), 80–86.

Bionanocomposites from Biofibers and Biopolymers

Muhammad Bilal, Tahir Rasheed, Faran Nabeel and Hafiz M. N. Iqbal

Abstract This particular chapter focuses on bionanocomposites, an emergent group of bio-hybrid materials at nanostructured level, as a concept of environmental, bioinspired, and functional hybrid materials. Bionanocomposites represents at least their one dimension on a nanometer scale and can be engineered using naturally occurring biofibers and/or biopolymers either in pristine form or the combination of both along with other inorganic elements. Nanoscale cues/constructs have now become a high requisite for new applications. Likewise, synthetic polymer-based nanocomposites, bionanocomposites (based on biofibers or biopolymers) also exhibit inherited or improved structural and multifunctional characteristics, such as renewability, recyclability, biocompatibility, biodegradability, (re)-generatability, high and efficient functionality against various substrates, induced turn-over, and overall cost-effectiveness are of high interest for numerous applications. Individually or collectively, all those properties of bionanocomposites open new and interesting perspectives with notable incidences in the environmental, biomedical, and biotechnological sector of the contemporary world. In this context, research is underway, around the globe, on the positioning of bionanocomposites as a new interdisciplinary area that could cover significant topics such as bioinspired biomaterials, green composites, bio-nanofabrication strategies and/or engineering processes, and biomimetic systems. Briefly, this chapter discusses various perspectives related to the biofibers and biopolymers, such as cellulose, chitosan, and polyhydroxyalkanoates, as building blocks of bionanocomposites, their sources, and classification along with the development of bionanocomposites using those fibers and polymers. Further to this end,

M. Bilal (✉)
School of Life Science and Food Engineering, Huaiyin Institute of Technology, Huaian 223003, China
e-mail: bilaluaf@hotmail.com

T. Rasheed · F. Nabeel
School of Chemistry and Chemical Engineering, State Key Laboratory of Metal Matrix Composites, Shanghai Jiao Tong University, 800 Dongchuan Road, Shanghai 200240, China

H. M. N. Iqbal (✉)
School of Engineering and Sciences, Tecnologico de Monterrey, Campus Monterrey, Ave. Eugenio Garza Sada 2501, CP 64849 Monterrey, NL, Mexico
e-mail: hafiz.iqbal@tec.mx

© Springer Nature Switzerland AG 2020
A. Khan et al. (eds.), *Biofibers and Biopolymers for Biocomposites*,
https://doi.org/10.1007/978-3-030-40301-0_6

the applied standpoints in relation to environmental and biomedical applications of bionanocomposites are also given with suitable examples.

1 Introduction

In recent years, the research directed to polymer-based nanocomposites plays a noteworthy role in a range of environmental, biomedical and engineering technology applications [1]. The biocomposite materials exhibit exceptional characteristics including thermal steadiness, permeability and mechanical properties that are generated from combining different materials. Generally, composite materials contained two matrix components and fillers. In order to develop composite materials from carbon fibers, Kevlar and natural fibers with various kinds of fillers such as nano- and micro-particles are widely employed that upgrade the mechanical attributes, isolation, conductivity, thermal stability, elasticity, and flame-retardant properties of the matrix. Use of petroleum-based polymers is highly advantageous and received a wider interest in many aspects of engineering applications depending on the requirements [2]. However, the polymers compounds are very detrimental and might significantly damage the natural environment. Three major drawbacks included (i) complicated reusing of composite materials because of the toxic release during the decomposing process, (ii) difficult decay of carbon fiber armored strong polymeric composites, and (iii) elevated cost of green and innovative composites for domestic bio-products [2]. Therefore, researchers and the scientific community is highly concerned to overcome these problems.

In this context, biopolymers or green polymers have been increasingly investigated as potential alternatives to synthetic polymers. Their decomposition circumvents the discharge of noxious compounds to the ecosystem. Biopolymers are applied as a structural matrix for the construction of bio-composite materials, for example, poly(vinyl alcohol), poly(L-lactide), poly(3-hydroxybutyrate-co-3-hydroxy valerate), etc. However, their application perspectives are quite narrow because of the high cost, poor thermal stability, and marginal mechanical features [3]. Nowadays, the tremendous focus has been paid to explore novel biocomposites to improve the material properties. A great variety of nanoparticle fillers is proposed for modifying the thermal and mechanical properties of biocomposites. Obviously, incorporation of some nanoparticles furnishes new or improve the existing properties the sample such as antimicrobial, water vapor permeability, UV protection, facilitating the biocomposite suitability for food packaging, biomedical, and therapeutic applications [4]. For instance, the introduction of blended silica nanostructured particles in polymer matrix composites improved the creep resistance owing to homogeneous nanoparticle distribution [5].

Nature offers an impressive range of biodegradable and biocompatible polymers with great potential to swap many existing polymers. Natural polymers i.e. starch, cellulose, and proteins are some notable examples of such kinds of polymers that illustrate potential alternative nanocomposites to non-biodegradable traditional polymers. Polymer-based nano-biocomposites are materials in which nanoscopic organic/inorganic particles, explicitly 10–1000 Å in at least one aspect, are distributed in an organic polymer matrix for dramatic improvement of the polymer properties. Because of the nanoscale length, nanocomposites are usually translucently exhibiting significantly improved properties than those of conventional polymers or composites. They present high strength, heat resistance, lower flammability, improved solvency, and exceptional barrier properties without any negative impacts on ductility.

2 Biofibers and Biopolymers—Building Blocks of Bionanocomposites

In nanobiocomposites, the biofibers are used as renewable and low-cost reinforcement by increasing the strength and toughness to the synthesized composites. The conventional fibers like aramid, carbon, glass, etc., can be manufactured with definite physicochemical characteristic, while the typical properties of natural fibers significantly vary depending on plant leaves or stem, and the quality and age of the plant species. Based on their origin, these fibers can be classified into seed, bast, leaf, core, fruit, and grass and reed fibers. The best recognized examples are: (i) Leaf (sisal, pineapple leaf fiber, and henequen) (ii) Bast (flax, kenaf/mesta, ramie, hemp and jute) (iii) Seed (cotton) and (iv) Fruit (coconut husk) [6]. Natural fibers exhibit a hollow structure, where each fibril possess an intricate and layered arrangement as depicted in Fig. 1 [7].

Biorenewable materials are essential for environmental sustainability and can be utilized as viable raw feedstocks in industrial sectors to substitute non-biodegradable petro-based products, which are expensive and undesirable leading to increase of carbon dioxide in the ecosystem and thus global warming [8]. Presently, biodegradable bio-based polymers play a pivotal role in all engineering aspects owing to their

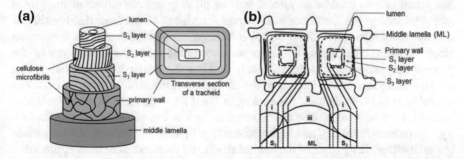

Fig. 1 **a** Three-dimensional structure of the secondary cell wall of a xylem cell and **b** the relative amounts of cellulose, hemicellulose, and lignin across a cross-section of two wood cells (i: cellulose; ii: lignin; and iii: hemicelluloses). Reprinted from Wei and McDonald [7], an open-access article distributed under the terms and conditions of the Creative Commons Attribution (CC-BY) license (https://creativecommons.org/licenses/by/4.0/)

ecological friendliness. Due to microbial decomposition, their consequence is environmentally friendly than that to the disintegration of petro-based materials. The contributing monomers to manufacture such polymers may be either bio-renewable i.e. based on agricultural plant and animal products or synthetic. Production of biopolymers from renewable resources is more desirable than the synthetic counterparts owing to their renewability, biocompatibility, biodegradability and carbon-neutrality characteristics. Therefore, the application of bio-based polymers would build a sustainable industry. When a bio-decomposable or degradable material, neat polymer, amalgamated product, or biocomposite is completely derived from sustainable resources, it can be referred to as a green polymeric material [9]. The nanocomposites fabrication from natural polymers, such as cellulose, starch, and chitin, and oriented investigation in this direction intended to increase the features of the bio-products. Similarly, polysaccharide polymers with natural abundant availability are also utilized to synthesize nanocomposites. Nevertheless, the associated drawbacks of biopolymer i.e. low strength, brittleness, and low stiffness need to be improved by incorporating some compounds in the matrix to constitute the biocomposite materials. Notably, the functionality of a biopolymer might be limited, but the design and life-cycle of biopolymers aid some important applications necessitating very fast degradation such as the medical industry. In this way, the biopolymers have become significant for such purposes [10].

2.1 Cellulose

The natural fibers are composed of lignocellulosic materials, which are the renewable and most plentiful organic resources on the planet. Lignocelluloses are broadly scattered in the atmosphere in the form of plants, crops, and trees. Cellulose, hemicellulose, and lignin are the major structural components of lignocellulosic bio-fibers. Among the components of lignocellulosic materials, only cellulose in its numerous arrangements establishes almost half of all polymer consumed in the global industrial sectors [6]. Cellulose is a hydrophilic glucan biopolymer that consists of a linear macromolecular chain of 1,4-β-bonded anhydroglucose units containing alcoholic hydroxyl groups [11]. The occurrence of multiple hydroxyl groups on the glucose ring promotes widespread intra- and intermolecular hydrogen bonds that hold them together tightly leading to the formation of insoluble crystalline cellulose microfibrils [11–13]. Indeed, the hydrogen bonding within and between cellulose chains are linked to its unique characteristics such as exceptional strength, durability, crystallinity, stiffness, and biocompatibility. In addition, these hydroxyl groups impart hydrophilicity to all of the natural fibers and their moisture level approached to 8–12.6%. Though the crystalline cellulose structure is identical in various natural fibers but varies with respect to their degree of polymerization. The mechanical traits of a lignocellulosic fiber are directly related to the degree of polymerization, and among the 10,000 different natural fibers, Bast fibers present the highest degree of polymerization [14]. Cellulose can be chemically modified to a range of cellulose esters and ether derivatives such as cellulose acetate/butyrate/ propionate, ethyl or methylcellulose, and hydroxypropyl cellulose. The reactivity of cellulose is based on the presence as well as locations of OH groups. For instance, the OH groups on the C6 position is 10-times more reactive during esterification compared to the OH groups attached on C2 and C3 [13]. Overall, OH-C6 is more reactive followed by OH-C2 and OH-C3 among the three types of OH groups. The mechanical properties of cellulose derived compounds are modified than pristine cellulose and the degree of replacements determine these properties [15].

2.2 Nanocellulose

Cellulose nanomaterials have recently received a great deal of attention. Two common forms of nanocellulose include CNCs and CNFs. CNCs, also recognized as cellulose whiskers, nanorods, or nanowhiskers, can be achieved by removal of amorphous or crystalline regions of the cellulose, hemicellulose, and lignin using a controlled acid hydrolysis treatment [7]. This treatment results in rod-shaped acid resistant CNCs with a reduced degree of polymerization in contrast to the native cellulose [16]. The The dimension extent of CNCs differs upon the cellulose origins ranging from 100 nm to several micrometers (celluloses of bacteria, algae, and tunicates) in length and 3–20 nm in diameter [17, 18]. CNCs typically exhibit an elevated specific surface area, crystallinity and Young's modulus of 150–170 m^2/g, 54–90% and 120–170 GPa, respectively, with highly reactive surfaces due to the OH moieties [17, 19, 20]. String-like CNFs particles can be fabricated by various methods, such as 2,2,6,6,-tetramethylpipelidine-1-oxyl radical (TEMPO)-assisted oxidation, enzymatic hydrolysis, ultrasonic methods, and high-pressure homogenization [21, 22]. These nanofibrils possess a size of 4–20 nm with a length of several micrometers. CNFs produced by TEMPO oxidation exhibit the magnitudes of 3–10 nm in diameter along with a length of few microns. CNFs encompass both amorphous and crystalline regions [23].

2.3 Lignin

Apart from cellulose, lignocellulosic fibers also consist of hemicellulose and phenolic compound lignin. Lignin is a complicated three-dimensional biochemical polymer containing both aromatic and aliphatic elements. It is characterized by three main monomers coniferyl alcohol, p-coumaryl alcohol and synapyl alcohol (Fig. 2) [24]. Notably, monolignols composition depends on classes and species of plants i.e., grass, hardwood, softwood. The main active functional group present in hardwood Kraft lignin include aliphatic hydroxyl, carboxylic acid, phenolic hydroxyl, and methoxy with a concentration of 1.7, 0.5, 4.3, and 580 mmol/g, respectively [25]. This non-polysaccharide matrix is generally a low-cost waste material of the pulp and paper

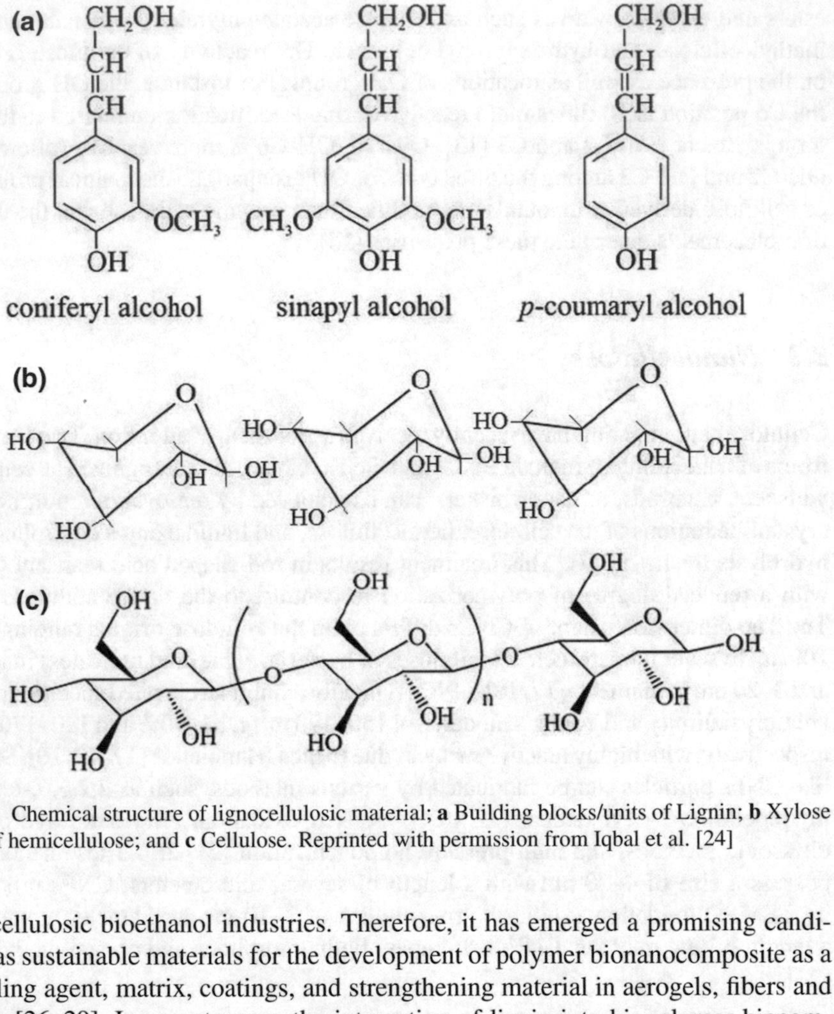

Fig. 2 Chemical structure of lignocellulosic material; **a** Building blocks/units of Lignin; **b** Xylose unit of hemicellulose; and **c** Cellulose. Reprinted with permission from Iqbal et al. [24]

and cellulosic bioethanol industries. Therefore, it has emerged a promising candidate as sustainable materials for the development of polymer bionanocomposite as a coupling agent, matrix, coatings, and strengthening material in aerogels, fibers and resins [26–29]. In recent years, the integration of lignin into biopolymer biocomposites as a reinforcement matrix has gained burgeoning interest owing to the rising consciousness of constructing a sustainable environment.

2.4 Chitin and Chitosan

Chitin is another important bio-fiber source, which can be prevalently found in the crustacean shells e.g., crab and shrimp, insect's cuticles, and cell walls of some fungi and microorganisms. It has a linear polymer containing β-(1–4)-linked 2-deoxy-2-acetamido-D-glucose units [30, 31]. Chitosan is a deacetylated chitin derivative with a degree of deacetylation more than 75%. Owing to admirable biocompatibility and biodegradability, chitosan finds increasing importance for biomedical purposes such as artificial skin, kidney membrane, and drug delivery systems. Like cellulose, it is of high interest to obtain chemically modified chitin and chitosan without any significant alterations in their physicochemical properties and backbone structure. Chitin nanofibers and chitin-protein blended composite nanofibers have been prepared and modified for diverse biomedical applications [32].

3 What are Bionanocomposites?

Bionanocomposite, also named as nano-biocomposites, bio-hybrids or green composites has become a common term to illustrate the nano-material blends of natural polymers with or without doping of inorganic moieties [33]. Bionanocomposite constitutes a fascinating interdisciplinary area that associates nanotechnology, biology and material sciences. Due to the systematic distribution of nano-sized particles in the polymeric matrix, the physical, optical and mechanical properties of nano-biocomposites are significantly upgraded [34]. Additionally, bionanocomposite displays improved biocompatibility and functional properties due to the occurrence of inorganic or biological entities. A synthetic or natural biodegradable polymer shows vitality in developing nano-sized materials incorporated bio-nanocomposites. Biopolymers such as lignin, cellulose, hemicellulose, chitin, chitosan, lignin, glucans, pullulan, protein, polyhydroxyalkanoates, and linear polyesters are often used for the synthesis of nanocomposites. Certain factors including nature of biopolymer, degree of cross-linking and the stoichiometric ratio of macromolecular skeletons and assembled primary constituents (naturally abundant biopolymers) affect the characteristics of bionanocomposites [35–38]. The ecological concerns and sustainability necessitate bio-constructs hybrid essentially synthesized from renewable sources apart from their real degradability and practical life cycle.

4 Biofibers and Biopolymers Based Bionanocomposites

Aiming to reduce/lower the environmental pollution burden caused by the excessive utilization of petroleum-based resources and their synthetic counterparts that results from large amounts of waste, current research is being refocused on the exploitation of green natural materials which are carbon–neutral and environmentally friendlier, in nature. Therefore, the development of bionanocomposites using bioinspired biofibers and biopolymers, such as cellulose, chitosan, protein, and polyhydroxyalkanoates, are the focus of many research groups, around the globe [39–42]. Moreover, petroleum-based resources and their synthetic counterparts are now being replaced with naturally occurring biofibers and/or biopolymers extracted from renewable natural resources, such as cellulose, chitosan, proteins, polyhydroxyalkanoates, and many others [43–50]. Additional to this, several inherited or improved characteristics of natural biofibers and biopolymers, such as renewability, recyclability, biocompatibility, biodegradability, and (re)-generatability, all signify a huge benefit to the environment and can potentially contribute to abridge the undesirable dependence on petroleum-based resources or their synthetic counterparts. Following subsections briefly discuss various bionanocomposites based on cellulose, chitin and chitosan, and polyhydroxyalkanoates with suitable examples.

4.1 Cellulose-Based Bionanocomposites

The chemistry and other highly requisite physiochemical characteristics of cellulose and other cellulose-based materials that make them best fit for the development of bionanocomposites are discussed earlier under the section of biofibers and biopolymers as building blocks of bionanocomposites. An array of diverse processing methods has been developed and introduced for different biofibers including cellulose. For instance, molding that includes injection and compression molding, and resin transfer molding, enzymatic grafting, and vacuum bagging, among others [39, 51]. Alemdar and Sain [52] used a solution casting method to develop biocomposites of cellulose-based wheat straw nanofibers and thermoplastic starch and characterized for morphology, mechanical and thermal properties. As an initial step, a combination of chemical and mechanical technique, so-called "chemi-mechanical technique" was implemented to isolate cellulose nanofibers from wheat straw as a raw material. As obtained cellulose nanofibers were found to have diameters in the range of 10–80 nm and lengths of several thousand nanometers. Very recently, Ayrilmis et al. [53] developed *Moringa oleifera* cellulose-based epoxy nanocomposites by acid hydrolysis and designated as cellulose nanowhiskers. A compared to neat cured epoxy resin, it was found that the epoxy composites filled with 0.18 wt% cellulose nanowhiskers displayed better mechanical characteristics of the composites, such as the tensile modulus (31.4%) and bending modulus (38.2%) [53]. In another study, one-pot in situ polymerization approach was exploited to engineer bio-nanocomposites of polylactide (PLA)-CNFs [54]. For a said purpose, bleached hardwood kraft pulp was first subjected to the enzymatic (*FiberCare R* type enzyme) and mechanical treatment to obtain CNF. Following that various amounts of CNF were used to fabricate bio-nanocomposites through ring opening polymerization (ROP) of $_L$-lactide. The end-product, i.e., bio-nanocomposites of PLA-CNF was speculated to be a composite mixture of both PLA chains coupled to CNF surface along with free PLA chains as well (Fig. 3) [54]. The development of cellulose acetate and lignin-rich rice straw nanofibers-based nanocomposite film is reported by Hassan et al. [55]. Unbleached neutral sulfite rice straw pulp was used to isolate nanofibers to formulate transparent films by casting cellulose acetate (CA) solution in acetone. The characterization profile revealed their transparency and light transmittance features as shown in Fig. 4 [55].

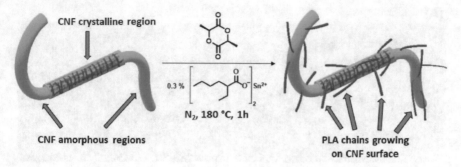

Fig. 3 Schematic representation of the "grafting-from" reaction resulting in PLA chain growing from the CNF surface. Reprinted from Gazzotti et al. [54], with permission from Elsevier. Copyright (2018) Elsevier Ltd.

4.2 Chitin and Chitosan-Based Bionanocomposites

Chitosan being the second most copious biopolymer available on the earth has been extensively studied and exploited for the development of robust bionanocomposites though using different methods/processes. Enormous inherited structural and bio-functional features of chitin and chitosan offer wide range possibilities for chemical, mechanical and biological modifications to achieve amendable requisite features. However, such modifications are not the focus of this work and have been reviewed elsewhere [56–59]. Regardless of the processing approach, various types of chitin and chitosan-based bionanocomposites have been developed in different geometries at the nanoscale. For instance, Peter et al. [60] prepared nanocomposite scaffolds of chitosan–gelatin/nanohydroxyapatite by combining gelatin and chitosan with nanophase hydroxyapatite (nHA). The as-synthesized composite frameworks were highly porous with a pore diameter of 150–300 µm. Recently, Enescu et al. [61] developed chitosan, nanoclays and cellulose nanocrystals-based bio-inspired films by applying a water-evaporation-induced self-assembly approach. However, this approach has considerable limitations such as unreacted chemical residues left behind on the casted surface that requires additional processing. Qiu et al. [62] constructed chitosan/ZnO nanocomposite film by a facile and green method via in-situ precipitation of nano-ZnO (nZnO) in the chitosan membrane. The characterization results revealed robust coordination interaction between chitosan matrix and Zn^{2+} for the good dispersal of nZnO in the chitosan membrane. The preparation process is shown in Fig. 5 [62].

4.3 Poly(hydroxyalkanoates)-Based Bionanocomposites

Poly(hydroxyalkanoates) (PHAs) are polymers produced by biological systems i.e. microorganisms through metabolic engineering reactions. Based on their nature of

(A)

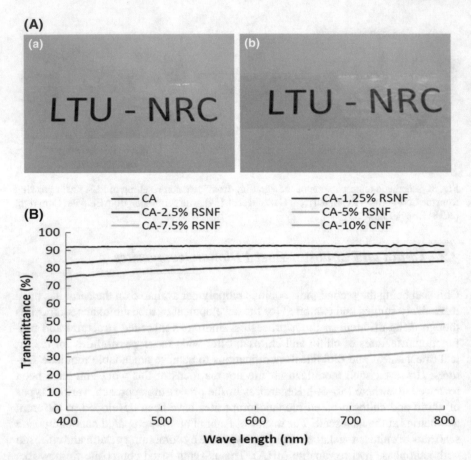

Fig. 4 (**A**) Photos of neat CA (**a**) and CA/10% RSNF films (**B**) over a printed paper, and **b** Visible light transmittance of CA films containing different ratios of RSNF. Transmittance values of 92%, 87%, and 80% were recorded for neat CA, CA/1.25% RSNF, and CA/2.5% RSNF films, respectively. Reprinted from Hassan et al. [55], an open-access article distributed under the terms and conditions of the Creative Commons Attribution (CC BY) license (https://creativecommons.org/licenses/by/4.0/). Copyright (2019) the authors. Licensee MDPI, Basel, Switzerland

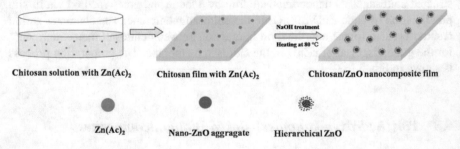

Fig. 5 Schematic representation of the preparation process. Reprinted from Qiu et al. [62], with permission from Elsevier. Copyright (2018) Elsevier B.V

origin, PHAs has been categorized or classified as natural or synthetic [39]. From the past few years, PHAs have gained substantial consideration as bio-sustainable candidates for the development of bionanocomposites in numerous geometries. Based on the literature evidence, more than 150 monomers have been known as components of PHAs [39]. Numerous PHAs offers a wide range of properties to developing novel constructs with multifarious attributes. Most studied PHAs are poly 3-hydroxybutyrate, 3-hydroxyvalerate copolymer, 3-hydroxyhexanoate copolymer, 3-hydroxybutyrate, poly 4-hydroxybutyrate, and poly 3-hydroxyoctanoate. Despite the use of PHAs in pristine form, various other materials have also been used in combination to achieve requisite properties. For instance, very recently, Relinque et al. [63] developed a surface coated PLA/PHA nanocomposites by mechanical mixing under gentle conditions and low loading contents (<0.10 wt %). As developed PLA/PHA nanocomposites were coated with silver nanoparticles or either Graphite NanoPlatelets (GNP). Iqbal et al. [64] synthesized poly(3-hydroxybutyrate) [(P(3HB)] grafted ethyl cellulose (EC) based green composites with multifunctional characteristics by laccase-mediated grafting. The newly grafted composites showed improved functionalities such as good mechanical and tensile strength [64]. In an earlier study, the same group has prepared a laccase-assisted grafting of gallic acid and thymol as functional entities onto the formerly constituted P(3HB)-g-EC composite [65, 66]. The surface dipping and incorporation technique were used to graft different concentrations of GA and T, each separately, onto the previously developed P(3HB)-g-EC composite [65, 67]. Evidently, it was recorded that as developed enzymatically grated bio-composites had strongest bacteriostatic and bactericidal activities against Gram + bacterial strains, i.e., *Bacillus subtilis* NCTC 3610 and *Staphylococcus aureus* NCTC 6571 and Gram- bacterial strains, i.e., *Escherichia coli* NTCT 10,418 and *Pseudomonas aeruginosa* NCTC 10,662. In addition to strong antibacterial activity, these grafted bio-composites maintained differential viability of human keratinocyte-like (HaCaT) skin cells. Figs. 6 and 7 shows the morphology of HaCaT cells grown on the GA-g-P(3HB)-g-EC and T-g-P(3HB)-g-EC bio-composites at 1, 3 and 5 days of seeding [66].

5 Applications of Bionanocomposites

Bio-nanocomposite exhibit good biocompatibility, biodegradability, and mechanical properties. Effective antimicrobial performance of bio-nanocomposite materials is attributed to the large surface to volume and high surface activity making them useful against microorganisms compared to other macro- or micro-scale counterparts [3]. These properties are of profound importance in the field of regenerative medicines and developing green and environmentally safe nanocomposite. One of the main features of bio-nanocomposite materials is the biocompatibility and the accurate functionality at the physiological condition of the human body for the required results without harmful effects [68]. Several progressive biomedical applications of bio-nanocomposite materials include drug delivery systems, tissue engineering, and

0GA-*g*-P(3HB)-EC

5GA-*g*-P(3HB)-EC

10GA-*g*-P(3HB)-EC

15GA-*g*-P(3HB)-EC

20GA-*g*-P(3HB)-EC

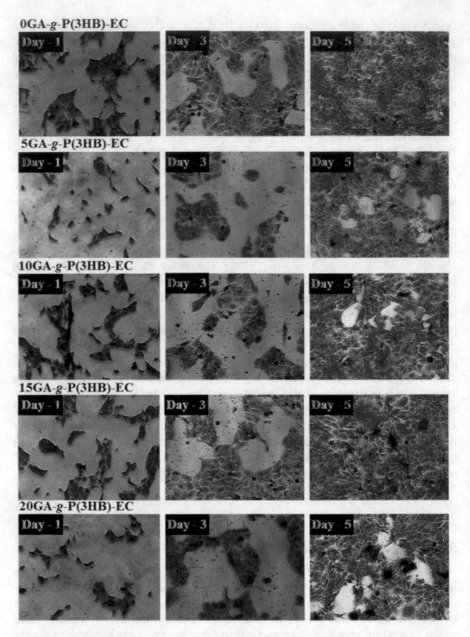

Fig. 6 Adherent morphology of HaCaT cells seeded on the GA-*g*-P(3HB)-*g*-EC bio-composites. All of the test samples were stained using neutral red dye (5 mg/mL) for 1 h followed by three consecutive washings with PBS at ambient temperature. Reprinted from Iqbal et al. [66], with permission from Elsevier. Copyright (2015) Elsevier Ltd.

0T-*g*-P(3HB)-EC

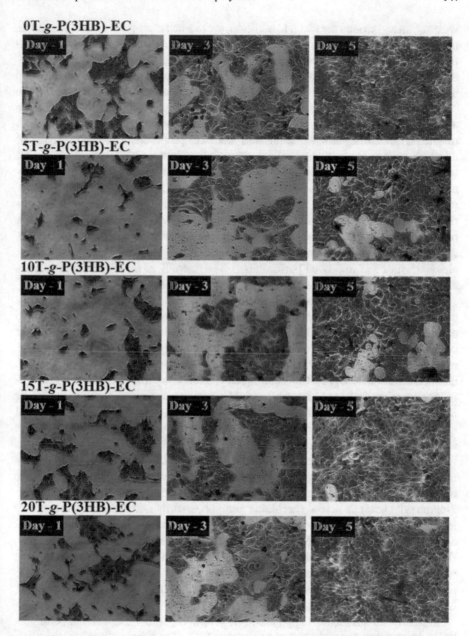

5T-*g*-P(3HB)-EC

10T-*g*-P(3HB)-EC

15T-*g*-P(3HB)-EC

20T-*g*-P(3HB)-EC

Fig. 7 Adherent morphology of HaCaT cells seeded on the T-*g*-P(3HB)-*g*-EC bio-composites. All of the test samples were stained using neutral red dye (5 mg/mL) for 1 h followed by three consecutive washings with PBS at ambient temperature. Reprinted from Iqbal et al. [66], with permission from Elsevier. Copyright (2015) Elsevier Ltd.

vaccination and wound dressing. In addition, the bio-nanocomposite materials can be used as bio-nanocomposite films for food packaging, which are renewable, environmentally friendly, cost-effective and manifest high antimicrobial activity. So far, bio-nanocomposite materials investigated for packaging applications include cellulose derivatives, starch, PLA, polyhydroxybutyrate, polycaprolactone and poly-(butylene succinate). The appropriate fillers are layered silicate nanoclays of montmorillonite and kaolinite [3]. Polymers armored with functional nanostructures can offer optical, magnetic, electromagnetic shielding and electrical properties to develop advanced materials such as sensors, solar cells, light emitting diodes, display panels, radiation shielding materials and medical devices [69]. The variation in the clay by biopolymers provides an alternative for pollutant removal [70]. For instance, chitosan bio-nanocomposites and hydrogels with adsorbent and biocatalytic properties were fabricated to remove heavy metals ions and azo dyes [71–73]. Bio-nanocomposites can be altered by using different nanofillers such as silver, metal matrix nanocomposite, ZnO, etc. with biopolymers, for example, starch and polylactic acid endows superior characteristics and improved performance. The nanofillers do affect the properties and the prospective performance of the bio-nanocomposites [74].

Chitosan matrix reinforced from nanostructures has been used for the development of a number of bionanocomposite materials such as chitin, cellulose nanofibers and clays [75, 76]. Acidic pesticides are mostly present in their anionic form in soil and contaminate ground and surface water. Therefore, the synthesis and engineering of materials for the decomposition of anionic pesticides to control the environment has attained much interest [77]. Celis et al. [78] prepared montmorillonite-chitosan bio-nanocomposites that could adsorb pesticides and remove anionic pesticides under mildly acidic conditions from water. Clay minerals can efficiently adsorb polar and anionic pesticides. Although, the clay minerals display some restrained attraction as of some repulsion from clay surfaces due to negative charge for anionic species from pesticides. The introduction of cationic polymer can reduce this repulsion by modifying the clay surface to hydrophobic from a hydrophilic character in case excessively adsorbed reversing the charge. The clopyralid (3,6-dichloropyridine-2-carboxylic acid) an anionic herbicide was successfully adsorbed from the soil by montmorillonite-chitosan. So, the introduction of organic cations could be a potential approach to increase the adsorption of anionic pesticides towards clay minerals. Bion-nanocomposite material based on chitosan matrix and others organic–inorganic substances such as organoclay, epichlorohydrin and Schiff's base@Fe_3O_4 have also been synthesized for eliminating toxic heavy metals such as Cr (VI), Cu (II), Ni (II), Cd (II), Pb (II), etc. [79–81].

Similarly, the biomedical application of several organic–inorganic nanocomposite hydrogels has been studied significantly [82]. The nanocomposite hydrogels developed for drug delivery purposes have shown superior performance in comparison to only polymer hydrogels. The nanoparticles enhance the efficacy of the drug by improving the release effect, reducing the bursting of the drug and progressive and slower drug releases [83]. Several studies have been reported using chitosan nanocomposites investigating the drug release behavior with different inorganic nanoparticles such as vermiculite, attapulgite, palygorskite, montmorillonite, hydroxyapatite, etc. [84, 85]. Metal nanoparticles, for example, silver nanoparticles (AgNPs) have shown promising results due to their excellent antimicrobial activity and low toxicity, are commercially used in medical devices, clothing, and pharmaceutics. Yadollahi et al. [82] have prepared chitosan/Ag nanocomposite hydrogel as shown in Fig. 8. The prepared hydrogel has shown higher swelling compared to hydrogel without AgNPs. The chitosan/Ag nanocomposite hydrogel also performed well against gram-positive and gram-negative microbial pathogens for the agar diffusion test. Similarly, the group has also developed chitosan/ZnO and cellulose/ZnO nanocomposite hydrogels for drug delivery applications [86, 87]. Therefore, metal nanoparticles could be potentially used for the production of bio-nanocomposite material for drug delivery application.

Encapsulation of bioactive molecules such as drug or high-quality supplement is important where a stabilized material is packed for specific estimated times [88]. These bioactive substances must be structurally aligned in such a manner without

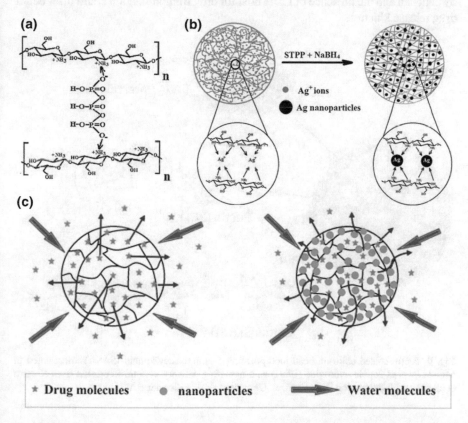

Fig. 8 Preparation of chitosan/Ag nanocomposite and drug release mechanism; **a** sodium tripolyphosphate (STPP) crosslinking with chitosan, **b** interaction of Ag ions and AgNPs with chitosan and **c** drug release mechanism of polymer (without NPs) and chitosan/Ag nanocomposite. Reprinted from Yadollahi et al. [82], with permission from Elsevier. Copyright (2015) Elsevier B.V

affecting their efficacy. An enzyme of firefly luciferase is little unstable at ambient situations can be employed in the pharmaceutics as well as food hygiene [89]. Therefore bio-nanocomposite materials were pleasantly developed to stabilize this enzyme. Liu et al. [90] prepared freeze-dried hydrogel bio-nanocomposite from chitosan/xanthan blended gum to firmly encapsulate firefly luciferase enzyme. The encapsulated enzyme was stable and could be discharged at suitable rates from rehydrated samples. Polymer protective coating has been employed for a number of drugs, for example, the synthetic polymer hydroxypropylmethylcellulose [91] and Eudragit®S [92] have been investigated as colonic drug delivery systems. These were more efficient and precise in drug liberation for specific site compared to conventional oral formulations of the same drug. Ribeiro et al. [93] have revealed a new drug delivery approach based on pectin coated chitosan microspheres with 5ASA and Mg–Al LDH intercalated as present in Fig. 9 [93]. This system was aimed to cover benefits of the pectin coated for low stomach pH, drug adsorption muco-adhesive pattern provided by chitosan and the presence of LDH host for drug immobilization could offer better drug release kinetics.

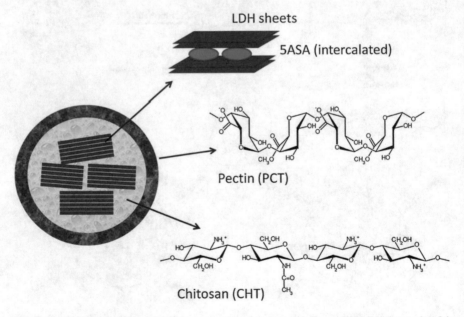

Fig. 9 Pectin-coated chitosan bead incorporating 5-aminosalicylic acid (5ASA) intercalated in Mg–Al layered double hydroxide (LDH) as a new drug delivery system. Reprinted from Ribeiro et al. [93], with permission from Elsevier. Copyright (2013) Elsevier B.V

6 Conclusions and Outlook

In summary, bionanocomposites based on naturally occurring biofibers and/or biopolymers have been the subject of intensive research in different sectors with a variety of applications, from environmental to biomedical at the nanoscale level. To signify the applied value of bionanocomposites, research is underway, around the globe, on the positioning of bionanocomposites as a new interdisciplinary area that could cover significant topics such as bioinspired biomaterials, green composites, bio-nanofabrication strategies and/or engineering processes, and biomimetic systems. More precisely, the area of bionanocomposites has considerable potential and can be integrated as a new field at the frontier of nanobioengineering, nanotechnology, and polymer science or macromolecular science. Clearly, based on literature evidence and published reports, the (re)-use or (re)-focus on naturally occurring biofibers and biopolymers for the development of multifunctional bionanocomposites for a variety of applications in the environmental and biomedical sector has revolutionized the materials science of this century.

In recent years, several types of bionanohybrid nano-constructs/nano-cues have been engineered by integrating naturally occurring charged biopolymers or proteins such as chitosan or keratin, respectively. Additional to this, the development of multi-materials based bionanocomposites which sometimes also terms as bionanohybrid materials have also been the interest of researchers from different fields. Considering the novel characteristics of multifunctional bionanocomposites, the buzz words like renewability, recyclability, biocompatibility, biodegradability, (re)-generatability, sustainability, "cradle-to-grave" design, environmentally friendly, and eco-efficiency are not just newly coined but also pave the way/principles or guidelines for the development of a new generation of "green materials" following the green chemistry agenda.

In future studies, novel pristine or multi-materials should be constructed following compact surface design and amendment of bionanocomposites. Besides, naturally occurring polymers should be utilized to design controllable degradable delivery systems. The use of natural biofibers and biopolymers will also reduce the environmental and pollution burden of petroleum-based resources. Additionally, such practices will also help to keep reserve the petroleum-based resources which are finite. The click chemistry mediated biomimetic bionanohybrid nano-constructs/nano-cues could encapsulate various other organic or inorganic solid components for industrially requisite applications. Innovative click chemistry-based techniques are simple, facile, controllable, and promising approaches to prepare cross-linked bionanocomposites. From the perspectives of the biomedical application, numerous materials-based cues with remarkable functionalities should be designed to manipulate the drug-loaded molecules and their structures to control indispensable features such as degradability, compatibility, toxicity, and controlled delivery. Moreover, to further strengthen the diagnostic are of the medical sector, numerous fluorescent imaging-based approaches should also be implemented to understand the drug-loaded nanocarriers-host interactions. This will further strengthen the requisite delivery insight and release mechanisms. Despite current advancement and technology innovation in the medical sector, significant challenges remain before such bionanocomposites can be used successfully and routinely for multi-purpose environmental and therapeutic applications.

Acknowledegment All listed authors are grateful to their representative universities/institutes for providing the literature facilities.

Conflict of Interest Authors declare no conflict of interest in any capacity including competition or financial.

References

1. Essabir, H., Raji, M., Laaziz, S. A., Rodrique, D., Bouhfid, R., & el kacem Qaiss, A. (2018). Thermo-mechanical performances of polypropylene biocomposites based on untreated, treated and compatibilized spent coffee grounds. *Composites Part B: Engineering, 149*, 1–11.
2. Lau, K. T., Hung, P. Y., Zhu, M. H., & Hui, D. (2018). Properties of natural fiber composites for structural engineering applications. *Composites Part B: Engineering, 136*, 222–233.
3. Rhim, J. W., Park, H. M., & Ha, C. S. (2013). Bio-nanocomposites for food packaging applications. *Progress in Polymer Science, 38*(10–11), 1629–1652.
4. Nafchi, A. M., Alias, A. K., Mahmud, S., & Robal, M. (2012). Antimicrobial, rheological, and physicochemical properties of sago starch films filled with nanorod-rich zinc oxide. *Journal of Food Engineering, 113*(4), 511–519.
5. Hassanzadeh-Aghdam, M. K., Ansari, R., Mahmoodi, M. J., & Darvizeh, A. (2018). Effect of nanoparticle aggregation on the creep behavior of polymer nanocomposites. *Composites Science and Technology, 162*, 93–100.
6. Mohanty, A. K., Misra, M. A., & Hinrichsen, G. I. (2000). Biofibres, biodegradable polymers and biocomposites: An overview. *Macromolecular Materials and Engineering, 276*(1), 1–24.
7. Wei, L., & McDonald, A. (2016). A review on grafting of biofibers for biocomposites. *Materials, 9*(4), 303.
8. Kargarzadeh, H., Mariano, M., Huang, J., Lin, N., Ahmad, I., Dufresne, A., et al. (2017). Recent developments on nanocellulose reinforced polymer nanocomposites: A review. *Polymer, 132*, 368–393.
9. Dufresne, A., Thomas, S., & Pothan, L. A. (2013). Bionanocomposites: State of the art, challenges, and opportunities. In *Biopolymer nanocomposites: Processing, properties, and applications* (pp. 1–10).
10. John, M. J., & Thomas, S. (2008). Biofibres and biocomposites. *Carbohydrate Polymers, 71*(3), 343–364.
11. Kadla, J. F., & Gilbert, R. D. (2000). Cellulose structure: A review. *Cellulose Chemistry and Technology, 34*(3–4), 197–216.
12. Klemm, D., Heublein, B., Fink, H. P., & Bohn, A. (2005). Cellulose: Fascinating biopolymer and sustainable raw material. *Angewandte Chemie International Edition, 44*(22), 3358–3393.
13. Roy, D., Semsarilar, M., Guthrie, J. T., & Perrier, S. (2009). Cellulose modification by polymer grafting: A review. *Chemical Society Reviews, 38*(7), 2046–2064.
14. Pearce, E. M. (1985). *Handbook of fiber science and technology: Fiber chemistry* (Vol. 4). Marcel Dekker Incorporated.

15. Pilla, S. (2011). *Handbook of bioplastics and biocomposites engineering applications* (Vol. 81). John Wiley & Sons.
16. Habibi, Y., Lucia, L. A., & Rojas, O. J. (2010). Cellulose nanocrystals: Chemistry, self-assembly, and applications. *Chemical Reviews, 110*(6), 3479–3500.
17. Šturcová, A., Davies, G. R., & Eichhorn, S. J. (2005). Elastic modulus and stress-transfer properties of tunicate cellulose whiskers. *Biomacromolecules, 6*(2), 1055–1061.
18. Li, M. C., Wu, Q., Song, K., Lee, S., Qing, Y., & Wu, Y. (2015). Cellulose nanoparticles: Structure–morphology–rheology relationships. *ACS Sustainable Chemistry & Engineering, 3*(5), 821–832.
19. Agarwal, U. P., Sabo, R., Reiner, R. S., Clemons, C. M., & Rudie, A. W. (2012). Spatially resolved characterization of cellulose nanocrystal–polypropylene composite by confocal Raman microscopy. *Applied Spectroscopy, 66*(7), 750–756.
20. Chen, L., Wang, Q., Hirth, K., Baez, C., Agarwal, U. P., & Zhu, J. Y. (2015). Tailoring the yield and characteristics of wood cellulose nanocrystals (CNC) using concentrated acid hydrolysis. *Cellulose, 22*(3), 1753–1762.
21. Abe, K., Iwamoto, S., & Yano, H. (2007). Obtaining cellulose nanofibers with a uniform width of 15 nm from wood. *Biomacromolecules, 8*(10), 3276–3278.
22. Xu, X., Liu, F., Jiang, L., Zhu, J. Y., Haagenson, D., & Wiesenborn, D. P. (2013). Cellulose nanocrystals vs. cellulose nanofibrils: A comparative study on their microstructures and effects as polymer reinforcing agents. *ACS Applied Materials & Interfaces, 5*(8), 2999–3009.
23. Saito, T., Kimura, S., Nishiyama, Y., & Isogai, A. (2007). Cellulose nanofibers prepared by TEMPO-mediated oxidation of native cellulose. *Biomacromolecules, 8*(8), 2485–2491.
24. Iqbal, H. M. N., Kyazze, G., & Keshavarz, T. (2013). Advances in the valorization of lignocellulosic materials by biotechnology: An overview. *BioResources, 8*(2), 3157–3176.
25. Brodin, I. (2009). *Chemical properties and thermal behaviour of kraft lignins*. Doctoral dissertation, KTH, Department of Fibre and Polymer Technology, KTH Royal Institute of Technology, Stockholm, Sweden.
26. Chung, Y. L., Olsson, J. V., Li, R. J., Frank, C. W., Waymouth, R. M., Billington, S. L., et al. (2013). A renewable lignin–lactide copolymer and application in bio-based composites. *ACS Sustainable Chemistry & Engineering, 1*(10), 1231–1238.
27. Pohjanlehto, H., Setälä, H. M., Kiely, D. E., & McDonald, A. G. (2014). Lignin-xylaric acid-polyurethane-based polymer network systems: Preparation and characterization. *Journal of Applied Polymer Science, 131*(1), 39714.
28. Liu, R., Peng, Y., Cao, J., & Chen, Y. (2014). Comparison on properties of lignocellulosic flour/polymer composites by using wood, cellulose, and lignin flours as fillers. *Composites Science and Technology, 103*, 1–7.
29. Thakur, V. K., Singha, A. S., & Thakur, M. K. (2014). Pressure induced synthesis of EA grafted *Saccaharum cilliare* fibers. *International Journal of Polymeric Materials and Polymeric Biomaterials, 63*(1), 17–22.
30. Arslan, H., Hazer, B., & Yoon, S. C. (2007). Grafting of poly (3-hydroxyalkanoate) and linoleic acid onto chitosan. *Journal of Applied Polymer Science, 103*(1), 81–89.
31. Kikkawa, Y., Fukuda, M., Kimura, T., Kashiwada, A., Matsuda, K., Kanesato, M., … Tanaka, T. (2014). Atomic force microscopic study of chitinase binding onto chitin and cellulose surfaces. *Biomacromolecules, 15*(3), 1074–1077.
32. Ifuku, S., & Saimoto, H. (2012). Chitin nanofibers: Preparations, modifications, and applications. *Nanoscale, 4*(11), 3308–3318.
33. Ponnamma, D., Sadasivuni, K. K., Grohens, Y., Guo, Q., & Thomas, S. (2014). Carbon nanotube based elastomer composites–an approach towards multifunctional materials. *Journal of Materials Chemistry C, 2*(40), 8446–8485.
34. Camargo, P. H. C., Satyanarayana, K. G., & Wypych, F. (2009). Nanocomposites: Synthesis, structure, properties and new application opportunities. *Materials Research, 12*(1), 1–39.
35. Byeon, J. H., & Kim, Y. W. (2013). Continuous gas-phase synthesis of graphene nanoflakes hybridized by gold nanocrystals for efficient water purification and gene transfection. *Chemical Engineering Journal, 229*, 540–546.

36. Sun, X. F., Qin, J., Xia, P. F., Guo, B. B., Yang, C. M., Song, C., et al. (2015). Graphene oxide–silver nanoparticle membrane for biofouling control and water purification. *Chemical Engineering Journal, 281*, 53–59.
37. Bedian, L., Villalba-Rodriguez, A. M., Hernandez-Vargas, G., Parra-Saldivar, R., & Iqbal, H. M. (2017). Bio-based materials with novel characteristics for tissue engineering applications–A review. *International Journal of Biological Macromolecules, 98*, 837–846.
38. Kolbasov, A., Sinha-Ray, S., Yarin, A. L., & Pourdeyhimi, B. (2017). Heavy metal adsorption on solution-blown biopolymer nanofiber membranes. *Journal of Membrane Science, 530*, 250–263.
39. Iqbal, H. M. N. (2015). *Development of bio-composites with novel characteristics through enzymatic grafting.* Doctoral dissertation, University of Westminster.
40. Iqbal, H. M., Kyazze, G., Locke, I. C., Tron, T., & Keshavarz, T. (2015). Poly (3-hydroxybutyrate)-ethyl cellulose based bio-composites with novel characteristics for infection free wound healing application. *International Journal of Biological Macromolecules, 81*, 552–559.
41. Bilal, M., Rasheed, T., Iqbal, H. M., Li, C., Hu, H., & Zhang, X. (2017). Development of silver nanoparticles loaded chitosan-alginate constructs with biomedical potentialities. *International Journal of Biological Macromolecules, 105*, 393–400.
42. Bilal, M., Zhao, Y., Rasheed, T., Ahmed, I., Hassan, S. T., Nawaz, M. Z., et al. (2019). Biogenic nanoparticle-chitosan conjugates with antimicrobial, antibiofilm, and anticancer potentialities: Development and characterization. *International Journal of Environmental Research and Public Health, 16*(4), 598.
43. Iqbal, H. M., Kyazze, G., Locke, I. C., Tron, T., & Keshavarz, T. (2015). In situ development of self-defensive antibacterial biomaterials: Phenol-g-keratin-EC based bio-composites with characteristics for biomedical applications. *Green Chemistry, 17*(7), 3858–3869.
44. Gallegos, A. M. A., Carrera, S. H., Parra, R., Keshavarz, T., & Iqbal, H. M. (2016). Bacterial cellulose: A sustainable source to develop value-added products–A review. *BioResources, 11*(2), 5641–5655.
45. Villalba-Rodriguez, A. M., Parra-Saldivar, R., Ahmed, I., Karthik, K., Malik, Y. S., Dhama, K., et al. (2017). Bio-inspired biomaterials and their drug delivery perspectives-A review. *Current Drug Metabolism, 18*(10), 893–904.
46. Bilal, M., & Iqbal, H. M. (2018). Bio-based biopolymers and their potential applications for bio-and non-bio sectors. In *Handbook of biopolymers: Advances and multifaceted applications* (p. 23).
47. Bilal, M., Rasheed, T., Ullah, A., & Iqbal, H. M. (2018). Valorization of green and sustainable advanced materials from a biomed perspective-potential applications. *Green and Sustainable Advanced Materials: Applications, 2*, 19–47.
48. Iqbal, H. M., Rasheed, T., & Bilal, M. (2018). Design and processing aspects of polymer and composite materials. *Green and Sustainable Advanced Materials: Processing and Characterization, 1*, 155–189.
49. Iqbal, H. M., & Keshavarz, T. (2018). Bioinspired polymeric carriers for drug delivery applications. In *Stimuli responsive polymeric nanocarriers for drug delivery applications* (Vol. 1, pp. 377–404). Woodhead Publishing.
50. Rasheed, T., Bilal, M., Abu-Thabit, N. Y., & Iqbal, H. M. (2018). The smart chemistry of stimuli-responsive polymeric carriers for target drug delivery applications. In *Stimuli responsive polymeric nanocarriers for drug delivery applications* (Vol. 1, pp. 61–99). Woodhead Publishing.
51. Kalia, S., Dufresne, A., Cherian, B. M., Kaith, B. S., Avérous, L., Njuguna, J., & Nassiopoulos, E. (2011). Cellulose-based bio-and nanocomposites: A review. *International Journal of Polymer Science*, Article ID 837875, 35 p.
52. Alemdar, A., & Sain, M. (2008). Biocomposites from wheat straw nanofibers: Morphology, thermal and mechanical properties. *Composites Science and Technology, 68*(2), 557–565.
53. Ayrilmis, N., Ozdemir, F., Nazarenko, O. B., & Visakh, P. M. (2019). Mechanical and thermal properties of *Moringa oleifera* cellulose-based epoxy nanocomposites. *Journal of Composite Materials, 53*(5), 669–675.

54. Gazzotti, S., Rampazzo, R., Hakkarainen, M., Bussini, D., Ortenzi, M. A., Farina, H., …
    Silvani, A. (2019). Cellulose nanofibrils as reinforcing agents for PLA-based nanocomposites:
    An in situ approach. *Composites Science and Technology*, *171*, 94–102.
55. Hassan, M., Berglund, L., Abou-Zeid, R., Hassan, E., Abou-Elseoud, W., & Oksman, K. (2019).
    Nanocomposite film based on cellulose acetate and lignin-rich rice straw nanofibers. *Materials*,
    12(4), 595.
56. Shukla, S. K., Mishra, A. K., Arotiba, O. A., & Mamba, B. B. (2013). Chitosan-based nano-
    materials: A state-of-the-art review. *International Journal of Biological Macromolecules, 59*,
    46–58.
57. Cheaburu-Yilmaz, C. N., Yilmaz, O., & Vasile, C. (2015). Eco-friendly chitosan-based
    nanocomposites: Chemistry and applications. In *Eco-friendly polymer nanocomposites*
    (pp. 341–386). New Delhi: Springer.
58. Yassue-Cordeiro, P. H., Severino, P., Souto, E. B., Gomes, E. L., Yoshida, C. M., de Moraes, M.
    A., & da Silva, C. F. (2018). Chitosan-based nanocomposites for drug delivery. In *Applications
    of nanocomposite materials in drug delivery* (pp. 1–26). Woodhead Publishing.
59. Ramachandran, S., Rajinipriya, M., Soulestin, J., & Nagalakshmaiah, M. (2019). Recent
    developments in chitosan-based nanocomposites. In *Bio-based polymers and nanocomposites*
    (pp. 183–215). Springer, Cham.
60. Peter, M., Ganesh, N., Selvamurugan, N., Nair, S. V., Furuike, T., Tamura, H., et al. (2010).
    Preparation and characterization of chitosan–gelatin/nanohydroxyapatite composite scaffolds
    for tissue engineering applications. *Carbohydrate Polymers, 80*(3), 687–694.
61. Enescu, D., Gardrat, C., Cramail, H., Le Coz, C., Sèbe, G., & Coma, V. (2019). Bio-inspired
    films based on chitosan, nanoclays and cellulose nanocrystals: Structuring and properties
    improvement by using water-evaporation-induced self-assembly. *Cellulose, 26*, 2389–2401.
62. Qiu, B., Xu, X. F., Deng, R. H., Xia, G. Q., Shang, X. F., & Zhou, P. H. (2019). Construction of
    chitosan/ZnO nanocomposite film by in situ precipitation. *International Journal of Biological
    Macromolecules, 122*, 82–87.
63. Relinque, J. J., de León, A. S., Hernández-Saz, J., García-Romero, M. G., Navas-
    Martos, F. J., Morales-Cid, G., et al. (2019). Development of surface-coated polylactic
    Acid/Polyhydroxyalkanoate (PLA/PHA) nanocomposites. *Polymers, 11*(3), 400.
64. Iqbal, H. M., Kyazze, G., Tron, T., & Keshavarz, T. (2018). Laccase from *Aspergillus niger*:
    A novel tool to graft multifunctional materials of interests and their characterization. *Saudi
    Journal of Biological Sciences, 25*(3), 545–550.
65. Iqbal, H. M., Kyazze, G., Tron, T., & Keshavarz, T. (2014). Laccase-assisted grafting of
    poly (3-hydroxybutyrate) onto the bacterial cellulose as backbone polymer: Development and
    characterization. *Carbohydrate Polymers, 113*, 131–137.
66. Iqbal, H. M., Kyazze, G., Locke, I. C., Tron, T., & Keshavarz, T. (2015). Development
    of bio-composites with novel characteristics: Evaluation of phenol-induced antibacterial,
    biocompatible and biodegradable behaviors. *Carbohydrate Polymers, 131*, 197–207.
67. Iqbal, H. M. N., Kyazze, G., Locke, I. C., Tron, T., & Keshavarz, T. (2015). Development
    of novel antibacterial active, HaCaT biocompatible and biodegradable CA-g-P(3HB)-EC
    biocomposites with caffeic acid as a functional entity. *Express Polymer Letters, 9*(9), 764–772.
68. Ratner, B. D., Hoffman, A. S., Schoen, F. J., & Lemons, J. E. (2004). *Biomaterials science:
    An introduction to materials in medicine*. Elsevier.
69. Kim, J. Y., Kim, M., Kim, H., Joo, J., & Choi, J. H. (2003). Electrical and optical studies of
    organic light emitting devices using SWCNTs-polymer nanocomposites. *Optical Materials,
    21*(1–3), 147–151.
70. Ruiz-Hitzky, E., Aranda, P., Darder, M., & Rytwo, G. (2010). Hybrid materials based on
    clays for environmental and biomedical applications. *Journal of Materials Chemistry, 20*(42),
    9306–9321.
71. Darder, M., Aranda, P., & Ruiz-Hitzky, E. (2012). Chitosan-clay bio-nanocomposites.
    In *Environmental silicate nano-biocomposites* (pp. 365–391). London: Springer.
72. Bilal, M., Iqbal, H. M., Hu, H., Wang, W., & Zhang, X. (2017). Enhanced bio-catalytic per-
    formance and dye degradation potential of chitosan-encapsulated horseradish peroxidase in a
    packed bed reactor system. *Science of the Total Environment, 575*, 1352–1360.

73. Bilal, M., Rasheed, T., Zhao, Y., & Iqbal, H. M. (2019). Agarose-chitosan hydrogel-immobilized horseradish peroxidase with sustainable bio-catalytic and dye degradation properties. *International Journal of Biological Macromolecules, 124*, 742–749.
74. Othman, S. H. (2014). Bio-nanocomposite materials for food packaging applications: Types of biopolymer and nano-sized filler. *Agriculture and Agricultural Science Procedia, 2*, 296–303.
75. de Moura, M. R., Aouada, F. A., Avena-Bustillos, R. J., McHugh, T. H., Krochta, J. M., & Mattoso, L. H. (2009). Improved barrier and mechanical properties of novel hydroxypropyl methylcellulose edible films with chitosan/tripolyphosphate nanoparticles. *Journal of Food Engineering, 92*(4), 448–453.
76. Mathew, A. P., Laborie, M. P. G., & Oksman, K. (2009). Cross-linked chitosan/chitin crystal nanocomposites with improved permeation selectivity and pH stability. *Biomacromolecules, 10*(6), 1627–1632.
77. Addorisio, V., Esposito, S., & Sannino, F. (2010). Sorption capacity of mesoporous metal oxides for the removal of MCPA from polluted waters. *Journal of Agricultural and Food Chemistry, 58*(8), 5011–5016.
78. Celis, R., Adelino, M. A., Hermosín, M. C., & Cornejo, J. (2012). Montmorillonite–chitosan bionanocomposites as adsorbents of the herbicide clopyralid in aqueous solution and soil/water suspensions. *Journal of Hazardous Materials, 209*, 67–76.
79. Tirtom, V. N., Dinçer, A., Becerik, S., Aydemir, T., & Çelik, A. (2012). Comparative adsorption of Ni (II) and Cd (II) ions on epichlorohydrin crosslinked chitosan–clay composite beads in aqueous solution. *Chemical Engineering Journal, 197*, 379–386.
80. Azzam, E. M., Eshaq, G. H., Rabie, A. M., Bakr, A. A., Abd-Elaal, A. A., El Metwally, A. E., et al. (2016). Preparation and characterization of chitosan-clay nanocomposites for the removal of Cu (II) from aqueous solution. *International Journal of Biological Macromolecules, 89*, 507–517.
81. Yan, Y., Yuvaraja, G., Liu, C., Kong, L., Guo, K., Reddy, G. M., et al. (2018). Removal of Pb (II) ions from aqueous media using epichlorohydrin crosslinked chitosan Schiff's base@ Fe_3O_4 (ECCSB@ Fe_3O_4). *International Journal of Biological Macromolecules, 117*, 1305–1313.
82. Yadollahi, M., Farhoudian, S., & Namazi, H. (2015). One-pot synthesis of antibacterial chitosan/silver bio-nanocomposite hydrogel beads as drug delivery systems. *International Journal of Biological Macromolecules, 79*, 37–43.
83. Zhang, J., Wang, Q., & Wang, A. (2010). In situ generation of sodium alginate/hydroxyapatite nanocomposite beads as drug-controlled release matrices. *Acta Biomaterialia, 6*(2), 445–454.
84. Venkatesan, P., Puvvada, N., Dash, R., Kumar, B. P., Sarkar, D., Azab, B., … Mandal, M. (2011). The potential of celecoxib-loaded hydroxyapatite-chitosan nanocomposite for the treatment of colon cancer. *Biomaterials, 32*(15), 3794–3806.
85. Wu, J., Ding, S., Chen, J., Zhou, S., & Ding, H. (2014). Preparation and drug release properties of chitosan/organomodified palygorskite microspheres. *International Journal of Biological Macromolecules, 68*, 107–112.
86. Yadollahi, M., Farhoudian, S., Barkhordari, S., Gholamali, I., Farhadnejad, H., & Motasadizadeh, H. (2016). Facile synthesis of chitosan/ZnO bio-nanocomposite hydrogel beads as drug delivery systems. *International Journal of Biological Macromolecules, 82*, 273–278.
87. Zare-Akbari, Z., Farhadnejad, H., Furughi-Nia, B., Abedin, S., Yadollahi, M., & Khorsand-Ghayeni, M. (2016). PH-sensitive bionanocomposite hydrogel beads based on carboxymethyl cellulose/ZnO nanoparticle as drug carrier. *International Journal of Biological Macromolecules, 93*, 1317–1327.
88. Madene, A., Jacquot, M., Scher, J., & Desobry, S. (2006). Flavor encapsulation and controlled release–a review. *International Journal of Food Science & Technology, 41*(1), 1–21.
89. Urata, M., Iwata, R., Noda, K., Murakami, Y., & Kuroda, A. (2009). Detection of living Salmonella cells using bioluminescence. *Biotechnology Letters, 31*(5), 737–741.
90. Liu, H., Nakagawa, K., Kato, D. I., Chaudhary, D., & Tadé, M. O. (2011). Enzyme encapsulation in freeze-dried bionanocomposites prepared from chitosan and xanthan gum blend. *Materials Chemistry and Physics, 129*(1–2), 488–494.

91. Gazzaniga, A., Iamartino, P., Maffione, G., & Sangalli, M. E. (1994). Oral delayed-release system for colonic specific delivery. *International Journal of Pharmaceutics, 108*(1), 77–83.
92. Ashford, M., Fell, J. T., Attwood, D., Sharma, H., & Woodhead, P. J. (1993). An in vivo investigation into the suitability of pH dependent polymers for colonic targeting. *International Journal of Pharmaceutics, 95*(1–3), 193–199.
93. Ribeiro, L. N., Alcântara, A. C., Darder, M., Aranda, P., Araújo-Moreira, F. M., & Ruiz-Hitzky, E. (2014). Pectin-coated chitosan–LDH bionanocomposite beads as potential systems for colon-targeted drug delivery. *International Journal of Pharmaceutics, 463*(1), 1–9.

Bamboo Strips with Nodes: Composites Viewpoint

Mohammad Irfan Iqbal, Rashed Al Mizan and Ayub Nabi Khan

Abstract A Chinese poet once wrote, "Man can live without meat, but he will die without bamboo" because of its multifunctional and ecofriendly in nature. It has recently entered the textile and composite sector with some attractive labels such as 'green'. The current commercial manufacturing methods of bamboo fibres and its reinforced composites are mainly based on removal of nodes portion of bamboo culm. This method generates a high amount of solid waste materials and hence the term 'green' becomes questionable. This chapter investigates the effects of culm nodes on strip properties from composite perspectives in order to seek their suitability for prospective fiber reinforced composites applications. In this work, strips of bamboo were collected and their mechanical properties; for instance, tensile strength, compression, flexural, impact and thermal properties were analyzed and compared. It was found that tensile, compression, flexural and impact properties as well as thermal properties of bamboo strips were comparable to that of wood material and far better than many other bio fibers due to their rigidity and durability.

1 Introduction

La Mantia and Morreale [1] state synthetic fibers like, fiberglass, carbon fiber, polyurethane lead the composite industries for their superior mechanical properties and lower preparatory cost. However, due to growing global environmental concern, synthetic materials are now losing appeal since most of them cause higher carbon footprint during production. In addition, they are not biodegradable after end use. So, the use of natural fibers as reinforcement like, jute, kenaf, sisal, hemp, coir, straw, bamboo, banana leaf is now a growing focal of attention of stakeholders throughout the globe. Scurlock et al. [2] mention that especially the appeal of bamboo fiber is

M. I. Iqbal (✉)
Department of Textile Engineering, Wuhan Textile University, Wuhan, Hubei, China
e-mail: irfan.iqbal@connect.polyu.hk

M. I. Iqbal · R. A. Mizan · A. N. Khan
Department of Textile Engineering, BGMEA University of Fashion & Technology, Dhaka, Bangladesh

© Springer Nature Switzerland AG 2020
A. Khan et al. (eds.), *Biofibers and Biopolymers for Biocomposites*,
https://doi.org/10.1007/978-3-030-40301-0_7

159

gradually being boosted in the world market for its high mechanical properties, rapid growing nature, low production cost, and environmentally friendly behavior.

Bamboo is the longest grass. It belongs to the Poaceae family, a sub-family of Bambusoideae with a large variety of 1250 spices and 75 genera. It is one of the oldest building materials. Now, it is more extensively used including household, handicraft, furniture, agriculture, architecture, paper-making and other industrial purposes due to advancement in processing technology and rising market demand [3, 4]. Ray et al. [5] compare the mechanical properties of bamboo with other mostly used fibers for composites viz. bamboo offers maximum mechanical strength and minimum density; 0.9 g/cm^3, whereas jute and fiberglass show 1.45 and 2.5 g/cm^3 respectively. Although mechanical properties of bamboo are relatively lower than fiberglass, they are approximately 10 times cheaper than fiberglass. La Mantia and Morreale [1] find bamboo have much higher aspect ratio than wood fibers originating from pine [1]. Bamboo is itself considered as a natural composite material in which cellulose fibers are implanted in lignin and hemicellulose matrix. It shows maximum strength along the fibers and minimum across the fibers. Thus, they provide maximum tensile strength than many other natural fibers. Cellulose (60%) is the main chemical constituent of bamboo. Other major constituents are hemicellulose and lignin [6]. The microfibrillar angle of bamboo is relatively small (10°–12°) [7]. Due to these outstanding specific properties, bamboo fiber is also known as 'natural glass fiber' [8]. These superior characteristics have made the bamboo suitable as reinforcement material for composite applications. But, the extraction of undamaged long technical bamboo fibers of about 25 cm is very challenging due to the nodes [9]. For this concern, recently an attempt has been made to develop composites from bamboo strips instead of fibers for its difficulty associated to extraction. This approach is appreciable as it enables the researchers to take the advantage of cohesiveness since the cohesive strength of individual fiber is significantly low [10].

There is a large volume of published studies describing the performance of bamboo strips as reinforcement for composites. Polyester resin composites can be prepared using bamboo stripes treated with Alkali (10–25%) as reinforcement [11]. Again, Kushwaha and Kumar [12] also treated bamboo strips with different concentrations (1–25%) of alkali at room temperature for 30 min long in order to enhance the mechanical properties in preparing bamboo fiber-reinforced plastic. A number of authors have considered bamboo strips to manufacture unidirectional bamboo–epoxy laminates by changing the numbers of laminae and assessing their mechanical properties [12, 13]. The moisture absorption properties of bamboo strips and their effects on the interfacial shear strength of bamboo–vinyl ester composites have been investigated [14]. To understand and compare the influence of different chemical treatments (silane, alkali, oxidation and acetylation) on moisture absorption performance of same composites have also been carried out [15]. Bamboo–polyester composites have been developed by hand lay-up method using alkali treated and untreated bamboo strips [16]. Furthermore, Das and Chakraborty have prepared novolac based bamboo composites after mercerizing the bamboo strips with NaOH of various concentrations (10, 15, 20 and 25%) [17]. In the following study, they report detailed information regarding dynamic mechanical and thermal properties of untreated and treated bamboo strip-reinforced Novolac composites [18]. Afterwards, they assess thermal and weathering properties of that composites as well [19]. Overall, all these studies highlight the applications of bamboo strips as a reinforcing material for composite applications. But there is a comparatively limited body of literature regarding the use of strips of bamboo along with nodes as reinforcement for composites.

However, available studies have hardly addressed the use of bamboo strips along with nodes for composite applications. In this work various physical, mechanical, and thermal properties of strips of bamboo determined having node and without having node with a view to assessing whether the strips of bamboo bear the properties of composite applications.

2 Materials

The bamboo strips were collected from mid of China with a typical length of 74 cm. The strips had an average length of 74 cm. The average distance between the nodes was 26.5 cm, 16.5 cm at the top and 20 cm at the bottom. The width and thickness of bamboo strips were 4.7 mm and 1.8 mm respectively. The strips were classified into two categories: without nodes or only internodes and with nodes in middle of strips. All specimens were tested in standard air-dried condition.

3 Measurements

In this study, four mechanical properties; for instance, Tensile, Compressive, Flexural and Impact as well as thermal properties (TGA) were evaluated as these properties have significant influence on ultimate performance of composite materials and their manufacturing process. For each mechanical property 20 samples were considered randomly, and their averages were reported.

4 Mechanical Properties Investigation

Tensile testing: Tensile testing was carried out using an Instron 2712 pneumatic grips machine, ISO 11566 was followed to determine tensile properties. The specimen geometry was 250 mm × 4.7 mm × 1.8 mm. The gauge length was 150 mm and the cross-head speed was 1 mm/min. The tests were carried out until the materials got broken. The samples with Jaw break were not taken into consideration for the analysis. Furthermore, tenacity has also been determined calculating linear density in decitex (dtex).

Compression testing: Tests were carried out on bamboo strip specimens with & without node to obtain the compressive strength. Data were collected using an electronic compression testing machine according to ASTM-D3410. The specimens were cut into samples with dimensions of 30 mm × 4.7 mm × 1.8 mm for compression test. For more accuracy, the samples were placed within the center of the cross head and perpendicular to the longitudinal axis. The cross-head speed was followed at 10 mm/min.

Flexural testing: Loading nose and its supports were arranged and finally three-point bend tests were performed with samples of 150 mm × 4.7 mm × 1.8 mm dimensions. Flexural strength was received from auto generated computer sheet through a software called WinWdw inbuilt in universal testing machine equipment as per procedure given in ASTM-D7164 standard. The ratio of support span to thickness was maintained as 20:1 so that breakage arises at the outer surface of specimens for higher bending moment. The similar speed of cross head was maintained like.

Impact testing: For impact behavior test specimens without notch were 60 mm long and the cross-section was 4 mm × 1.2 mm. A 7.5 J pendulum was used to break the specimens. The impact energy was note down. The final impact strength was obtained from dividing the impact energy by the cross-sectional area. The unit of the impact strength was KJ/m^2.

5 Thermal Properties Investigation

The thermal degradation of bamboo strips occurs during the processing of composite that decreases the effect of mechanical reinforcement. Thus, to analyze the degradation behavior of bamboo strips, TGA analysis was conducted. The thermal decomposition of bamboo strip was evaluated by thermogravimetric analysis (TGA) using a NETZSCH instrument of TG 209 F1 instruments. Roughly 5 mg of sample was heated under air from room temperature to 600 °C at a rate of 10 °C/min to yield the onset temperature of decomposition, mass loss and maximum decomposition peak.

6 Surface Morphology Investigation

A scanning electron microscope (JSM-6510LV, voltage: 20 kV) was used to investigate the surface morphology of the fibers. Before observation, the samples were coated in gold by ion sputtering. Figure 1a and b show the structure of bamboo strip nodes and internodes fibers respectively. It was observed that internodes of bamboo strips had smoother and compact structure compared to nodes. Again, bamboo strips including nodes had more cracks compared to internodes.

(a)   (b)

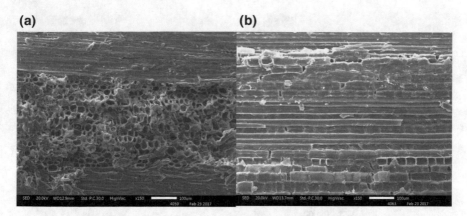

Fig. 1 **a** SEM photograph of bamboo strip with node **b** SEM photograph of bamboo strip without node

7 Statistical Analysis

The average of the values and the comparisons of different properties between nodes and internodes of the strips have been calculated using SPSS software at 5% level of significance.

8 Tensile Properties of Bamboo Strips

Tensile properties of the bamboo strips with node and without node are shown in Table 1. The average value of Tensile Strength, Tensile Modules and Strain at failure% of bamboo strips with node and without node are 183 MPa, 14 GPa, 1.6% and 235 MPa, 16GPa, 2% respectively. The difference in Tensile Strength (P = 0.0002) and Tensile Modules (P = 0.0005) has been found statistically significant between nodes and internodes. These results seem to be consistent with data of other research where the tensile strength, tensile modulus and strain at failure% are 140–230 Mpa, 11–17 GPa and 1.1% respectively [20–22].

Table 1 Tensile properties of bamboo strips

Name of the specimen	Tensile strength (MPa) ± STD	Tensile modulus (GPa) ± STD	Strain to failure% ± STD
Bamboo strip with node	183 ± 26	14 ± 2	1.6 ± 0.4
Bamboo strip without node	235 ± 14	16 ± 0.9	2 ± 0.2

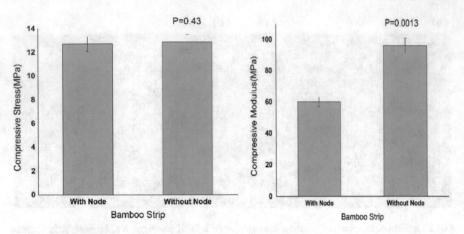

Fig. 2 Compressive properties of bamboo strip with and without node

From the Table 1, it is also observed that the tensile properties of internodes are significantly greater than the nodes. These results are in agreement with the study of Taylor et al. [23]. A possible explanation for this might be that the fiber cells in the bamboo internodes arranged directly and continuously, while node portions are arranged discontinuously. Thus, it reduces tensile properties of bamboo strips in node portions. Therefore, the significant difference in the tensile properties between the bamboo strips with nodes and internodes might be due to the differences in fiber morphology and fiber alignment in nodes and internodes [24]. Another possible explanation could be the number of vascular bundles that also determines the mechanical properties of bamboo. The vascular bundles have a positive influence on the tensile strength and modulus for the bamboo [25]. Figure 2b shows that the amount of fibers in the vascular bundles of internodes is higher than the nodes, which contributes to higher tensile strength and modulus of internodes. During testing tensile properties, it is also observed shear failures occurs first before tensile failure in most of the internode specimens. On the other hand, in case of strips with nodes failure due to tension occurs first. Consequently, shear failure specimens, internodes, have higher tensile strength than tension failure specimens, nodes.

9 Compressive Properties of Bamboo Strips

The variation in maximum compressive stress and compressive modulus is presented in Fig. 2. The compressive stress and compressive moduli of internodes and nodes are 13 and 96 MPa respectively. In case of nodes, the corresponding values are 12 and 60 MPa respectively. The values of compressive stress are almost supported by Li's [3] finding which is 16.1 MPa for one-year-old bamboo. It appears that the difference in compressive moduli between nodes and internodes is statistically

Fig. 3 Flexural properties of
bamboo strip with and
without node

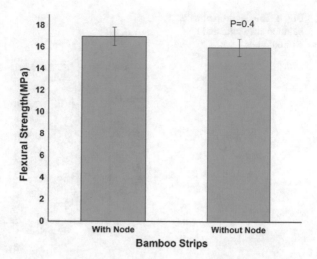

significant (P = 0.0013). In addition, surprisingly, there were no significant differences (P = 0.43) in compressive strength between strips with node and without node. On the other hand, compressive moduli have been found statistically significant between nodes and internodes. In contrast to earlier findings, no evidence of compressive strength of bamboo strip is found. This result may be explained by the fact of unique morphological and anatomical characteristics of nodes [24] that allows that contribute to the almost similar compressive strength.

10   Flexural Properties of Bamboo Strips

Figure 3 shows the results of the flexural tests conducted on bamboo strips of nodes and internodes. Paired Student's t-tests were also used to determine statistical significance. It is observed that nodes have little effect on the bending strength. Nodes have 6% higher strength than internodes, and this is not statistically significant. This may be because of Specific Gravity (SG) as there is a strong relationship between bending properties and SG of bamboo. The SG varied with age and height location of bamboo [3]. However, these results are not very encouraging. A further study with more focus on flexural properties is therefore suggested.

11   Impact Properties of Bamboo Strips

Impact strength is defined as "the ability of a material to resist fracture under stress applied at high speed" [26]. The Pendulum type impact test provides a record of the impact event. Figure 4 shows the results of impact tests conducted on specimens

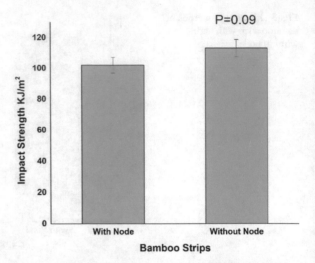

of nodes and internodes. This finding is far better than the previous research which reveals that the impact strength of bamboo along the fibers is 63.54 (±4.63) [8]. Paired Students t-tests were used to determine statistical significance. It is apparent from the Fig. 4 nodes cause a small decrease in strength from 113 kJ/m^2 (SD = ± 5) to 103 kJ/m^2 (SD = ± 8) which is not statistically significant (P = 0.09). It is probably due to different shapes and sizes of samples. These results suggest that the nodes have only a minor effect on the impact properties of the strips. The significant effect could be obtained for high frequency of the nodes along the culm and more compact structure in the nodes [23].

12 Thermal Behavior of Bamboo Strips

The thermo-gravimetric curves of waste bamboo strip at a heating rate 10 °C/min are shown in Fig. 5 There is a peak at approximately 100 °C indicating the removal of moisture from the strips. Bamboo strips show thermal decomposition within the temperature range of 200 and 360 °C which are in line with those of previous studies. However, the maximum decomposiion peak has been observed at 343 °C. Thus, the bamboo strips had higher thermal stability as compared to other natural fibers commonly used in the polymer processing industries [27].

13 Concluding Remarks

Bamboo in the form of strips including nodes and internodes have been characterized in terms of mechano-physical and thermal properties from composite perspective.

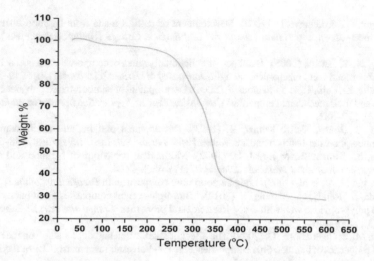

Fig. 5 TGA results of bamboo strip

This study has shown that all the criteria of mechanical and thermal properties vary with nodes and internodes. Although the both mechanical and thermal properties of bamboo strips do not possess as high as those of synthetic materials (Glass, Carbon) which have been commonly and widely accepted materials for composite applications, they do have better performance than other bio fibers like wood, alifa, coir, feather, pineapple etc. [20, 21, 28]. The second major finding, the contribution of nodes to the mechanical properties of bamboo strips, is not appreciable.

However, the properties revealed from the characterization imply that materials could be applied for structural and non-structural applications. Consequently, future research can be conducted by developing composites from bamboo strip materials to understand their performance and behavior in the composites and other high-performance materials. As the material is lower in cost and easily available, it could be a good replacement for traditional wood in terms of indoor and outdoor applications for example; furniture, housing, packaging, transport automobile body etc. along with daily life applications like, decking, fencing, dustbin etc.

References

1. La Mantia, F., & Morreale, M. (2011). Green composites: A brief review. *Composites Part A: Applied Science and Manufacturing, 42*(6), 579–588.
2. Scurlock, J., Dayton, D., & Hames, B. (2000). Bamboo: An overlooked biomass resource? *Biomass and Bioenergy, 19*(4), 229–244.
3. Li, X. (2004). *Physical, chemical, and mechanical properties of bamboo and its utilization potential for fiberboard manufacturing.* Beijing Forestry University.

4. Taijun, C., & Gangyi, L. (2005). Development of textiles made from bamboo fiber and its prospect. *Journal of Hunan Liberal Art and Science College* (Natural Science Ed.), *17*(1), 57–59.
5. Ray, A. K., et al. (2005). Bamboo—a functionally graded composite-correlation between microstructure and mechanical strength. *Journal of Materials Science, 40*(19), 5249–5253.
6. Okubo, K., Fujii, T., & Yamamoto, Y. (2004). Development of bamboo-based polymer composites and their mechanical properties. *Composites Part A: Applied Science and Manufacturing, 35*(3), 377–383.
7. Jain, S., Jindal, U., & Kumar, R. (1993). Development and fracture mechanism of the bamboo/polyester resin composite. *Journal of Materials Science Letters, 12*(8), 558–560.
8. Jain, S., Kumar, R., & Jindal, U. (1992). Mechanical behaviour of bamboo and bamboo composite. *Journal of Materials Science, 27*(17), 4598–4604.
9. Van Vuure, A., et al. (2009). Long bamboo fibre composites. In *Proceedings of the ICCM*.
10. Huda, S., Reddy, N., & Yang, Y. (2012). Ultra-light-weight composites from bamboo strips and polypropylene web with exceptional flexural properties. *Composites Part B: Engineering, 43*(3), 1658–1664.
11. Das, M., & Chakraborty, D. (2009). The effect of alkalization and fiber loading on the mechanical properties of bamboo fiber composites, Part 1:—Polyester resin matrix. *Journal of Applied Polymer Science, 112*(1), 489–495.
12. Kushwaha, P., & Kumar, R. (2009). Enhanced mechanical strength of BFRP composite using modified bamboos. *Journal of Reinforced Plastics and Composites, 28*(23), 2851–2859.
13. Corradi, S., et al. (2009). Composite boat hulls with bamboo natural fibres. *International Journal of Materials and Product Technology, 36*(1–4), 73–89.
14. Chen, H., Miao, M., & Ding, X. (2009). Influence of moisture absorption on the interfacial strength of bamboo/vinyl ester composites. *Composites Part A: Applied Science and Manufacturing, 40*(12), 2013–2019.
15. Chen, H., Miao, M., & Ding, X. (2011). Chemical treatments of bamboo to modify its moisture absorption and adhesion to vinyl ester resin in humid environment. *Journal of Composite Materials, 45*(14), 1533–1542.
16. Das, M., & Chakraborty, D. (2009). Effects of alkalization and fiber loading on the mechanical properties and morphology of bamboo fiber composites. II. Resol matrix. *Journal of Applied Polymer Science, 112*(1), 447–453.
17. Das, M., & Chakraborty, D. (2007). Role of mercerization of the bamboo strips on the impact properties and morphology of unidirectional bamboo strips–novolac composites. *Polymer Composites, 28*(1), 57–60.
18. Das, M., & Chakraborty, D. (2008). Processing of the Uni-directional powdered phenolic resin–bamboo fiber composites and resulting dynamic mechanical properties. *Journal of Reinforced Plastics and Composites*.
19. Das, M., Prasad, V., & Chakrabarty, D. (2009). Thermogravimetric and weathering study of novolac resin composites reinforced with mercerized bamboo fiber. *Polymer Composites, 30*(10), 1408–1416.
20. Faruk, O., et al. (2012). Biocomposites reinforced with natural fibers: 2000–2010. *Progress in Polymer Science, 37*(11), 1552–1596.
21. Yan, L., Chouw, N., & Jayaraman, K. (2014). Flax fibre and its composites–a review. *Composites Part B: Engineering, 56*, 296–317.
22. Liu, L., et al. (2011). Modification of natural bamboo fibers for textile applications. *Fibers and Polymers, 12*(1), 95–103.
23. Taylor, D., et al. (2015). The biomechanics of bamboo: Investigating the role of the nodes. *Wood Science and Technology, 49*(2), 345–357.
24. Qi, J., et al. (2015). Effects of characteristic inhomogeneity of bamboo culm nodes on mechanical properties of bamboo fiber reinforced composite. *Journal of Forestry Research, 26*(4), 1057–1060.
25. Tong, J.K., et al. (2015). Infrared-transparent visible-opaque fabrics for wearable personal thermal management, *2*(6), 769–778.

26. Srinivasa, C., & Bharath, K. (2013). Effect of alkali treatment on impact behavior of areca fibers reinforced polymer composites. *Fiber Composites, 1*(2), 8.
27. Yao, F., et al. (2008). Thermal decomposition kinetics of natural fibers: Activation energy with dynamic thermogravimetric analysis. *Polymer Degradation and Stability, 93*(1), 90–98.
28. Okubo, K., & Fujii, T. (2002). Eco-composites using natural fibres and their mechanical properties. *WIT Transactions on The Built Environment, 59.*

Water Hyacinth for Biocomposites—An Overview

A. Ajithram, J. T. Winowlin Jappes, Thiagamani Senthil Muthu Kumar,
Nagarajan Rajini, Anumakonda Varada Rajulu,
Sanjay Mavinkere Rangappa and Suchart Siengchin

Abstract In recent years, there is a mounting interest in the utilization of natural fibers in composite materials due to their abundancy, low density and weight, low cost, recyclability and biodegradable properties. It is well known that these plant fibers are rich in cellulose and have the greater potential as reinforcements in polymeric materials to form polymer composites. Natural fibers were already proved as a better alternative for high cost synthetic fibers such as glass, carbon, kevlar and basalt etc. This article presents an overview on the environmental impact of aquatic weed water hyacinth (*Eichhornea crassipe*). Furthermore, emphasis is given on the extraction of fibers from water hyacinth, fabrication of composites and the effective utilization of the extracted natural fiber in composite materials for various applications.

1 Introduction

Natural fibers have gained more attraction due to their excellent characteristic properties and are considered as potential replacement for the traditional synthetic fibers. Scientists all over the world are exploring the development of new natural fibers as reinforcement materials in composites for different applications. One of the potential sources of natural fiber is the water hyacinth (*Eichhornea crassipe*) which is an invasive and resistant plant that has infested the water bodies of most tropical countries [1, 2]. Water hyacinth causes many problems and pose serious threat to the

A. Ajithram · J. T. Winowlin Jappes · T. Senthil Muthu Kumar (✉)
Department of Mechanical Engineering, Kalasalingam Academy of Research and Education,
Anand Nagar, Krishnankoil, Tamil Nadu 626126, India
e-mail: tsmkumar@klu.ac.in

N. Rajini · A. Varada Rajulu
Centre for Composite Materials, Department of Mechanical Engineering, Kalasalingam Academy
of Research and Education, Anand Nagar, Krishnankoil, Tamil Nadu 626126, India

T. Senthil Muthu Kumar · S. M. Rangappa · S. Siengchin
Department of Mechanical and Process Engineering, The Sirindhorn International Thai-German
Graduate School of Engineering (TGGS), King Mongkut's University of Technology North
Bangkok, 1518 Wongsawang Road, Bangsue, Bangkok 10800, Thailand

© Springer Nature Switzerland AG 2020
A. Khan et al. (eds.), *Biofibers and Biopolymers for Biocomposites*,
https://doi.org/10.1007/978-3-030-40301-0_8

171

biodiversity. This is mainly because of its rapid rate of spreading which causes invasions over large parts of water bodies. This results in hindrance to water transport, blockage of water for irrigation, hydropower and water supply systems, obstruction of canals and rivers causing flooding, micro-habitat for a variety of disease vectors, increased evapotranspiration, problems related to fishing [3, 4]. But its high growth rate makes it a potential renewable source of fibers. Further, with relatively high cellulose contents, water hyacinth shows greater potential as a reinforcement in composite materials. When compared to other natural fibers, water hyacinth has a high percentage of holocellulose, which makes it a potential reinforcing candidate [5, 6]. Indeed, water hyacinth is successfully used in sewage water treatments for nutrient removal and retention of particles [2]. Many researchers have developed composites with water hyacinth fiber as a reinforcement. Flores Ramirez et al. [7] studied the properties of water hyacinth polyester composites. Similarly, Abral et al. [8] examined the mechanical properties of water hyacinth polyester composites with and without treatment of fibers. In another study the effect of chemical treatment of water hyacinth on the mechanical properties of poly (vinyl alcohol) (PVC) and Low-density polyethylene (LDPE) composites were investigated [9]. The chemical treatments of the fibers enhanced the fiber matrix interaction and therefore presented better functional properties of the composites [5]. The water hyacinth fiber in woven form is useful in domestic applications such as carry bags and engineering application purposes, like spinning sewing and indusial waste water treatment and some other useful industrial applications. It is also used in the production of biomass. Recently, a number of results have been reported with respect to thermoplastic composites. It was also reported that the addition of organic fillers can also improve the functional properties of water hyacinth-based composites. A study revealed that the addition of organic fillers with polypropylene and poly methyl methacrylate improved the mechanical, thermal and absorption properties [10]. Due to the negative effects of water hyacinth and the large amount of money spent for their removal it becomes very important to find better solutions for the utilization of this plant. Hence, this article presents an overview of the physio mechanical properties of the fibers and their potential as a reinforcement in composite materials.

2 Properties of Water Hyacinth Fiber

The physical and chemical properties of water hyacinth fiber are presented in Tables 1 and 2.

It can be seen from Table 2 that; the water hyacinth contains a comparable amount of cellulose and hemicellulose when compared with other natural fibers used.

Table 1 Physical properties of water hyacinth fiber [11–14]

Property	Value
Moisture content (%)	24.63
Ash content (%)	2.17
Volatile acidity (%)	37.22
Particle density	1.12
Porosity (%)	61.45
Pore volume (cm^2 g^{-1})	0.89
Surface area (m^2 g^{-1})	102.6
Surface acidity ($mmol\ g^{-1}$)	0.13

Table 2 Comparison of chemical composition of water hyacinth [3, 15, 16]

Fiber	Cellulose (wt. %)	Lignin (wt. %)	Hemi cellulose (wt. %)	Pectin (wt. %)	Wax (wt. %)	Moisture (wt. %)
Jute	61–71.5	12–13	13.6–20.4	0.4	0.5	12.6
Hemp	70–74.4	3.7–5.7	17.4–22.7	0.9	0.8	10
Kenaf	31–39	15–19	21.5	–	–	–
Flax	71	2.2	18.6–20.6	2.3	1.7	10
Sisal	67–78	8–11	10–14.2	10	2	11
Coir	36–43	41–45	10–20	3–4	–	–
Banana	63 67	5	19	–	–	8.7
Water hyacinth	61.63	3.78	16.26	–	–	11.8

3 Issues and Opportunities in Extraction

The water hyacinth fibers were extracted from the fresh water hyacinth plant. The fiber extraction process includes the drying of the fibers until the required weight percentage were obtained with no damage. After drying, the fibers were stored in vacuum desiccators [17, 18]. The process of extraction is shown in Fig. 1. First, the plant leaves and roots were removed. Only the stem of the plant was considered the fiber extraction process. The average length of the stem of water hyacinth plant would be around 50 mm. After removing the leaves and roots, the stem was rinsed through the flow fresh water to separate water hyacinth fibers from their skins. Then the water hyacinth fiber was dried by the help of dehumidifier with the gradual relative humidity of 40% at the 40 °C. After drying the water hyacinth fibers were obtained which are stored in desiccators to maintain the relative humidity value [19–21].

Fig. 1 Extraction of fibers from water hyacinth plant

4 Treatment of the Water Hyacinth Fiber

The chemical treatment of water hyacinth fibers was done by immersing the fiber in different alkali concentration of 2.5, 7.5 and 10%. The fibers were soaked for about 1 h after the alkalization process for the neutralization of cellulosic fibers. Another method of treatment is done by boiling the water hyacinth fibers in fresh water for about 3 h to achieve gradual reduction in their weight. Further, these fibers were neutralized by 1% acetic acid in distilled water. After this the normal alkali treatment process would be carried out. This is called as the before immersion and after immersion processes in the fiber extraction processes [22, 23].

5 Composite Preparations

Composites were prepared by using some advance and traditional methods. Composite pipes and circular shaped composites can be fabricated using the z—blade mixer method. The composite laminates were prepared using compression molding, vacuum bagging and hand layup methods. Most of the researchers used the compression molding method because of outcome of defect free laminates and quick process time compared with the other traditional methods [24].

6 Compression Molding

Many researchers used the compression molding method to produce the water hyacinth-based composites laminates. The process parameters for compression molding technique were as follows: Both the top and bottom plates of the mold were maintained at 180 C. Then before placing a specimen the mold was heated up to 2 min. After that the specimen were subjected into the mold, they were preheated for first 2.5 min and compressed for another 2.5 min. After, the compression processes the composites were cooled for about 2 min. Further, complex shapes and parts can also be fabricated by processes like hand lay-up and vacuum bagging. The main advantages of these methods were the dimensional control, stability and less expensive finished parts compared to other methods. The temperature of the mold plates plays a main role in the shape and strength of the composites made using natural fibers as reinforcing materials [25–29].

7 Properties of Water Hyacinth Composites

It can be seen from the above Table 3 that the flexural strength of water hyacinth fiber is more than double that of other natural fibers such as sisal, banana and coir. Furthermore, the tensile strength of the water hyacinth fiber is also comparable with sisal fiber but is higher than that of the banana and coir fibers. These strengths clearly

Table 3 Comparison of different composites with water hyacinth

Fiber	Tensile strength (MPa)	Flexural strength (MPa)	Impact strength (MPa)
Sisal [30]	34.35	40.12	30.15
Banana [31]	16.62	57.22	13.44
Coir [23]	17.867	31.08	11.49
Water hyacinth [32]	30.76	128.0	–

indicate the bonding characteristics of the matrix and the hyacinth fiber. It is well known that most of the natural fibers are combined with thermosetting resins and especially the epoxy resin [33, 34].

8 Biomass Production

Reddy discussed the water hyacinth fiber and the potential use for biomass production from wastewaters. The water hyacinth is one of the worst aquatic plant in all around the world. But this all were biomass and are waiting to convert it to the biofuel. Renewability of this plant was maddening and it can be converted in the form of profitable uses like biofuel and biomass. The stem of the water hyacinth is the usable part, cut and kept separately and it was well dried to make crush. The materials were kept in an anaerobic digester and maintained at a temperature of 32 °C for about 28 days. During the process, the pH value was measured both the initial and final stages. The gas properties were analyzed using digital flame photometer and gas chromatography. The water hyacinth plant was used for reducing the oxygen depletion from the water bodies and from these plant bio gas was used instead of LPG [35, 36]. Verma and sSingh collected the various water hyacinth plants located from India. These water hyacinth plants were does not grow up in electro plating industry effluent and diluted brass because of its inability property. These plants were under the several percentages of concentration like 20 and 40 and 60% concentration for 30 days and then they were chopped to maintain 20 mm length and finally oven dried at 60 °C for 48 h [30, 31, 37, 38].

9 Domestic Application

These water hyacinth fiber plants were well growth in the water and pretty useful on urban and village faction like water hyacinth spinning and sewing (Fig. 2). It is low in economic values of the people and the growth rate is vast in majority compare to other fibers and plants.

Textile application of water hyacinth fiber   Water hyacinth used in paper production

Fig. 2 Applications of water hyacinth

This plant can also be used in day today life. Recent trends about the fiber is utilization in wastewater treatment in industries and some other factories application. The water hyacinth fiber stem was used in to the three different styles. The first way of the water hyacinth stem is utilized directly in to the straw material. Second one the stem was slit and the pith is removed and finally the water hyacinth was used in looms [23, 32].

10 Engineering Application

The water hyacinth plant was used in phytoremediation process to clean up the several number of pollutants in industrial and domestic waste water. In 1970s the waste water treatment of the water hyacinth fibers was found to be useful to treat various several types of sugar factory use and textile and paper mill production. The organic and inorganic content removal process from the waste water the water hyacinth plant was effectively done so now all the world exclusively researchers focused the water hyacinth plant and fiber and composite materials. Especially water hyacinth petioles were performed most useful construction materials and make a particle board. These boards were proved they were good thermal conductivity values [39–41].

11 Conclusion

This review study explained the mechanical properties, extraction, and application of the water hyacinth fiber. In these studies, first these fibers were extracted from the fresh water and then they were cut into small sized pieces. Then some chemical treatments and alkali process were done for making the composites with some suitable manufacturing methods like z—blade mixture and compression molding. It is well known that that some chemical modifications and the alkali treatment of the water hyacinth fibers have more and better properties and good adhesion bonding compare to normal untreated fibers. These chemical compounds and some of the treatments increase the composite strength. And these fibers also have better interfacial adhesion between the fibers and the matrix. And the water hyacinth fiber has more mechanical properties like flexural strength compare to various other types of natural fibers. These fibers were mainly used some domestic applications like textile application and paper production and some furniture materials and sewing and spinning. In engineering applications like industrial waste water treatment these water hyacinth fibers were used and they also used in successful construction materials. However, the use of water hyacinth fiber as a potential reinforcement in composite materials for engineering and structural applications are less explored. Hence, there is a large scope for the further exploration of the fiber for composite fabrication targeting structural and other applications.

References

1. Herrera-Franco, P. J., & Valadez-Gonzalez, A. (2005). A study of the mechanical properties of short natural—fiber reinforced composites. *Composites Part B: Engineering, 36*(8), 597–608.
2. Tan, S.J., Supri, A.J., & Chong, K.M. (2007). Properties of recycled high-density polyethylene/water hyacinth fiber composites: the effect of different concentration of compatibilizer. *Polymer Bulletin*, 1387–1393.
3. Bhattacharya, A., & Kumar, P. (2010). Water hyacinth as a potential biofuel crop. *Electronic journal of Environmental, Agricultural and Food Chemistry, 9*(1), 112–122.
4. Hairul, A., Hendri, P., Sapuan, S.M., & Ishak, M.R. (2013). Effect of alkalization on mechanical properties of water hyacinth fibers-unsaturated polyester composites, *Polymer-Plastics Technology and Engineering*, *52*, 446–451.
5. Malik, A. (2007). Environmental challenge vis a vis opportunity: The case of water hyacinth. *Environment International, 33*(1), 122–138.
6. Mansour, O., Abdel-Hady, B., Ibrahem, S.K., & Goda, M. (2011). *Polymer Plastic Tech Engineering, 40*, 311.
7. Flores Ramirez, N., Sanchez Hernandez, Y., Cruz de Leon, J., Vasquez Garcia, S. R., Domratcheva, L., & Garcia Gonzalez, L. (2015). Composites from water hyacinth and polyester resin. *Fibers and Polymers, 16*(1), 196–200.
8. Abral, H., Kadriadi, D., Rodianus, A., Mastariyanto, P., Ilhamdi, Arief, S., Sapuan, SM., & Ihak, R. (2014). Mechanical properties of water hyacinth fibers—polyester composites before and after immersion in water. *Materials and Design, 58*, 125–129
9. Supri, A. G., Tan, S. J., Ismail, H., & Teh, P. L. (2013). Enhancing Interfacial Adhesion performance by using Poly (vinyl alcohol) in (low-density polyethylene)/(natural rubber)/(water hyacinth fiber) composites. *Journal of Vinyl & Additive Technology, 19*, 47–54.
10. Phenology, T. H. E., Hyacinth, W., Lake, F., & Spencer, N. R. (1981). *Aquatic Botany, 10*(1–32), 10.
11. Kgser, H.J.K., & Schmalstieg, G. (1982). Densification of water hyacinth basic data, *61*, 791–798.
12. Solms, M. (2018). the resource utilization of water hyacinth. *Journal of Environmental Management, 87*.
13. Tumolva, T., Ortenero, J., Kubouchi, M., & City. (2019). Characterization and treatment of water. *International Journal of Engineering and Technology, 8*(1.9), 1–11.
14. Sahari, J., Sapuan, S. M., Zainudin, E. S., & Maleque, M. A. (2013). Mechanical and thermal properties of environmentally friendly composites derived from sugar palm tree. *Materials and Design, 49*, 285–289.
15. Singha, A.S., & Thakur, V.K. (2014). Physical, Chemical and Mechanical Properties of Hibiscus sabdariffa Fiber/Polymer Composite. *International Journal of Polymeric Materials and Polymeric Biomaterials, 37–41*.
16. Sundari, M. T., & Ramesh, A. (2012). Isolation and characterization of cellulose nanofibers from the aquatic weed water hyacinth—*Eichhornia crassipes. Carbohydrate Polymers, 87*(2), 1701–1705.
17. Ghani, S.A., & Lim, B.Y. (2009). Effect of treated and untreated filler loading on the mechanical, morphological, and water absorption properties of water hyacinth fibers-low density polyethylene composites. *Journal of Physical Science, 20*(2), 85–96
18. Supri, A.G., Tan, S.J., & Teh, P.L. (2011). Effect of poly (methyl Methacrylate) modified water hyacinth fiber on properties of low density Polyethylene/Natural Rubber/Water Hyacinth Fiber Composites. *Polymer Plastics Technology and Engineering, 2559*(2016).
19. Reddy, K.R., & Sutton, D.L. (1984). Reviews and analyses Waterhyacinths for Water Quality Improvement. *Journal of Environmental Quality, 13*(4868).
20. Singhal, V., & Rai, J. P. N. (2003). Biogas production from water hyacinth and channel grass used for phytoremediation of industrial effluents. *Biosource Technology, 86*, 221–225.
21. Verma, V. K., Singh, Y. P., & Rai, J. P. N. (2007). Biogas production from plant biomass used for phytoremediation of industrial wastes. *Biosource Technology, 98*, 1664–1669.

22. Bernard, P., Lhote, A., & Legube, B. (2019). Principal component analysis : an appropriate tool for water quality evaluation and management—application to a tropical lake system. *Ecological Modelling, 178*, 295–311
23. Harish, S., Michael, D. P., Bensely, A., Lal, D. M., & Rajadurai, A. (2008). Mechanical property evaluation of natural fiber coir composite. *Materials Characterization, 60*(1), 44–49.
24. Asrofi, M., Abral, H., Kasim, A., Pratoto, A., Mahardika, M., & Hafizulhaq, F. (2018). Mechanical properties of a water hyacinth nanofiber cellulose reinforced thermoplastic starch bionanocomposite: Effect of ultrasonic vibration during processing. *Fibers, 6*(2), 1–9.
25. Abral, H., Lawrensius, V., Handayani, D., & Sugiarti, E. (2018). Preparation of nano-sized particles from bacterial cellulose using ultrasonication and their characterization. *Carbohydrate Polymers, 191*(September 2017), 161–167.
26. Bledzki, A. K., Reihmane, S., & Gassan, J. (1998). Thermoplastics reinforced with wood fillers: A literature review. *Polymer—Plastics Technology and Engineering, 37*(4), 451–468.
27. Kalia, S., Dufresne, A., Cherian, B.M., Kaith, B.S., Avérous, L., Njuguna, J., & Nassiopoulos, E. (2011). Cellulose-based bio- and nanocomposites: A review. *International Journal of Polymer Science, 2011*.
28. Moorhead, K. K., Reddy, K. R., & Graetz, D. A. (1988). Water hyacinth productivity and detritus accumulation. *Hydrobiologia, 157*(2), 179–185.
29. Patel, V., Desai, M., & Madamwar, D. (1993). Thermochemical pretreatment of water hyacinth for improved biomethanation. *Applied Biochemistry and Biotechnology, 42*(1), 67–74.
30. Jarukumjorn, K., & Suppakarn, N. (2009). Effect of glass fiber hybridization on properties of sisal fiber-polypropylene composites. *Composites Part B: Engineering, 40*(7), 623–627.
31. Reed, K. E. (1980). Dynamic mechanical analysis of fiber reinforced composites. *Polymer Composites, 1*(1), 44–49.
32. Taylor, P., Van Wyk, E., & Van Wilgen, B.W. (2002). The cost of water hyacinth control in South Africa. *African Journal of Aqatic Science, 37*–41.
33. Sanjay, M.R., Arpitha, G.R., Naik, L.L., Gopalakrishna, K., & Yogesha, B. (2016). Applications of natural fibers and its composites : An overview. *Natural Resources*, 108–114.
34. Sanjay, M R , Madhu, P., Jawaid, M., Senthamaraikannan, P., Senthil, S., & Pradeep, S. (2018). Characterization and properties of natural fiber polymer composites: A comprehensive review. *Journal of Cleaner Production, 172*.
35. Zhou, W., Zhu, D., Tan, L., Liao, S., Hu, Z., & Hamilton, D. (2007). Extraction and retrieval of potassium from water hyacinth (*Eichhornia crassipes*). *Bioresource Technology, 98*(1), 226–231.
36. Abral, H., Kadriadi, D., Rodianus, A., Mastariyanto, P., Arief, S., Sapuan, S. M., et al. (2014). Mechanical properties of water hyacinth fibers—polyester composites before and after immersion in water. *Materials and Design, 58*, 125–129.
37. Adhikary, K. B., Pang, S., & Staiger, M. P. (2008). Dimensional stability and mechanical behaviour of wood—plastic composites based on recycled and virgin high-density polyethylene (HDPE). *Composite Part B: Engineering, 39*, 807–815.
38. Goswami, T., & Saikia, C. N. (1995). Water hyacinth—a potential source of raw material for greaseproof paper. *50*(1994), 235–238.
39. Temi, T., & Michael, H. Jr. (2007). Adsorption of methyl red by water-hyacinth (*Eichornia crassipes*). *Biomass Chemistry and Biodiversity, 4*.
40. Mishima, D. (2008). Ethanol production from candidate energy crops: Water hyacinth (*Eichhornia crassipes*) and water lettuce. *Bio resource Technology, 99*, 2495–2500.
41. Rezania, S., Fadhil, M., & Fatimah, S. (2016). Evaluation of water hyacinth (*Eichornia crassipes*) as a potential raw material source for briquette production. *Energy, 111*, 768–773.

Ionic Liquids Based Processing of Renewable and Sustainable Biopolymers

Sadia Naz and Maliha Uroos

Abstract In view of immense potential of sustainable and renewable biopolymers for future biorefineries, development of green and carbon economic methods for their processing are highly demanding. Despite of numerous protocols established so far, innovations leading to sustainable methods for integration of multi-step volarization of low value biopolymeric feedstock are still highly concerned. One of such innovations is the ionic liquids based biorefinery concept for various advanced biofuels, valuable chemicals and other bio-products. Superiority of ionic liquids is due to their green, non-degradative, non-toxic, nono-volatile and chemically and thermally stable profile for upgrading renewable biopolymers based biorefinery. Some processing applications of ionic liquids for biofuels and fine chemicals production are covered in this chapter.

Keywords Ionic liquids · Renewable and sustainable biopolymers · Cellulose · Hemicellulose · Lignin · Closed loop biorefinery

1 Introduction

Future energy requirements intensely entertain the use of ionic liquids (ILs) as a platform media for processing of renewable and sustainable biopolymers. The growing attention towards these renewable biopolymers is due to the limited non-renewable fossil resources and alarming climate concerns associated to their processing. In contrast, renewable biopolymers are widely distributed in nature and meet the green and sustainable biorefinery concepts for advanced biofuels and valuable chemicals thus retaining low carbon economy. It is predicted that a decade after, chemical industry will rely on these renewable resources for 30% of their total raw material [1]. Thus, the development of 'integrated biorefineries' based on sustainable and renewable biopolymers is the topmost med-century targets for critical 'Bioenergy' demands to welcome the dreamt golden era reveling the green and sustainable environment.

S. Naz · M. Uroos (✉)
Institute of Chemistry, University of the Punjab, Lahore 54000, Pakistan
e-mail: malihauroos.chem@pu.edu.pk

© Springer Nature Switzerland AG 2020
A. Khan et al. (eds.), *Biofibers and Biopolymers for Biocomposites*,
https://doi.org/10.1007/978-3-030-40301-0_9

Traditionally, various chemical, physical, physiochemical and biological methods or their combination have been employed for the processing of renewable biopolymers but these methods are associated with harsh conditions such as high temperature, pressure, use of environmentally toxic chemicals making the process uneconomical. The continuous search for green processing strategies underpinned the use of ILs as highly appreciable media.

1.1 Ionic Liquids (ILs)

The term 'Ionic liquid' adverts a special class of molten organic salts having melting points less than 100 °C, comprising of unending possible combinations of organic cations and organic or inorganic anions. They are highly designed solvents exhibiting vast adjustable striking features like high thermal stability, insignificant vapor pressure and wide electrochemical assortment. Some other discrete properties that make them distinct from other organic solvents include polarity, hygroscopicity, viscosity as well as high solvation power even for polymeric compounds. These properties can be tuned accordingly by sensible selection of cation or anion (Fig. 1) [2].

ILs are considered as best media for processing of renewable biopolymers in view of promptly emerging green chemistry and clean technology. They reduce the consumption of energy, solvent loss, chances of carbon dioxide evolution and by-products formation. Thus these are highlighted as promising 'green', 'designed' and 'degradative' solvents for future sustainable biorefinery. Most potent classes of ILs for the purpose involve aromatic ILs, deep eutectic solvents (DES) and salt hydrates

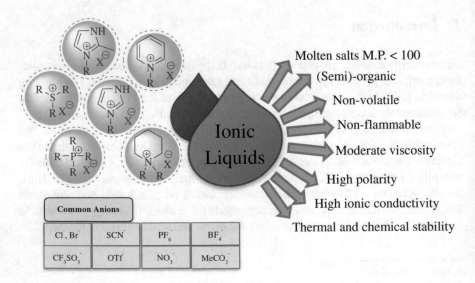

Fig. 1 General profile of ionic liquids (ILs)

[3]. Some most commonly employed ILs for different biopolymers processing are piled up in Table 1 while their detailed description is given in succeeding sections.

1.2 Renewable and Sustainable Biopolymers

General and most abundant renewable and sustainable biopolymers involve carbohydrates and polyaromatic lignin. Former one contains polysaccharides such as semi-crystalline cellulose and amorphous multicomponent hemicellulose while the later one is recognized as phenylpropanoid lignin. Both of these biopolymers exhibit high potential for bioenergy production and are thus promising feedstock for future biorefinery. Most imperative cost-effective natural source of these biopolymers is lignocellulosic biomass that is light weight, porous, carbon–neutral composite material present in cell walls of woody plants like trees, shrubs and grasses. It contains 35–50 wt% cellulose, 23–32 wt% hemicellulose and 15–25 wt% of lignin (Fig. 2). Other minor constituents of lignocellulose involve traces of proteins, pectins, inorganic compounds and some extractives such as waxes and lipids. Nonetheless, the precise lignocellulose composition is variable with respect to plant tissue type, species and growth conditions [4]. Percentage chemical composition of biopolymers in some plant derived biomass materials is listed in Table 2 and their brief overview is given in next sections.

1.2.1 Carbohydrate Biopolymers

Carbohydrates mainly contain cellulose and hemicellulose; among which cellulose is widely distributed in nature having wide renewability potential. It is a linear polymer having flat sheet or stretched chain conformation containing varying number of glucopyranosyl monomeric units linked to each other via β-1,4-glycosidic linkages (Fig. 3). Linearity of the structure is due to the β-configuration and hydrogen bonding associated to anomeric carbons of glucopyranosyl backbone [7]. Each monomer of natural cellulose holds back three hydrogen bonds of which two are intramolecular and third one is intermolecular to the adjacent polymeric cellulose in the same sheet. These repeating interactions are responsible for uniform well-organized packing of countless cellulosic elements to form crystalline fibrils that are stabilized via van der Waals interactions [8, 9]. Nearly 10,000 or higher monomeric units are present in a single polymeric strand depending upon polymer origin and its treatment protocol (Table 3) [10]. Certainly, this is the highest degree of polymerization (DP) amongst all other renewable biopolymers.

Despite the potential of glucose monomeric units and other short oligomers for water, cellulose is completely insoluble in water and even in organic solvents. The reason lies in its comparatively low flexibility, high molecular weight as well as systematic close packing due to hydrogen-bonding and van der Waals interactions resulting in hydrophobic flat upper and lower sides.

Table 1 An outline of some ILs used in processing of renewable and sustainable biopolymers

Dissolution of biopolymers	
Cellulose dissolution	 X = Cl, OAc, PO₄, HCOO⁻, OH R = H, C₁ R = C₄, Bn R₁ = H, Allyl, C₁, C₂, C₄, .. R₁ = H, C₁ X = Cl, Br, OAc
Lignin dissolution	 R = C₁ R₁ = H, Allyl, C₁, C₂, C₄, .. X = MeSO₄, CF₃SO₃, Cl, Br

Processing of biopolymers	
Cellulose hydrolysis and further processing	 X = Cl, OAc, PO₄, X = Cl, OAc, X = Cl, HSO₄, MeSO₄, HCOO⁻, OH⁻ Lys, Asp CF₃SO₃, OAc R = H, C₁ R = C₄, Bn R₁ = H, Allyl, C₁, C₂, C₄, .. R₁ = H, C₁ X = Cl, Br, OAc, HSO₄
Lignin degradation	 X = Cl, Me₂PO₄ R₁ = C₁, C₂, C₄, .. X = Cl, Br, Me₂PO₄ X = Cl, Br, BF₄, HSO₄, Me₂PO₄

Characterization of biopolymeric biomass	
Solution state NMR spectroscopy	 (MIMCl-d₆)-DMSO-d₆ (PyCl-d₆)-DMSO-d₆
2D HSQC NMR analysis	 ([EMIM]OAc-d₆)-DMSO-d₆
GPC analysis of cellulose biopolymer	

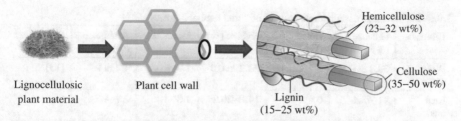

Fig. 2 Composition of lignocellulosic biomass

Table 2 Percentage of renewable and sustainable biopolymers in some lignocellulosics

Biomass	Cellulose (%)	Hemicellulose (%)	Lignin (%)	Proteins (%)	Ash (%)	References
Wheat straw	35–45	20–30	8–15	3.1	10.1	[3]
Rice husk	32–47	19–27	5–24	–	12.4	[5]
Corn stover	42.6	21.3	8.2	5.1	4.3	[3]
Bagasse	65 (Net sugars)		18.4	3	2.4	[6]

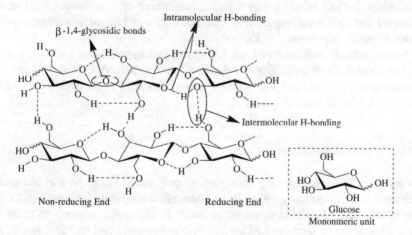

Fig. 3 Flat sheet linear conformation of cellulose biopolymer

Table 3 Polymerization degree of different cellulose based on origin

Source	Degree of polymerization (DP)
Cotton	800–1700
Wood	300–1700
Microcrystalline	150–300
Regenerated	250–500
Bacterial	6500–10,000

Table 4 Percentage carbohydrate content of some lignocellulosics

Biomass	Glucose (%)	Galactose (%)	Xylose (%)	Mannose (%)	Arabinose (%)	References
Wheat straw	38.8 ± 0.5	2.7 ± 0.1	22.2 ± 0.3	1.7 ± 0.2	4.7 ± 0.1	[13]
Rice husk	41–43.4	0.4	14.8–20.2	1.8	2.7–4.5	[5]
Corn stover	39	0.8	14.8	0.3	3.2	[14]
Bagasse	38.1	1.1	23.3	–	2.5	[14]

After cellulose, the second most eminent carbohydrate natural biopolymer is hemicellulose having amorphous structure with multicomponent composition. Contrarily to cellulose, polysaccharides present in hemicellulose are of low molecular weight with only 100–200 monomeric units per polymeric structure [11]. It shows wide structural diversity with respect to branching and type of five or six membered monomeric carbohydrate units having versatile functional groups such as methyl, acetyl or carboxylic moieties like cinnamic acid, galacturonic acid or glucuronic acid. Mannose and xylose are two major monomers of hemicellulose present in softwood and hardwood respectively. Hemicellulose carbohydrate content of some lignocellulosics is presented in Table 4.

An example of hemicellulose is galactoglucomannan present in softwood exhibiting branched chain structure due to α-1,6-D-galactopyranosyl units as well as β-1,4-D-mannopyranosyl and β-1,4-D-glucopyranosyl monomeric linkages with C–2 and C–3 acetyl substitutions (Fig. 4) [12].

1.2.2 Lignin Biopolymer

Second most frequent naturally occurring biopolymer is lignin present along with carbohydrates in secondary cell walls of lignocellulosic plant biomass (Fig. 2). The role of this water insoluble aromatic polymer is to provide strength to plant cell wall, water proofing property and flexibility for biological and physical attack. It is biosynthesized in plants after ceasing their growth. Three monomeric units taking part in biosynthetic pathway are sinapyl, p-coumaryl and coniferyl alcohol that upon polymerization in lignin give rise to syringyl, p-hydroxyphenyl and guaiacyl units respectively (Fig. 5). These monomeric units are varied in different types of plant biomasses; softwood contains only guaiacyl units, guaiacyl and syringyl both are present in hardwood while grasses also contain p-hydroxyphenyl as minor lignin contributors. Various linkages are present all over the polymer like C–C cross-linking [15] responsible for rigidness and β-O-4 ether bonds [16] responsible for linear elongation of polymer as it is present in 50% inter-subunit bonds.

The binding of lignin with carbohydrates in composite is facilitated by its covalent cross-linking with hemicellulose in soft and hardwood. While in grasses binding is

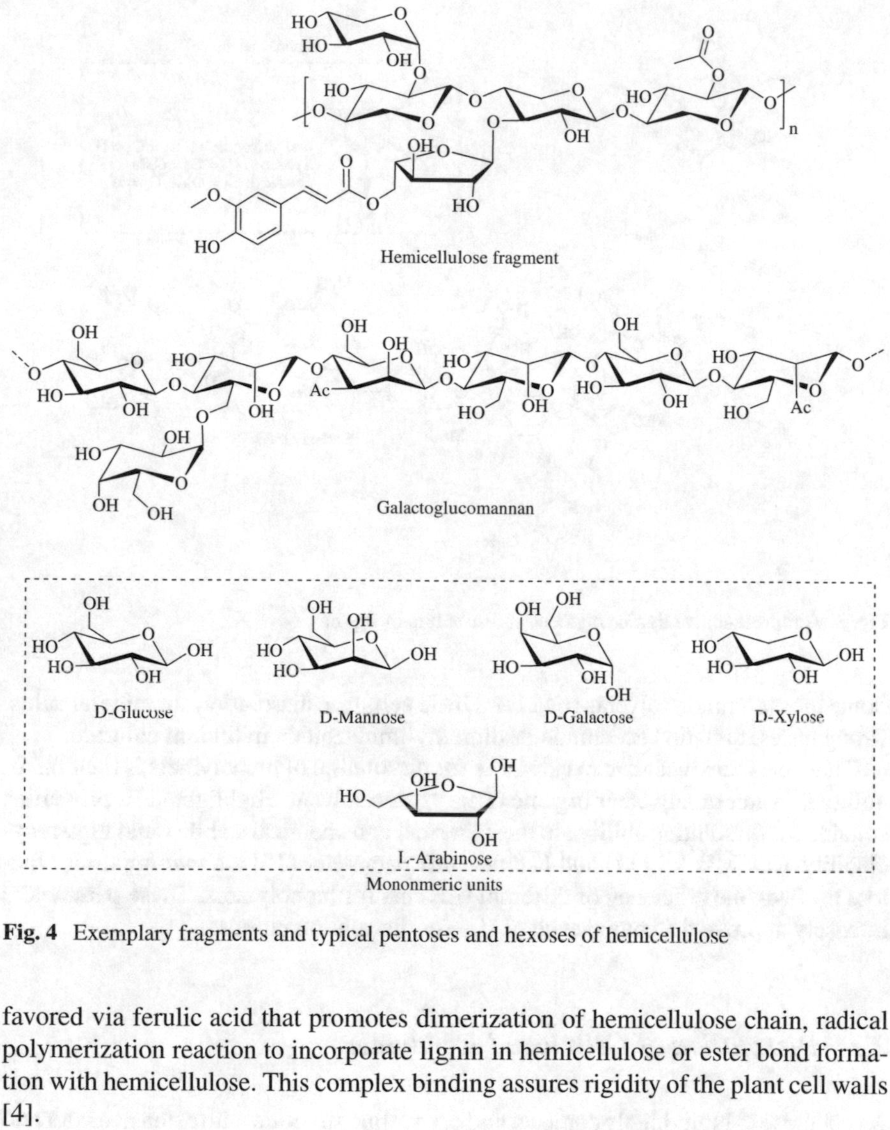

Fig. 4 Exemplary fragments and typical pentoses and hexoses of hemicellulose

favored via ferulic acid that promotes dimerization of hemicellulose chain, radical polymerization reaction to incorporate lignin in hemicellulose or ester bond formation with hemicellulose. This complex binding assures rigidity of the plant cell walls [4].

2 Dissolution of Biopolymers in ILs

Dissolution of renewable and sustainable biopolymers in any appropriate media is essential for their effective processing. But it remains a challenge due to the compact and rigid structure and extensive inter and intramolecular forces present in them.

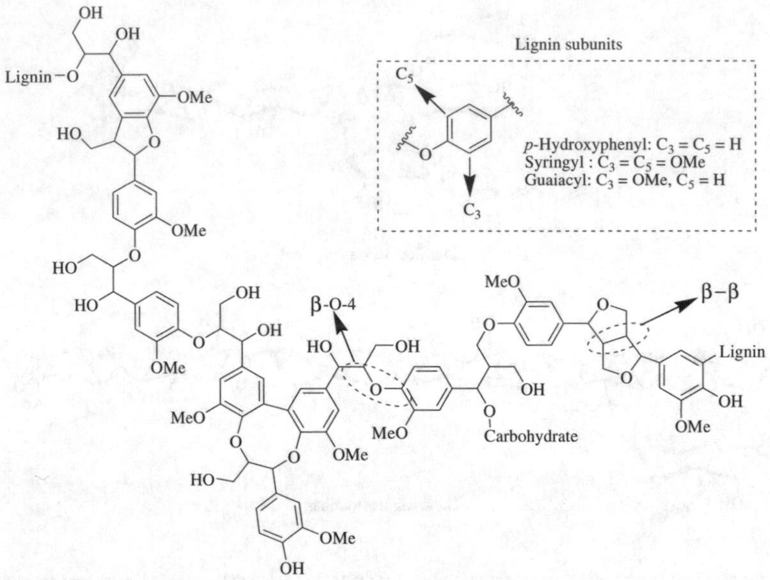

Fig. 5 A representative lignin fragment and subunits of lignin

Some most common solvents tried for a little cellulose dissolution are mineral acids, strong bases, dimethyl acetamide or dimethylimidazolone in lithium chloride.

Thus, the main incentive in using ILs for dissolution of biopolymers is their insolubility in water or any other organic or inorganic solvent. Highlighted IL properties suitable for dissolution ability are their thermal and chemical stability and high recyclability. COSMO RS [17] and Kamlet Taft parameter [18] are two important factors to check the efficiency of different solvents for biopolymers. These parameters intensely approve the competence of ILs for dissolution purpose.

2.1 Dissolution of Cellulose Biopolymer

As cellulose exhibits highly compact and crystalline structure with extensive inter and intramolecular hydrogen bonding network; its dissolution in any solvent remained a question mark for a long time. The first report on IL dissolution ability dates back to a US patent filed in 1934 when N-ethylpyridinium chloride [EtPy]Cl was used for dissolution in the presence of a base [19]. But it gained no care at that time due to high melting point of [EtPy]Cl. Attention has been given to ILs about one and half decade ago when 25 wt% cellulose dissolution was reported with 1-butyl-3-methylimidazolium chloride [BMIM]Cl without any derivatization [20]. After that, a number of ILs have been tested for efficient dissolution of individual cellulose or lignocellulose.

ILs tend to penetrate inside the crystalline structure, disrupting the hydrogen bonding network and thus facilitating the dissolution. Process is accompanied by regeneration of dissoluted material with appropriate anti-solvents [21]. Analysis of regenerated cellulose shows that dissolution causes alteration of its symmetrical structure and crystalline cellulose having high polymerization degree is converted to cellulose type II exhibiting reduced crystallinity and lesser degree of polymerization and is easy to be converted to fuels and other targeted products. Various factors affecting the dissolution process in ILs involve cellulose crystallinity and its polymerization degree, initial cellulose concentration, nature of selected IL and operating parameters like time and temperature [22].

For effective dissolution, careful designing of IL by selecting effective ion pairs is most important. Anion is most crucial for dissolution as it forms hydrogen bonding with the biopolymers facilitating the process. Thus the size, concentration and hydrogen bond basicity or electron donating ability of anions is most important for the process. In Kamlet Taft parameter when finding the dissolution ability of ILs, α and β factors are most important. α regards to hydrogen bond acidity of IL while β is the hydrogen bond basicity that is mostly related to the anion. Most widely used IL anions having highest hydrogen bond accepting abilities are chloride and acetate having high β values. Contrarily, dicyanamide $N(CN)_2$, bis(trifluromethylsulfonyl)imide (NTf_2), hexafluorophopsphate (PF_6) and tetrafluoroborate (BF_4) having less hydrogen bonding abilities are less effective for dissolution [23].

Protons of cation must be highly acidic and it should contain short side chain to reduce the steric hindrance between biopolymers and IL [2]. Thus acidic ILs are proven to be more powerful for dissolution and further processing of renewable biopolymers. Substituting the cation with unsaturated functional groups like cyanide or allyl also enhances the process efficiency due to their increased interactions with that of cellulose [24]. For example, replacing butyl of [BMIM]Cl with allyl group increases the dissolution up to 14.5%. In addition, this IL [AMIM]Cl exhibits a lesser viscosity than many others having extended side chains thus positively contributing in dissolution.

2.2 Dissolution of Lignin

Just like cellulose, lignin dissolution is also of profound investigation to enhance its further processing to obtain fuels and targeted products. Dissolution causes exposure of monomeric phenylpropane connecting points thus facilitating the degradation process [25]. ILs have high dissolution power for these biopolymers as compared to other solvents. A maximum of 500 g lignin have been reported to dissolve per kg of 1-butyl-3-methylimidazolium trifluormethanesulfonate [BMIM]CF_3SO_3 and 1,3-dimethylimidazolium methylsulfate [MMIM]MeSO$_4$ at 90 °C when incubated for 24 h [26]. Alkylmethylimidazolium ILs like [HMIM]CF_3SO_3, [MMIM]MeSO$_4$

and [BMIM]MeSO$_4$ also have significant lignin dissolution powers. Again, the dissolution is highly dependent upon anion type and trend is almost similar to cellulose [27][B].

Ji et al. studied the mechanism of dissolution process by natural bond orbital (NBO) analysis, atoms in molecules (AIM) theory, density functional theory (DFT) and Wiberg bond index (WBI) method. They modeled the lignin with 1-(4-methoxyphenyl)-2-methoxyethanol (LigOH) and employed [AMIM]Cl as dissolution media. Theoretical studies revealed that interaction of [AMIM]Cl and LigOH happens via hydrogen bonding and it is stronger than interactions present in LigOH itself. Successive regeneration process is favored by weakening or destroying this developed hydrogen bonding between IL and LigOH by addition of water as anti-solvent [28].

3 Processing of Biopolymers in ILs

All-encompassing profile of ILs towards future green biorefinery concepts demands the establishment of efficient protocols for obtainment and processing of renewable and sustainable biopolymers. Generally, two methods namely ionosolv and dissolution are in practice for obtaining the renewable biopolymers from natural lignocellulosic biopolymeric composites. By ionosolv method, considerable amount of lignin is selectively dissolved in IL while keeping the carbohydrates intact. Solid cellulosic content is filtered and dissolved lignin is isolated via re-precipitation using water. Significant delignification efficiency of ionosolv process depends upon nucleophilic, neutral or acidic IL anions employed in the process [29]. On the other hand, dissolution method uses such ILs that have high selectivity towards cellulose. Lignocellulose is processed in ILs for suitable time and temperature for all possible dissolution. Anti-solvents mixture is used to obtain cellulose and lignin. For example, in mostly used water–acetone mixture, water precipitates cellulose enriched pulp while acetone helps in lignin recovery. Generally, the deconstruction of lignocellulose is not complete in this process due to the recovery of hemicellulose traces with lignin and also a little fraction with that of cellulose [4].

3.1 Processing of Carbohydrate Biopolymers

Carbohydrate biopolymers possess wide range of applications for integrated biorefineries. From all carbohydrate contents, cellulose is hard to process or depolymerize due to its highly compact crystalline structure [4]. Hydrolysis of cellulose into monomeric sugars is the fundamental stairway to various biofuels and targeted platform chemicals. It cleaves biopolymeric β-1,4-glycosidic bonds thus unlocks the potential of cellulose for further processing; hydrolytic products are more prone to chemical conversion and fermentation. A variety of products exhibiting one to six carbon skeletons can be obtained depending upon the mode of processing (Fig. 6). Chemical conversion is generally accomplished via hydrolysis, dehydration, isomerization, hydrogenation, aldol condensation, oxidation and reforming [30]. Remarkable products of these methods include sorbitol, 5-hydroxymethylfurfural (5-HMF), levulinic acid (LVA), 2,5-furan dicarboxylic acid, gluconic acid, glucaric acid and many more. Highlighted products of fermentation involve bioethanol, glycerol, 1,3-propanediol, glutamic acid, fumaric acid, lactic acid, 3-hydroxypropanoic acid, itaconic acid, ascorbic acid, fumaric acid, succinic acid, lactic acid, citric acid, malic acid, penicillin, lysine and riboflavin [31–33].

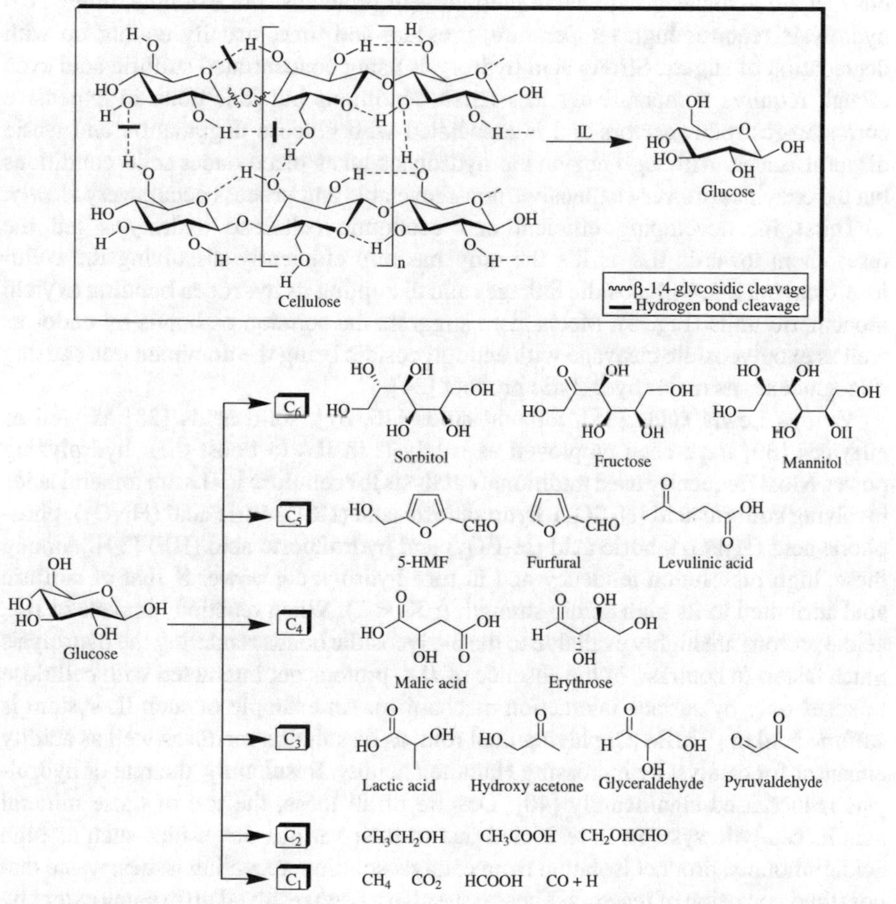

Fig. 6 Hydrolysis of cellulose in ILs and important cellulose conversion products

3.1.1 Hydrolysis of Carbohydrates

As stated earlier, hydrolysis of carbohydrates into simplest reducing sugars serves as the starting point for carbohydrate-based biorefineries approach; cellulose is usually converted to glucose simplest units that further leads towards wide spectrum biorefinery products. On the other hand, hydrolysis of hemicellulose ends up in xylose monomers that in turn are hydrolyzed to give xylitol, xylal, dithioacetal, furfural, hydroxy-xylal esters, levulinic acid, and aza-heterocycles such as pyrazole and imidazole [32].

Traditionally the hydrolysis is achieved via alkaline, acidic or enzymatic methods that are associated with harsh pretreatment protocols. For example, dilute acid hydrolysis requires high temperature, pressure and time, usually ending up with degradation of sugars. Strong acid hydrolysis using concentrated sulfuric acid even though requires comparatively less harsh conditions but it is done in expensive corrosion-resistant reactors and is associated with glucose degradation and waste disposal issues. Although enzymatic hydrolysis takes place under mild conditions but the enzymes are very expensive, non-recyclable and process occurs very slowly.

Thirst for developing efficient and economic cellulose hydrolysis led the researchers towards ILs as it's the only medium efficiently dissolving the cellulose, cleaving β-1,4-glycosidic linkages and disrupting its hydrogen bonding to yield monomeric units (Fig. 6). Mechanism suggests the scission of bonds by endo- as well as exoglycosidic cleavage with endoglycosidic being the dominant one causing oligoglucoses as major hydrolytic product [34].

Various Lewis acids [35], mineral acids [36, 37], solid acids [38] as well as enzymes [39] have been employed as catalysts in ILs to boost their hydrolyzing power. Most frequently used traditional catalysts for cellulose in ILs are mineral acids involving sulfuric acid (H_2SO_4), hydrochloric acid (HCl), nitric acid (HNO_3), phosphoric acid (H_3PO_4), boric acid (H_3BO_3) and hydrofluoric acid (HF) [34]. Among these, high dissolution tendency and in turn hydrolyzing power is that of sulfuric acid attributed to its high acidic strength ($pK_a < 1$). When reaction happens in ILs, acidic protons are highly available to the β-glycosidic bonds rendering the hydrolysis much faster. In contrast, in the absence of ILs, protons get interacted with cellulose linkages only by surface interaction mechanism. An example of such IL system is sulfuric acid in [BMIM]Cl playing dual role; as dissolution media as well as acidity enhancer for catalyst by increasing Hammett acidity. Resultantly, the rate of hydrolysis is increased significantly [40]. Despite of all these, the use of these mineral acid/IL catalytic systems is restricted due to their various downsides such as high acidic amounts, product isolation from catalyst solution, recycling issues, waste disposal and corrosion of reactors. These issues have been resolved up to some extent by replacing mineral acids with solid acids. Highlighted solid acid/IL catalytic systems involve supported metals, metal oxides, zeolites, functionalized silicas, acidic resins, magnetic acids and carbonaceous acids [41].

3.1.2 Dehydration of Carbohydrates

Dehydration of simplest five and six carbon carbohydrates results in 5-hydroxmethyl furfural (5-HMF) that is the most important chemical derived from renewable and sustainable biopolymeric composites. It is documented as a promising renewable platform chemical by U.S. Department of Energy [42]. The reason lies in its structural versatility as a 'sleeping giant' containing multiple functional groups; it's an aromatic aldehyde, aromatic alcohol and also contains a furan ring system. Thus it serves as an important intermediate in petroleum-based industrial chemistry as well as biomass derived carbohydrate chemistry. Some highlighted platform chemicals obtained from 5-HMF involve functional polymers, biomass fuels, drugs and other important compounds like 2,5-diformylfuran (2,5-DFF), 5-ethoxymethylfurfural (EMF), 2,5-dihydroxymethylfurfural (DHMF), 2,5-furandicarboxylic acid, 5-hydroxymethyl furanoic acid (HMFA), 2,5 furandicarboxylic acid (FDCA), 5-formyl-2-furancarboxylic acid (FFCA), formic acid, levulinic acid and humins [43].

Lewis acids in ILs are the most suitable catalysts for transformation of carbohydrates to 5-HMF. The role of these Lewis acids is to enhance the cellulose hydrolysis to produce glucose monomers, isomerization of glucose and finally the dehydration of resultant fructose to produce 5-HMF. This mechanism is not as much favored by other systems like mineral or solid acid catalysts due to their lack of competence for glucose isomerization [44]. A wide range of Lewis acids have been tested till date with the best ever results with halides of chromium, aluminium and iron. Mechanistically, these halides first get coordinated with the IL to ensure IL–Lewis acid complex whose anion facilitates the mutarotation of glucose thus easing its isomerization to fructose (Fig. 7) [35].

Rehydration of 5-HMF results in levulinic acid (LVA) that is the imperative building block for a number of organic compounds and various biofuels as identified by National Renewable Energy Laboratory (NREL) of US. Starting from carbohydrate biopolymers of lignocellulosic biomass, LVA is synthesized via deep hydrolysis method. Just like 5-HMF, it also exhibits structural versatility and diverse transformational capabilities due to its dual functional group moieties; carboxylic acid and ketone (Fig. 8). Again, deep hydrolysis of renewable carbohydrate biopolymers using ILs is superior to traditional mineral acids and other heterogeneous catalysts due to dual catalyst-solvent behavior of IL, being green, mild reaction times, high recycling and so on. Various Lewis as well as acidic ILs mostly with imidazolium based cations have been reported for said transformation [45].

3.2 Processing of Lignin

Despite of being a primary natural renewable biopolymer, a little work has been done on lignin with regard to its volarization, degradation or chemical transformation as compared to carbohydrate biopolymers. Now-a-days, its practical applications

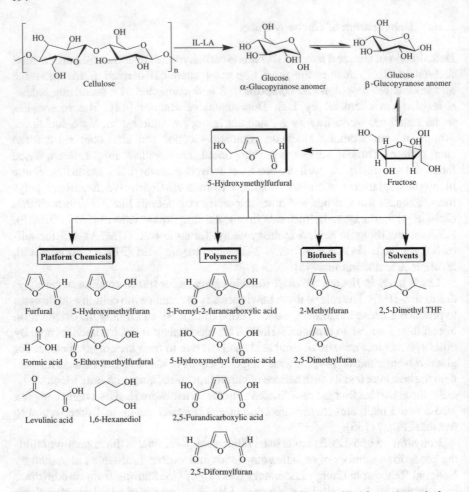

Fig. 7 Synthesis of 5-HMF via IL-Lewis acid catalysis and its versatility for various platform chemicals

are confined only to low value products such as binding, dispersing or emulsifying agents, carbon fibers, wood panel products, phenolic resins, low-grade fuels, automotive brakes, polyurethane foams and printed circuit boards [46]. But yes, it can serve as a promising biopolymer providing efficient routes towards economic production of valuable chemicals and biofuels sustaining the future bioeconomy. Due to its unique structural and chemical properties, a broad range of bulk and fine renewable aromatic compounds can be obtained by designing its effective processing. It's a potent source for various petrofuels like benzene, xylene and toluene [47], countless bio-products [48] and efficiently imparts anti-oxidant, anti-bacterial and anti-UV properties to other composites to enhance their polymeric properties [49]. It also serves as natural source of inhibition compounds such as syringyl aldehyde and vanillic acid for fermentative organisms and hydrolase enzymes [50]. Thus exploration of selective

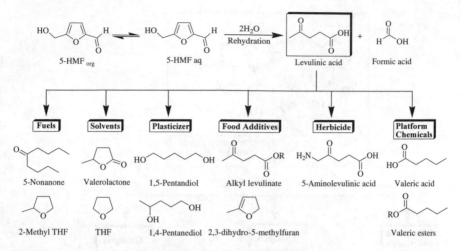

Fig. 8 Formation of levulinic acid and its synthetic transformations

and robust catalytic systems for lignin depolymerization is most suggested one by a recent review on lignin role in biomass refineries by U.S. Department of Energy [51]. Patently, lignin is well thought-out to be the chief, economic and well recognized aromatic resource of the bio-based economy (Fig. 9) approving the myth that says: "You can make anything out of lignin… except money" [52].

In view of vast applicability account of lignin, various methods have been established to deconstruct the lignocellulosic biomass; the most abundant natural source of lignin. Mostly it is obtained via pulping methods like Organosolv, Kraft and sulfite pulping that tend to remove the lignin from pulp [4]. The deconstruction or delignification competence of various lignocellulosics depends upon the cross-linking of monomeric units and complex binding of entire composite. Grasses and hardwood can be delignified relatively easily than softwoods. The reason is the C–C cross-linking present in softwood at C–5 of guaiacyl monomeric unit [15] that does not hydrolyze easily by acid or base. Various deconstruction methods have also been reported for lignin. Key method is its modification via hydrolysis of ether bonds.

Currently, biorefineries are taking on two-step strategy for biofuels production from lignin biopolymer. Firstly, the lignin is depolymerized and then fuels are produced from depolymerization products preferably by hydrodeoxygenation into alkanes. Depolymerization is mainly done to convert the lignin into simple aromatic products and it's a difficult task due to recalcitrant nature and complex structure of lignin, different extraction techniques and variability of sources. Different tools used for the purpose include hydrogenation, fast pyrolysis or acid/alkaline hydrolysis [53] while oxidative method is the most widely used. Oxidation may be carried out via peroxides, mesoporous materials, vanadium-based catalysis, photocatalysis and electrocatalysis done either via conventional method or by processing in ILs. Leading fragmentation products of lignin oxidation involve low molecular weight phenolic compounds (LMWPC) and dicarboxylic acids (DCAs) such as quinones,

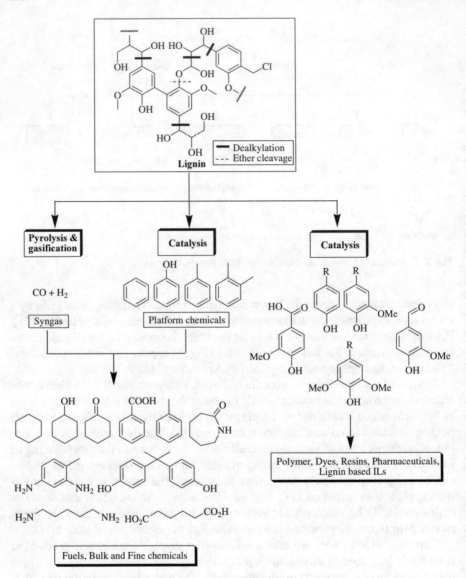

Fig. 9 Lignin biopolymer as chief aromatic resource for the bio-based economy

malonic acid, succinic acid, maleic acid, muconic acid and muconolactone that serve as efficient precursor of high value chemicals important for pharmaceutical, food and polymer industries. Quinones themselves are used as catalyst for depolymerization of lignin. They can also be used for energy storage purposes in battery technology via benzoquinone/hydroquinone redox coupling [52]. Careful conversion of lignin also leads towards aromatic aldehydes such as vanillin, *p*-hydroxybenzaldehyde and syringaldehyde extensively useful in flavoring, agricultural pesticides as well as chemical intermediates for therapeutic drugs [54]. However, there is a need to explore more about mechanism, selectivity and control of these depolymerizations to generate valuable end products. Because complex mixture of aromatic compounds or chemically modified lignin is generally obtained due to obstinate C–C linkages between propylphenol monomers present in it [55]. Also, the selectivity and reaction separation of oxidative degradation needs to be explored for aromatic aldehydes [54].

Although this first step has been made somewhat more improved and established, second fuel forming step is still a challenging one. Traditional hydrodeoxygenation methods for converting phenols to alkanes involve NiMo and CoMo sulfite catalysts having drawbacks of sulfur contamination, accumulation of coke and deactivation of catalysts due to the presence of water. To overcome this, aqueous-phase catalytic systems are used to carry out a series of hydrogenation and dehydration reactions. Although this catalytic system is superior to traditional ones, some limitations are also associated with it like dehydration reaction taking place in water, high temperature and energy consumption. In this scenario, ILs are an important medium having plus points of being green, high efficiency, easy phase separation and water-based system acting both as catalyst as well as solvents. Different catalytic systems in ILs have been developed for volarization and biofuel production from lignin; some highlighted ones comprise metal nanoparticles (NPs) [53, 56], metal coupling with various oxidants such as oxygen [54], hydrogen peroxide [55, 57] or titanium oxide [58], metal chlorides [59, 60] and nitrates [61], palladium supported on activated charcoal [62], biomimetic catalysts such as porphyrin [63] and electrocatalysis by using IL [64] or catalytically active ruthenium–vanadium–titanium mixed oxide as electrodes [65]. Various ILs used along with these catalysts are alkylimidazolium based Bronsted acidic ionic liquids (BAILs) [53], trimethyl phosphite anion containing varying imidazolium, pyridinium, ammonium and morpholinium cations [54], hydrogen sulfate anion with triethylammonium and butylimidazolium cations [55], triethylammonium hydrogen sulfate [58], alkylsulfonates based imidazolium ILs [61], choline based ILs [62] and triethylammonium methanesulfonate [65].

Besides these, some reports reveal the dual behavior of ILs for lignin depolymerization; as solvent as well as catalyst. For example 3-methylimidazolium chloride [MIM]Cl has been reported for depolymerization of lignin obtained from oak wood [66]. Other ILs used for this purpose include butylmethylimidazolium tetrafluoroborate [BMIM]BF$_4$, butylimidazolium bromide [BIM]Br, butylpyridinium bromide [BPy]Br [67], butyl-1,8-diazobicyclo[5.4.0]undec-7-enium chloride (DBUB]Cl [68] and many others.

In addition to fuel production, lignin also exhibits remarkable potential for synthesis of micro/nanoparticles [69], functionalized lignin hybrid materials [70] as well as various ionic liquids as patented recently [71, 72].

Due to wide spectrum bio-refinery applications of lignin, International Lignin Institute (ILI) association has been made to provide a platform to academia and industrial researchers having interest in volarization of lignin for future technology. For example, PureVision Technology, Inc., succeeded in production of low molecular weight lignin serving as fuel as well as a co-product to the cellulose stream [25].

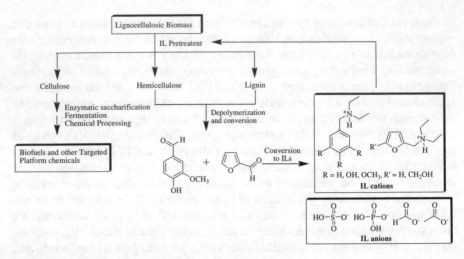

Fig. 10 Diagrammatic representation of closed loop biorefinery approach

4 Closed Loop Biorefinery

The concept of closed loop biorefinery is set forth to make the process more green and viable. According to this hypothetical statement, the hemicellulose and lignin biopolymeric contents obtained after pretreatment of lignocellulosic biomass can serve as potential raw materials for synthesis of ILs. For this purpose, important thing is the controlled depolymerization of these biopolymers to get low cost IL precursors such as alcohols, aromatic aldehydes and acids. Pretreatment of lignocellulosic biomass is also accompanied with lignin depolymerization to small aromatics stream that could potentially be used for formation of renewable ILs (Fig. 10) [73]. Thus, an ecofriendly closed loop biorefinery concept was provided and somewhat tried to be implemented by research group of Singh [74] who pretreated and deconstructed lignocellulose using room temperature ILs; and re-entered hemicellulose and lignin biopolymers into the biorefinery route by converting them into renewable ILs while cellulose was converted to biofuels and other platform chemicals in successive pathways via enzymatic saccharification, fermentation and some other processes.

5 ILs for Characterization of Renewable and Sustainable Biopolymers

Structure elucidation of natural biopolymeric composites namely lignocellulose containing three major types of renewable and sustainable biopolymers is important to explore in order to use them in better way. Normally, the plant cell wall of concerned lignocellulosic biomass is analyzed by first isolating its different components that is time taking job and also can change their native structure. Thus, to preserve the structure and properties of lignocellulosic biomass during analysis, effective dissolution of biomass should be done to get better spectroscopic characterization from homogeneous solutions. ILs are only best non-degradative solvents for this purpose that exhibit effective dissolution ability for all the biopolymers and even for plant cell walls of lignocellulosic biomass under mild conditions. Certain ILs has been successfully used as solvents in several 1D and 2D high-resolution nuclear magnetic resonance (NMR) spectroscopy. For instance, cellulose, hemicellulose, lignin and its inter-unit linkages as well as its complexes with cellulose and hemicellulose are characterized by solution state NMR spectroscopy using a mixture of 1-methylimidazolium chloride (MIMCl-d_6)-DMSO-d_6 and pyridinium chloride (PyCl-d_6)-DMSO-d_6 [75, 76]. In the same way, 2D HSQC NMR analysis of total plant cell walls is done by 1-ethyl-3-methylimidazolium acetate ([EMIM]OAc-d_6)-DMSO-d_6 [77].

In addition, ILs are also used in biomass characterization via gel permeation chromatography (GPC) to understand the bonding of carbohydrates–lignin complexes. For this purpose, biomass is first dissolved in 1-allyl-3-methylimidazolium chloride [AMIM]Cl, derivatized using benzoyl chloride in pyridine and is then checked for distribution of cellulose molecular weight and/or entire biomass cell walls [78]. Another advanced and more potent method that proficiently checks the molecular weight distribution of whole biomass and lignin uses [EMIM]OAc in DMF and do not require pre-swelling, activation or derivatization treatment [79]. These inventive gel permeation chromatographic analysis methods are well thought-out to be more substantial for thorough study of cellulose hydrolysis using enzymatic assay.

6 Challenges for ILs Based Processing of Biopolymers

Irrespective of continuously emerging research for the development and advancements in virtuous practicality of ILs for processing of renewable and sustainable biopolymers in integrated biorefineries, some economic and environmental issues associated to sustainability still persist. These issues make the scaling up of process a little bit difficult. This section covers a short overview of some of these.

6.1 Reducing the Particle Size of Renewable Lignocellulosic Composites

Processing and pretreatment of lignocellulosic biomass into ILs requires a pronounced particle size so that their surface area should be enhanced causing high solubility in ILs. For this purpose, lignocellulose is thermally deconstructed to obtain reduced size via grinding or milling consuming so much mechanical energy [80]. Different types of lignocellulosics require different energies as grasses need less energy than woody biomass [81]. According to an energy calculation report, grinding the corn stover to reduce its particle size utilizes 1.1% of its energy content [82]. The solution to this problem is the addition of IL to lignocellulosic biomass prior to grinding which will lower the energy consumption due to their lubricating properties [83].

6.2 Stability and Recycling Issues of ILs

Generally, the ILs are considered to have high temperature tolerating properties. But in real, this statement is not true for all types as different ILs possess different temperature sensitivities. Typically, the processing of biopolymers in ILs is done at high temperatures mostly above 100 °C. High temperature causes decomposition of organic cation of some ILs favored by their counter anion at certain temperatures. Therefore, most important thing to be considered is the stability of IL that is often been ignored in most of the studies. At present, the obtained compounds as well as ILs are checked for stability via thermogravimetric analysis (TGA) that measures the percentage of sample's weight loss with respect to temperature [84].

Another major issue of ILs based biorefineries is moisture sensitivity of ILs [20]. Lignocellulosic biomass containing renewable biopolymers itself holds significant moisture content that should be removed prior to pretreatment with ILs. According to a report, 0.15% of water, if present in IL, highly affects the dissolution by precipitating the cellulose [85]. Water content of IL renders recycling process most problematic. To stamp down this issue, IL as well as biomass both are well dried before processing that requires so much energy rendering process a little bit malign. So, the time demands discovery of temperature as well as moisture sensitive ILs having high recycling ability to efficiently process biopolymers [19].

6.3 Product Isolation from IL Post-Reaction Phase

Separation of hydrolytic sugar products from IL post-reaction liquid is a major challenge in carbohydrate biopolymers based biorefineries to make the process economically viable. Due to strong interactions between IL and monosaccharides, separation process should be such that do not affect the properties of IL. So, the separation is highly reliant on molecular weight, polarity and some other physicochemical properties of IL and products.

An easy-to-do method reported for sugars isolation is by using alkaline solutions that develop a biphasic system when added to imidazolium ILs thus helping in sugars isolation by phase separation [86]. Preparative chromatography was also employed for separation of xylose from [EMIM]HSO$_4$ by sensibly tuning the stationary phase as well as polarity of eluent [87].

6.4 Toxicity and Eco-Protection Hazards

Growing fame of ILs towards biorefineries has questioned their widespread applications with regard to the possible threats to human life and environment. At present, a number of reports have been presented on toxicity of ILs [88–92]. Usually, the ILs are considered as highly green solvents imparting high environmental protection but their (eco)toxicological and biodegradation studies by current European Union Environmental Legislation for safety assessment revealed some of their (eco)toxic effects. Studies show aprotic ILs to be more toxic as compared to protic ones. In protic ionic liquids (PILs) further lessening of toxicity is also achieved by lessening of alkyl chain [93] as well as substitution of some polar group in to the cation [94]. The anion of IL is also crucial for the same [95].

Talking about their direct entrance into the environment, they do not directly go into the air due to their non-volatility but can cause harm when entering to water bodies as these are highly water miscible. The PILs proved to have by far lower aquatic toxicity than the other acidic ionic liquids (AILs) and are proved to be potentially biodegradable in water, unlike the AILs [96].

6.5 Cost Effectiveness of ILs

Expensiveness of ILs is the main problem for scaling up the ILs based processing of renewable biopolymers. The reagents and precursors used to synthesize ILs are highly priced and the demand of searching low cost ILs is going up day by day. Luckily, the problem is being resolved by discovering alternate methods and employing low cost starting materials (chloine, mineral acids and many more) for the synthesis. An option is to prefer PILs over aprotic ones due to their ease of synthesis as well as low price. Recycling of ILs after the complete process also dramatically contributes to the cost efficiency of IL used for processing of renewable biopolymers [97].

Cost calculation by techno-economic model of ILs by bearing in mind the synthetic expense, loading and recycling, a practical design of low cost ILs has been developed [98] and a number of ILs having varying cation and anions with reasonable cost are known today. Thus, the ILs could pave the way to economical integrated biorefinery approaches to meet the future energy requirements.

7 Conclusion

Ionic liquids serve as highly appreciable media for processing of renewable and sustainable biopolymers. Dissolution, pretreatment, chemical conversion into valuable fine chemicals and characterization analyses of natural biopolymeric lignocellulosic composites are highly facilitated in ILs. Due to high process efficiencies as well as green chemistry and clean technologies associated with these IL based approaches over traditional ones, need of the time is to develop highly designed, cost effective and more benign methods that can be scaled up to meet the integrated biorefineries demand.

8 Future Perspectives

In view of appreciable results of ILs based processing of renewable and sustainable biopolymers, it is suggested to further explore the process for scaling it up to industrial level. Cheap and economic ILs should be developed for the purpose and optimized their specific features regarding this. Natural sources should be considered as efficient precursors of ILs in this regard. Attention must also be paid to product yields and purity and also to reduce the consumption of time.

References

1. Alvira, P., Tomas-Pejo, E., Ballesteros, M. J., & Negro, M. J. (2010). Pretreatment technologies for an efficient bioethanol production process based on enzymatic hydrolysis: A review. *Bioresource Technology,101*, 4851–4861.
2. Moyer, P., Smith, M. D., Abdoulmoumine, N., Chmely, S. C., Smith, J. C., Petridis, L., et al. (2018). Relationship between lignocellulosic biomass dissolution and physicochemical properties of ionic liquids composed of 3-methylimidazolium cations and carboxylate anions. *Physical Chemistry Chemical Physics,20*, 2508–2516.
3. Saha, B. C., & Cotta, M. A. (2006). Ethanol production from alkaline peroxide pretreated enzymatically saccharified wheat straw. *Biotechnology Progress,22*, 449–453.
4. Brandt, A., Grasvik, J., Hallett, J. P., & Welton, T. (2013). Deconstruction of lignocellulosic biomass with ionic liquids. *Green Chemistry,15*, 550–583.
5. Karimi, K., Kheradmandinia, S., & Taherzadeh, M. J. (2006). Conversion of rice straw to sugars by dilute-acid hydrolysis. *Biomass and Bioenergy,30*, 247–253.
6. Georgieva, T. I., Mikkelsen, M. J., & Ahring, B. K. (2008). Ethanol production from wet-exploded wheat straw hydrolysate by thermophilic anaerobic bacterium Thermoanaerobacter BG1L1 in a continuous immobilized reactor. *Applied Biochemistry and Biotechnology,145*, 99–110.
7. O'sullivan, A.C. (1997). Cellulose: The structure slowly unravels. *Cellulose, 4*, 173–207.
8. Nishiyama, Y., Langan, P., & Chanzy, H. (2002). Crystal structure and hydrogen-bonding system in cellulose Iβ from synchrotron X-ray and neutron fiber diffraction. *Journal of the American Chemical Society,124*, 9074–9082.
9. Qian, X., Ding, S.-Y., Nimlos, M. R., Johnson, D. K., & Himmel, M. E. (2005). Atomic and electronic structures of molecular crystalline cellulose Iβ: a first-principles investigation. *Macromolecules,38*, 10580–10589.
10. Jimenez de la Parra, C., Navarrete, A., Dolores Bermejo, M., & Jose Coccro, M. (2012). Patents review on lignocellulosic biomass processing using ionic liquids. *Recent Patents on Engineering,6*, 159–181.
11. Timell, T. E. (1967). Recent progress in the chemistry of wood hemicelluloses. *Wood Science and Technology,1*, 45–70.

12. Willfor, S., Sundberg, K., Tenkanen, M., & Holmbom, B. (2008). Spruce-derived mannans–A potential raw material for hydrocolloids and novel advanced natural materials. *Carbohydrate Polymers,72*, 197–210.
13. Erdei, B., Barta, Z., Sipos, B., Reczey, K., Galbe, M., & Zacchi, G. (2010). Ethanol production from mixtures of wheat straw and wheat meal. *Biotechnology for Biofuels,3*, 16.
14. Lee, J. (1997). Biological conversion of lignocellulosic biomass to ethanol. *Journal of Biotechnology,56*, 1–24.
15. Boerjan, W., Ralph, J., & Baucher, M. (2003). Lignin biosynthesis. *Annual Review of Plant Biology,54*, 519–546.
16. El Hage, R., Brosse, N., Chrusciel, L., Sanchez, C., Sannigrahi, P., & Ragauskas, A. (2009). Characterization of milled wood lignin and ethanol organosolv lignin from miscanthus. *Polymer Degradation and Stability,94*, 1632–1638.
17. Klamt, A. (1995). Conductor-like screening model for real solvents: a new approach to the quantitative calculation of solvation phenomena. *The Journal of Physical Chemistry,99*, 2224–2235.
18. Spange, S., Keutel, D., & Simon, F. (1992). Approaches to empirical donor-acceptor and polarity-parameters of polymers in solution and at interfaces. *Journal de Chimie Physique,89*, 1615–1622.
19. Ogura, K., Ninomiya, K., Takahashi, K., Ogino, C., & Kondo, A. (2014). Pretreatment of Japanese cedar by ionic liquid solutions in combination with acid and metal ion and its application to high solid loading. *Biotechnology for Biofuels,7*, 120.
20. Swatloski, R. P., Spear, S. K., Holbrey, J. D., & Rogers, R. D. (2002). Dissolution of cellulose with ionic liquids. *Journal of the American Chemical Society,124*, 4974–4975.
21. Fort, D. A., Remsing, R. C., Swatloski, R. P., Moyna, P., Moyna, G., & Rogers, R. D. (2007). Can ionic liquids dissolve wood? Processing and analysis of lignocellulosic materials with 1-n-butyl-3-methylimidazolium chloride. *Green Chemistry,9*, 63–69.
22. Vitz, J., Erdmenger, T., Haensch, C., & Schubert, U. S. (2009). Extended dissolution studies of cellulose in imidazolium based ionic liquids. *Green Chemistry,11*, 417–424.
23. Zhao, H., Baker, G. A., Song, Z., Olubajo, O., Crittle, T., & Peters, D. (2008). Designing enzyme-compatible ionic liquids that can dissolve carbohydrates. *Green Chemistry,10*, 696–705.
24. Khan, A. S., Nasrullah, A., Ullah, Z., Bhat, A. H., Ghanem, O. B., Muhammad, N., et al. (2018). Thermophysical properties and ecotoxicity of new nitrile functionalised protic ionic liquids. *Journal of Molecular Liquids,249*, 583–590.
25. Zakzeski, J., Bruijnincx, P. C., Jongerius, A. L., & Weckhuysen, B. M. (2010). The catalytic valorization of lignin for the production of renewable chemicals. *Chemical Reviews,110*, 3552–3599.
26. Lee, S. H., Doherty, T. V., Linhardt, R. J., & Dordick, J. S. (2009). Ionic liquid-mediated selective extraction of lignin from wood leading to enhanced enzymatic cellulose hydrolysis. *Biotechnology and bioengineering,102*, 1368–1376.
27. Pu, Y., Jiang, N., & Ragauskas, A. J. (2007). Ionic liquid as a green solvent for lignin. *Journal of Wood Chemistry and Technology,27*, 23–33.
28. Ji, W., Ding, Z., Liu, J., Song, Q., Xia, X., Gao, H., et al. (2012). Mechanism of lignin dissolution and regeneration in ionic liquid. *Energy & Fuels,26*, 6393–6403.
29. Brandt, A., Ray, M. J., To, T. Q., Leak, D. J., Murphy, R. J., & Welton, T. (2011). Ionic liquid pretreatment of lignocellulosic biomass with ionic liquid–water mixtures. *Green Chemistry,13*, 2489–2499.
30. Eminov, S., Filippousi, P., Brandt, A., Wilton-Ely, J. D., & Hallett, J. P. (2016). Direct catalytic conversion of cellulose to 5-hydroxymethylfurfural using ionic liquids. *Inorganics,4*, 32–47.
31. Werpy, T., Petersen, G., Aden, A., Bozell, J., Holladay, J., White, J., et al. (2004). *Top value added chemicals from biomass. Volume 1-Results of screening for potential candidates from sugars and synthesis gas*. Washington DC: Department of Energy.
32. Tundo, P., Perosa, A., & Zecchini, F. (2007). *Methods and reagents for green chemistry*. Hoboken: Wiley.

33. Carole, T.M., Pellegrino, J., & Paster, M.D. (2004). Opportunities in the industrial biobased products industry. In *Proceedings of the Twenty-Fifth Symposium on Biotechnology for Fuels and Chemicals Held May 4–7, 2003* (pp. 871–885) Breckenridge, CO: Springer.
34. Li, C., & Zhao, Z. K. (2007). Efficient acid-catalyzed hydrolysis of cellulose in ionic liquid. *Advanced Synthesis & Catalysis,349*, 1847–1850.
35. Zhao, H., Holladay, J. E., Brown, H., & Zhang, Z. C. (2007). Metal chlorides in ionic liquid solvents convert sugars to 5-hydroxymethylfurfural. *Science,316*, 1597–1600.
36. Muranaka, Y., Suzuki, T., Sawanishi, H., Hasegawa, I., & Mae, K. (2014). Effective production of levulinic acid from biomass through pretreatment using phosphoric acid, hydrochloric acid, or ionic liquid. *Industrial & Engineering Chemistry Research,53*, 11611–11621.
37. Sievers, C., Musin, I., Marzialetti, T., Valenzuela Olarte, M.B., Agrawal, P.K., & Jones, C.W. (2009). Acid-catalyzed conversion of sugars and furfurals in an ionic-liquid phase. *ChemSusChem: Chemistry & Sustainability Energy & Materials, 2*, 665–671.
38. Chen, T., Xiong, C., & Tao, Y. (2018). Enhanced hydrolysis of cellulose in ionic liquid using mesoporous ZSM-5. *Molecules,23*, 529–539.
39. Bose, S., Armstrong, D. W., & Petrich, J. W. (2010). Enzyme-catalyzed hydrolysis of cellulose in ionic liquids: A green approach toward the production of biofuels. *The Journal of Physical Chemistry B,114*, 8221–8227.
40. de Oliveira, H. F. N., Fares, C., & Rinaldi, R. (2015). Beyond a solvent: the roles of 1-butyl-3-methylimidazolium chloride in the acid-catalysis for cellulose depolymerisation. *Chemical Science,6*, 5215–5224.
41. Hu, L., Lin, L., Wu, Z., Zhou, S., & Liu, S. (2015). Chemocatalytic hydrolysis of cellulose into glucose over solid acid catalysts. *Applied Catalysis B: Environmental,174*, 225–243.
42. Qu, Y., Li, L., Wei, Q., Huang, C., Oleskowicz-Popiel, P., & Xu, J. (2016). One-pot conversion of disaccharide into 5-hydroxymethylfurfural catalyzed by imidazole ionic liquid. *Scientific Reports,6*, 26067.
43. Mukherjee, A., Dumont, M.-J., & Raghavan, V. (2015). Sustainable production of hydroxymethylfurfural and levulinic acid: Challenges and opportunities. *Biomass and Bioenergy,72*, 143–183.
44. Wang, P., Yu, H., Zhan, S., & Wang, S. (2011). Catalytic hydrolysis of lignocellulosic biomass into 5-hydroxymethylfurfural in ionic liquid. *Bioresource Technology,102*, 4179–4183.
45. Tiong, Y. W., Yap, C. L., Gan, S., & Yap, W. S. P. (2018). Conversion of biomass and its derivatives to levulinic acid and levulinate esters via ionic liquids. *Industrial & Engineering Chemistry Research,57*, 4749–4766.
46. Chatel, G., & Rogers, R. D. (2013). Oxidation of lignin using ionic liquids—an innovative strategy to produce renewable chemicals. *ACS Sustainable Chemistry & Engineering,2*, 322–339.
47. Rinaldi, R., Jastrzebski, R., Clough, M. T., Ralph, J., Kennema, M., Bruijnincx, P. C., et al. (2016). Paving the way for lignin valorisation: recent advances in bioengineering, biorefining and catalysis. *Angewandte Chemie International Edition,55*, 8164–8215.
48. Mainka, H., Tager, O., Korner, E., Hilfert, L., Busse, S., Edelmann, F. T., et al. (2015). Lignin—an alternative precursor for sustainable and cost-effective automotive carbon fiber. *Journal of Materials Research and Technology,4*, 283–296.
49. Kadla, J. F., & Kubo, S. (2004). Lignin-based polymer blends: analysis of intermolecular interactions in lignin–synthetic polymer blends. *Composites Part A: Applied Science and Manufacturing,35*, 395–400.
50. Berlin, A., Balakshin, M., Gilkes, N., Kadla, J., Maximenko, V., Kubo, S., et al. (2006). Inhibition of cellulase, xylanase and β-glucosidase activities by softwood lignin preparations. *Journal of Biotechnology,125*, 198–209.
51. Holladay, J.E., Bozell, J.J., White, J.F., Johnson, D. (2007). *Top value added chemicals from biomass: results of screening for potential candidate from sugars and synthesis gas*, vol. 2. Pacific Northwest National Laboratory: US Department of Energy.
52. Ma, R., Xu, Y., & Zhang, X. (2015). Catalytic oxidation of biorefinery lignin to value-added chemicals to support sustainable biofuel production. *Chemsuschem,8*, 24–51.

53. Yan, N., Yuan, Y., Dykeman, R., Kou, Y., & Dyson, P. J. (2010). Hydrodeoxygenation of lignin-derived phenols into alkanes by using nanoparticle catalysts combined with bronsted acidic ionic liquids. *Angewandte Chemie,122*, 5681–5685.
54. Liu, S., Shi, Z., Li, L., Yu, S., Xie, C., & Song, Z. (2013). Process of lignin oxidation in an ionic liquid coupled with separation. *RSC Advances,3*, 5789–5793.
55. Prado, R., Brandt, A., Erdocia, X., Hallet, J., Welton, T., & Labidi, J. (2016). Lignin oxidation and depolymerisation in ionic liquids. *Green Chemistry,18*, 834–841.
56. Scott, M., Deuss, P. J., de Vries, J. G., Prechtl, M. H., & Barta, K. (2016). New insights into the catalytic cleavage of the lignin β-O-4 linkage in multifunctional ionic liquid media. *Catalysis Science & Technology,6*, 1882–1891.
57. Wiermans, L., Schumacher, H., Klaaßen, C.-M., & de Maria, P. D. (2015). Unprecedented catalyst-free lignin dearomatization with hydrogen peroxide and dimethyl carbonate. *RSC Advances,5*, 4009–4018.
58. Prado, R., Erdocia, X., De Gregorio, G. F., Labidi, J., & Welton, T. (2016). Willow lignin oxidation and depolymerization under low cost ionic liquid. *ACS Sustainable Chemistry & Engineering,4*, 5277–5288.
59. Zakzeski, J., Jongerius, A. L., & Weckhuysen, B. M. (2010). Transition metal catalyzed oxidation of Alcell lignin, soda lignin, and lignin model compounds in ionic liquids. *Green Chemistry,12*, 1225–1236.
60. Das, L., Xu, S., & Shi, J. (2017). Catalytic oxidation and depolymerization of Lignin in Aqueous Ionic Liquid. *Frontiers in Energy Research,5*, 21.
61. Stark, K., Taccardi, N., Bosmann, A., & Wasserscheid, P. (2010). Oxidative depolymerization of lignin in ionic liquids. *Chemsuschem,3*, 719–723.
62. Liu, F., Liu, Q., Wang, A., & Zhang, T. (2016). Direct catalytic hydrogenolysis of kraft lignin to phenols in choline-derived ionic liquids. *ACS Sustainable Chemistry & Engineering,4*, 3850–3856.
63. Lange, H., Decina, S., & Crestini, C. (2013). Oxidative upgrade of lignin recent routes reviewed. *European Polymer Journal,49*, 1151–1173.
64. Dier, T. K., Rauber, D., Durneata, D., Hempelmann, R., & Volmer, D. A. (2017). Sustainable electrochemical depolymerization of lignin in reusable ionic liquids. *Scientific Reports,7*, 5041.
65. Reichert, E., Wintringer, R., Volmer, D. A., & Hempelmann, R. (2012). Electro-catalytic oxidative cleavage of lignin in a protic ionic liquid. *Physical Chemistry Chemical Physics,14*, 5214–5221.
66. Cox, B. J., & Ekerdt, J. G. (2012). Depolymerization of oak wood lignin under mild conditions using the acidic ionic liquid 1-H-3-methylimidazolium chloride as both solvent and catalyst. *Bioresource Technology,118*, 584–588.
67. Thierry, M., Majira, A., Pegot, B., Cezard, L., Bourdreux, F., Clement, G., et al. (2018). Imidazolium-based Ionic Liquids as efficient reagents for the C−O bond cleavage of Lignin. *Chemsuschem,11*, 439–448.
68. Diop, A., Jradi, K., Daneault, C., & Montplaisir, D. (2015). Kraft lignin depolymerization in an ionic liquid without a catalyst. *BioResources,10*, 4933–4946.
69. Liu, C., Li, Y. M., & Hou, Y. (2018). Preparation and structural characterization of lignin micro/nano-particles with ionic liquid treatment by self-assembly. *Express Polymer Letters,12*, 946–956.
70. Szalaty, T. J., Klapiszewski, L., Kurc, B., Skrzypczak, A., & Jesionowski, T. (2018). A comparison of protic and aprotic ionic liquids as effective activating agents of kraft lignin. Developing functional MnO_2/lignin hybrid materials. *Journal of Molecular Liquids,261*, 456–467.
71. Dutta, T., Sun, J., Simmons, B.A., & Singh, S. (2017). Conversion of lignin to ionic liquids.
72. Socha, A., Singh, S., Simmons, B.A., & Bergeron, M. (2014). Synthesis of novel ionic liquids from lignin-derived compounds.
73. Varanasi, P., Singh, P., Auer, M., Adams, P. D., Simmons, B. A., & Singh, S. (2013). Survey of renewable chemicals produced from lignocellulosic biomass during ionic liquid pretreatment. *Biotechnology for Biofuels,6*, 14.

74. Socha, A. M., Parthasarathi, R., Shi, J., Pattathil, S., Whyte, D., Bergeron, M., et al. (2014). Efficient biomass pretreatment using ionic liquids derived from lignin and hemicellulose. *Proceedings of the National Academy of Sciences,111*, E3587–E3595.
75. Jiang, N., Pu, Y., Samuel, R., & Ragauskas, A. J. (2009). Perdeuterated pyridinium molten salt (ionic liquid) for direct dissolution and NMR analysis of plant cell walls. *Green Chemistry,11*, 1762–1766.
76. Yelle, D. J., Ralph, J., & Frihart, C. R. (2008). Characterization of nonderivatized plant cell walls using high-resolution solution-state NMR spectroscopy. *Magnetic Resonance in Chemistry,46*, 508–517.
77. Cheng, K., Sorek, H., Zimmermann, H., Wemmer, D. E., & Pauly, M. (2013). Solution-state 2D NMR spectroscopy of plant cell walls enabled by a dimethylsulfoxide-d 6/1-ethyl-3-methylimidazolium acetate solvent. *Analytical Chemistry,85*, 3213–3221.
78. Zoia, L., King, A. W., & Argyropoulos, D. S. (2011). Molecular weight distributions and linkages in lignocellulosic materials derivatized from ionic liquid media. *Journal of Agricultural and Food Chemistry,59*, 829–838.
79. Engel, P., Hein, L., & Spiess, A. C. (2012). Derivatization-free gel permeation chromatography elucidates enzymatic cellulose hydrolysis. *Biotechnology for Biofuels,5*, 77.
80. Cadoche, L., & Lopez, G. D. (1989). Assessment of size reduction as a preliminary step in the production of ethanol from lignocellulosic wastes. *Biological Wastes,30*, 153–157.
81. Miao, Z., Grift, T. E., Hansen, A. C., & Ting, K. C. (2011). Energy requirement for comminution of biomass in relation to particle physical properties. *Industrial Crops and Products,33*, 504–513.
82. Sokhansanj, S., Kumar, A., & Turhollow, A. F. (2006). Development and implementation of integrated biomass supply analysis and logistics model (IBSAL). *Biomass and Bioenergy,30*, 838–847.
83. Brandt, A., Erickson, J. K., Hallett, J. P., Murphy, R. J., Potthast, A., Ray, M. J., et al. (2012). Soaking of pine wood chips with ionic liquids for reduced energy input during grinding. *Green Chemistry,14*, 1079–1085.
84. King, A. W., Parviainen, A., Karhunen, P., Matikainen, J., Hauru, L. K., Sixta, H., et al. (2012). Relative and inherent reactivities of imidazolium-based ionic liquids: the implications for lignocellulose processing applications. *RSC Advances,2*, 8020–8026.
85. Mazza, M., Catana, D.-A., Vaca-Garcia, C., & Cecutti, C. (2009). Influence of water on the dissolution of cellulose in selected ionic liquids. *Cellulose,16*, 207–215.
86. Sun, N., Liu, H., Sathitsuksanoh, N., Stavila, V., Sawant, M., Bonito, A., et al. (2013). Production and extraction of sugars from switchgrass hydrolyzed in ionic liquids. *Biotechnology for Biofuels,6*, 39.
87. da Costa Lopes, A. M., & Lukasik, R. M. (2018). Separation and Recovery of a Hemicellulose-Derived Sugar Produced from the Hydrolysis of Biomass by an Acidic Ionic Liquid. *Chemsuschem,11*, 1099–1107.
88. Amde, M., Liu, J.-F., & Pang, L. (2015). Environmental application, fate, effects, and concerns of ionic liquids: a review. *Environmental Science & Technology,49*, 12611–12627.
89. Reid, J. E., Prydderch, H., Spulak, M., Shimizu, S., Walker, A. J., & Gathergood, N. (2018). Green profiling of aprotic versus protic ionic liquids: Synthesis and microbial toxicity of analogous structures. *Sustainable Chemistry and Pharmacy,7*, 17–26.
90. Egorova, K. S., & Ananikov, V. P. (2014). Toxicity of ionic liquids: Eco (cyto) activity as complicated, but unavoidable parameter for task-specific optimization. *Chemsuschem,7*, 336–360.
91. Petkovic, M., Seddon, K. R., Rebelo, L. P. N., & Pereira, C. S. (2011). Ionic liquids: A pathway to environmental acceptability. *Chemical Society Reviews,40*, 1383–1403.
92. Pham, T. P. T., Cho, C.-W., & Yun, Y.-S. (2010). Environmental fate and toxicity of ionic liquids: A review. *Water Research,44*, 352–372.
93. Oliveira, M.V., Vidal, B.T., Melo, C.M., de Miranda, R.D.C., Soares, C.M., Coutinho, J., et al. (2016). (Eco) toxicity and biodegradability of protic ionic liquids. *Chemosphere, 147*, 460–466.

94. Grzonkowska, M., Sosnowska, A., Barycki, M., Rybinska, A., & Puzyn, T. (2016). How the structure of ionic liquid affects its toxicity to Vibrio fischeri? *Chemosphere,159*, 199–207.
95. Biczak, R., Pawlowska, B., Balczewski, P., & Rychter, P. (2014). The role of the anion in the toxicity of imidazolium ionic liquids. *Journal of Hazardous Materials,274*, 181–190.
96. Peric, B., Sierra, J., Marti, E., Cruanas, R., Garau, M. A., Arning, J., et al. (2013). (Eco) toxicity and biodegradability of selected protic and aprotic ionic liquids. *Journal of Hazardous Materials,261*, 99–105.
97. Wiredu, B., & Amarasekara, A. S. (2014). Synthesis of a silica-immobilized Bronsted acidic ionic liquid catalyst and hydrolysis of cellulose in water under mild conditions. *Catalysis Communications,48*, 41–44.
98. Klein-Marcuschamer, D., Simmons, B. A., & Blanch, H. W. (2011). Techno-economic analysis of a lignocellulosic ethanol biorefinery with ionic liquid pre-treatment. *Biofuels, Bioproducts and Biorefining, 5*, 562–569.

Development of Porous Bio-Nano-Composites Using Microwave Processing

Nishant Verma, Manoj Kumar Singh and Sunny Zafar

Abstract The current scenario of every polymer processing industry is focussed towards use of high strength material at a low cost. After discarding of polymer component, it creates problem for environment due to its non-biodegradability. It takes several years to degradation and reduces fertility of soil. So researchers focussed toward maximum utilization of biodegradable polymer for sustainable development. The biodegradable polymers have capability to replace non-biodegradable polymers. This chapter includes various fabrication techniques to process biodegradable polymers.

1 Introduction

The current demand of material for every structural application is lightweight with high strength [1]. The available categories of materials are metals, ceramics and polymers. The metals were very popular in various structural applications such as automobiles, constructions and aircraft. Due to the heavier weight of metals, researchers are focussed toward the development of alternatives for metals. Applications of ceramics are limited due to low fracture toughness. Polymers are lightweight, flexible and corrosion-resistant but due to low mechanical strengths, neat polymer not gain much attention from automobile and aerospace industry. However, polymer composites are gaining much attention due to better performance in aircraft industry. Carbon fibre reinforced polymer is a very popular material in aircraft for their good performance [2]. During construction of Boeing 787, replacement of aluminium structure was done with carbon fibre reinforced polymer. This attempt saves 20% of the average weight of aircraft [3]. Polymers composite are not only popular in the aircraft industry, but it is also popular in the implant industry due to its excellent biocompatibility. The various biocompatible polymers are ultra-high molecular weight polyethylene (UHMWPE), Polyether ether ketone (PEEK), Polylactide (PLA) and polycaprolactone (PCL) [4]. The porous structures of polymers are also popular in acoustic and

N. Verma · M. K. Singh · S. Zafar (✉)
School of Engineering, Indian Institute of Technology (IIT), Mandi, Himachal Pradesh, India
e-mail: sunnyzafar@iitmandi.ac.in

© Springer Nature Switzerland AG 2020
A. Khan et al. (eds.), *Biofibers and Biopolymers for Biocomposites*,
https://doi.org/10.1007/978-3-030-40301-0_10

bone scaffold applications. Polymeric foam is a great example of porous material. Porous polymers (foam) was invented in 1931, before the second world war [5]. Traditionally porous polymers were manufactured by polyurethane and polystyrene [6]. The traditional porous polymer materials were non-biodegradable. Biodegradability of polymers is not only important to the environment but it is also important for self-degradation in the body after growing a tissue (bone scaffold) [7]. This limitation of biodegradability stimulates the researchers to focussed towards development of porous biodegradable polymeric materials. Biodegradable polymers are the polymers which can be degraded by microbial action and produce end products, such as carbon dioxide and water in a reasonable interval. The complete decomposition of materials depends upon material, conditions of environment such as moisture, temperature and decomposition location.

2 Mechanism of Degradation

The mechanism involved in the biodegradation of polymers is shown in Fig. 1. The biodegradation involves two processes.

2.1 Fragmentation

It is a natural process involved in the degradation of a biodegradable polymer. This process occurs when biodegradable polymer exposed under the external environment. In this process, long chains of carbon–carbon bonds shortened by moisture, heat and humidity effect. In this stage, the polymer becomes weaker and polymer transform into smaller groups known as polymer fragments.

2.2 Biodegradation

This process involves the passing of carbon chains through cell walls of microbes. During passing the molecular chains of polymers are completely broken into biomass, CO_2, methane and water depend on anaerobic or aerobic conditions.

3 Classification of Biodegradable Polymers

The biodegradable polymer materials are classified in Fig. 2.

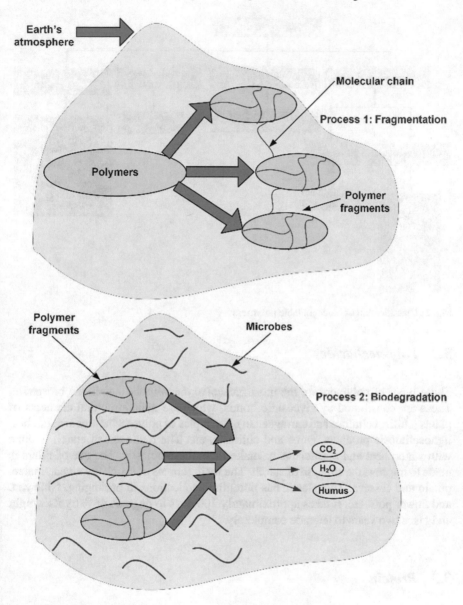

Fig. 1 Mechanism of biodegradation

3.1 Agro-Polymers

These types of polymers are obtained from renewable sources of various kinds. Saccharides, carbohydrates and oil can be transformed into polymer materials. The properties of these polymers are comparable to polymers derived from mineral oil.

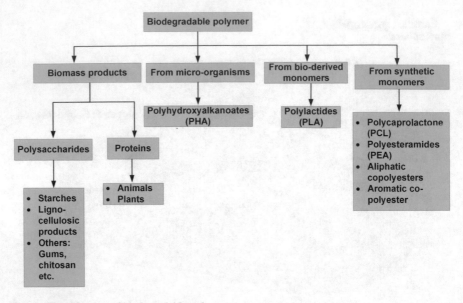

Fig. 2 Classification of biodegradable polymers

3.2 Polysaccharides

These types of polymers are the most presented macromolecules in the biosphere. These are constituted by glycosidic bonds, which are main structural elements of plants (chitin, cellulose and carrageenan). Examples of polysaccharides are starches, lignocellulosic products, gums and chitosan, etc. The treatment of starch is done with water, heat and plasticizers to make them thermoplastic. The use of fillers is made to improve the strength of starch. The main sources of starch are wheat, maize, potato and cassava. This plastic has potential applications in packaging, tableware and flower pots, etc. It takes approximately 100 days to degrade 46% by its weight and takes two years to degrade completely.

3.3 Protein

These type of materials obtained from plants, animal and bacteria. The potential examples of plant protein for the transformation of polymeric materials are corn protein, soy protein and wheat proteins. The available animal proteins are and keratin and gelatine. The bacterial proteins are chymotrypsin, lactate dehydrogenase and fumarase.

3.4 Micro-organism Derived

These types of polymers derived from micro-organisms. Polyhydroxyalkanoate (PHA) is an example of micro-organism derived polymer. It is produced by bacterial and genetically modified organisms. Recently researchers are focussed to derive PHA from food waste. PHA is expensive material and a limited amount can be derived from bacteria. These are used as cups, food wraps, coating for cardboard, and plates. Apart from it, this polymer has potential applications in medical, which includes gauzes, sutures, and medicines coating. It is replacing most of the fossil fuel-based polymers such as polyethylene, polystyrene and polyvinyl chloride. It is compostable in soil completely in the presence of microbes and fungi. The microbe presents in soil breakdown the PHA with the help of enzymes. The degradation depends on the concentration of presented microbes in the soil. As per bio-based press, PHA takes 2–3 months for decomposition. The decomposition rate is very slower in marine waters As per ASTM 7081, 30% decomposition was recorded in six months.

3.5 Bio-Derived Monomers

It is a thermoplastic polymer and derived through fermentation by bacteria, which is part of biotechnology, therefore, it is also known as bio-derived polymers. Polylactide Acid (PLA) is a kind of bio-derived monomers. The molecules of PLA are joined by a long chain of lactic acid. The production cost of PLA is quite lesser than the PHA. But applications of PLA are restricted due to its brittleness. It is used to make food packaging, grocery bags, cups bottles and plates. It is well decomposable material in the presence of acids. It has potential applications in medical plates and sutures. PLA is not easily decomposable in the backyard because water and temperature levels required for decomposing the PLA are not available in the environment. PLA generally takes 6–12 months for degradation in soil. When decompositions of PLA is done in the presence of oxygen, CO_2 and H_2O formed as the end product. When degradation occurs in the absence of oxygen, methane gas will form as the end product. PLA degrades only 3% in the marine water as per ASTM D7081 standard.

3.6 Petroleum Derived Monomers

These types of polymers obtained chemically from synthetic monomers. PCL and PEA are a the major examples of petroleum-derived polymers. These types of polymers are soft at room temperature. The production of these polymers depends on petroleum resources.

3.6.1 PCL

It is obtained by polymerization of e-caprolactone in the presence of metal alkoxides. It is widely used as a solid plasticizer, polyurethane applications. The biodegradability property of PCL leads to increase its applications in biomedicine such as controlled drug delivery and in environmental aspects, it is used to make soft compostable packaging. These polymers can be decomposed by yeast. The degradation time of PCL is 15 days.

3.6.2 PBS

It is resin obtained from petroleum. The mechanical performance of neat PBS is not good. So, bio-based polyester or fibres such as jute and coir may be used as an additive to improve its mechanical performance. It is used to make service ware, food packaging, plant pots, agricultural sheets and fishing nets.

The properties of various biodegradable materials are present in Table 1.

Table 1 Properties of various biodegradable materials [8]

Material	Ultimate tensile strength (MPa)	Modulus of elasticity (GPa)	Melting point (°C)	Degradation
Polylactic acid (PLA)	14.0–114	0.0850–13.8	90.0–180	12 months in soil and less than 3% in marine water as per ASTM D7081
Polyhydroxyalkanoate (PHA)	40	3.5	120–177	2 months in Backyards and less than 50% in marine water at the time period of 6 months
Polycaprolectone (PCL)	10	1.2	60	90% of film degrade in 15 days
Polybutylene Succinate(PBS)	34	–	114	–
Polybutyrate adipate terephthalate (PBAT)	>17	–	110–120	Biodegradable as well as compostable in soil
Polyvinyl alcohol (PVA)	151.68	–	200	–

4 Composite Fabrication

The applications of polymers are limited due to higher flexibility, which leads to lower Young's modulus and other properties. The reinforcing of polymers may be required to developed high strength polymer composite foam. The purpose of reinforcement is to provide strength and stiffness to composite. The properties of the fabricated composite are not only dependent on the properties of reinforcement, but it also depends on compatibility and bonding between reinforcement and matrix. When the load acts on composite, various phenomena may happen i.e. fracture on matrix, delamination of fibre, fracture of fibre. The fracture in the matrix may occur due to failure of load transfer from matrix to reinforcement. Delamination occurs due to poor compatibility and interfacial bonding between matrix and reinforcement. The reinforcement may be present in two forms.

(i)   Fibre reinforcement
(ii) Particulate Reinforcement.

4.1 Fibre Reinforcement

Fibres are systematic arrangement of threads or strands in one or two directions. If the arrangement of fibres in one direction, it is known as unidirectional fibres. When the arrangement of fibres is in two directions is known as bidirectional fibres. The fibres are categorized as natural fibres and synthetic fibres. Natural fibres are obtained from nature while synthetic fibres are man-made fibres with better mechanical properties than natural fibres. The energy required to obtain synthetic fibres is higher than natural fibres. Using natural fibre is advantageous because of its biodegradable nature and economical as compared to synthetic fibre.

4.2 Particulate Reinforcement

Composite contains reinforcement of particle size less than 0.25 μm is known as particulate reinforced composite. The enhanced mechanical behaviour of composite depends upon wettability between matrix and reinforced particles. Particulate reinforced composites are more popular in the case of MMC. For example, Graphite, Molybdenum disulphide and tungsten disulphide PEEK reinforced composite may be used in wear applications due to its better lubrication. Hydroxyapatite reinforced composite may also be used as a reinforcement in UHMWPE, PLA and PCL. The biocompatibility of polymers remains unaffected due to the use of hydroxyapatite. For better mechanical properties, nano-hydroxyapatite (nHA) may be used as reinforcement. Nano reinforcement leads to uniform dispersion due to the high surface area

Table 2 Properties of nHA [8]

Compressive strength (MPa)	Tensile strength (MPa)	Elastic modulus (GPa)	Fracture toughness (MPa$\sqrt{m}$)
>400	40	100	1.0

to volume ratio. The presence of nHA in composite leads to better tensile strength, flexural strength, compression strength and shore hardness. However, the presence of nHA leads to decrease impact behaviour of the composite due to brittle behaviour of nHA. The properties of nHA is present in Table 2.

5 Introduction to Microwave-Assisted Heating

The fabrication of porous composite is a challenge due to its porosity, pore size distributions, longer process cycle, high power consumption and interconnectivity. The efficient and industrial accepted techniques to make porous biodegradable polymer composite are solid-state foaming, fluid foaming, phase separation and particulate leaching. The comparative data of above discussed technique is presented in Table 3.

The mentioned techniques have the limitation of temperature gradient, residual stresses and overheating at the localised region. Residual stresses lead to laminates warping, fibre waviness and failure of the specimen at low loads. Microwave may be an alternative to develop porous biodegradable polymer composite. Researcher observed better mechanical properties of microwave processed polymer composite.

Table 3 Comparison of various fabrication technique used in fabrication of porous composite [8]

Fabrication route	Advantages	Disadvantages
Phase separation	• High porosity • High interconnectivity • Pore size can be controlled by varying processing condition	• It takes longer time to sublimation of solvent • Shrinkage issue • Small scale production
Particle leaching	• Controlled porosity • Controlled interconnectivity	• Use of organic solvent
Solid free-form	• The porous structure can be tailored	• Use of organic solvent
Scaffold coating	• Easy to process	• Pores clogging • Adhesion of coating with the surface is poor
Microwave processing	• Rapid technique • Controlled porosity • Better interconnectivity • Pores size can be varied by wt.% of NaCl	• Not suitable for mass production

Microwave radiations are highly penetrating in nature, which leads to rapid and uniform heating characteristics. Processing of materials using microwave energy offers to save in processing time as well as cost. It is an environment-friendly procedure to process polymers as well as metals. In microwave processing, there is the conversion of energy into heat inside the material rather than the transfer of heat from the source to the target material. The frequency of microwaves varies between 0.3 and 300 GHz. Microwaves are polarised and coherent, which can be absorbed, reflected and transmitted depends on the type of material exposed under the microwave. Traditionally microwaves were only useful for the purpose of telecommunication, radar detection and non-destructive testing. Application of microwaves in the heating of material was invented by Percy Spencer in 1946. From that time microwaves find potential applications in material processing due to its unique properties such as rapid, selected and volumetric heating. From 1946–2000, the ceramics and polymers composites were successfully processed with microwave energy due to their good microwave absorbing properties at room temperature. In the conventional mode of material processing, the heat transfers by conduction, radiation and convection, creating the temperature difference between source and target.

6 Microwave Material Interaction Mechanism

In the case of microwave processing of materials, energy is directly delivered to materials by molecular interaction and transformation of energy into heat occurs by oscillating of molecules under electromagnetic field and dielectric heating takes place. Dielectric constant (ε^*) can be mathematically expressed in term of electrical energy stored by a material (ε') and microwave energy dissipated by a heated material (ε'') in Eq. (1).

$$\varepsilon^* = \varepsilon' - i\varepsilon'' \tag{1}$$

The microwave energy absorbed per unit volume may be evaluated by Eq. (2).

$$P = \omega\left(\varepsilon_\circ \varepsilon'' E_{rms}^2 + \mu_\circ \mu'' H_{rms}^2\right) \tag{2}$$

where,
ε_0: Permittivity of free space (8.854×10^{-12} F/m).
ε'': Ability of heated material to dissipate energy.
E_{rms}: Root mean square value of the applied electric field (V/m).
H_{rms}: Root mean square value of the applied magnetic field (A/m).

In case of polymers (non-magnetic materials), the value of μ'' is considered as negligible, hence no contribution of the magnetic field in absorption of microwaves so it can be neglected from Eq. (2) and written as

$$P = \omega\left(\varepsilon_0 \varepsilon'' E_{rms}^2\right) \tag{3}$$

The difference in way of heating may lead to influence in resultant properties of the material. However, bulk metals reflect microwaves at a frequency of 2.45 GHz during processing at room temperature due to low skin depth (d_s). The skin depth of material is defined as the distance from the surface of material to depth of material, where magnitude of microwave decreased by factor of 1/e (36.8%).

The skin depth of material may be evaluated by Eq. (4).

$$d_s = \frac{1}{\sqrt{\pi \mu f \sigma}} \tag{4}$$

where,

μ: Imaginary part of magnetic permeability (H/m).

σ: Electrical conductivity (S/m).

f: Operated frequency of microwaves (1/s).

In microwave material processing, microwaves absorption by materials is volumetric and transform the amount of energy into heat by two modes known as ionic conduction and polarisation. Polarisation refers to a short-range displacement of the charged particle, occurs at high frequency (2.45 GHz). The presented dipoles in the materials rotate with the alternating field. Ionic conduction refers to the transport of charge, which increase oscillation in the molecules. Figure 3a–d shows microwave transparent material, microwave reflecting material, microwave absorbing material and mixed microwave absorber respectively.

However, processing of poor microwave absorber in the microwave is challenging, in this case, technique of Microwave hybrid heating (MHH) may be used for effective utilisation of microwave. In this technique, susceptor is used to enhance the microwave interaction, as a result, the heating rate of material increased. Susceptors refers to the material, which is highly microwave absorber. The susceptor has ability to couples with microwaves even at room temperature. Coupling leads to rapid heating, which raises the temperature of subsequent material, which is poor microwave absorber. When poor absorbing material reaches the critical temperature value through conventional heat mode then it will start acting as microwave absorbing material.

7 Case Study: Development of Biodegradable Porous Composite of Hydroxyapatite Reinforced PCL and PLA Composites

The applications of microwave energy for the development of composite are receiving huge interest since last decade. Researchers observed the better mechanical performance of microwave processed polymer composite [9–14]. Zhang et al. used

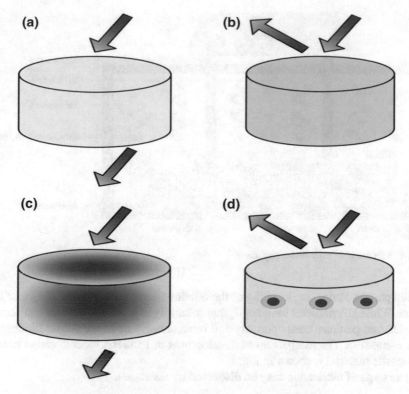

Fig. 3 Microwave material interaction: **a** transparent material, **b** reflecting material, **c** absorbing material, **d** mixed microwave absorber

microwave energy for synthesis of shape memory polycaprolactone foam [15]. The schematic used to fabricate porous composite of PLA and HA is shown in Fig. 4. To apply compaction pressure of 0.02 MPa, deadweight is used to apply the load on the alumina plate, which is put on the alumina mould. Hole of 7 mm was made on the surface of the alumina pressure plate for monitoring of temperature inside the mould using optical pyrometer.

The mechanism involved in the fabrication of PLA/HA composite is shown in Fig. 5. The PLA, HA and NaCl are exposed under microwaves and directly couples NaCl as well as HA. This leads to rise in temperature of HA as well as NaCl due to conversion of microwave energy into heat. The further exposure leads to a saturation point in the microwave heat energy conversion after that point heat transfer will take place. The heat starts to transfer from HA and NaCl to PLA particles. This heat transfer is due to the high thermal conductivity of NaCl. The temperature of PLA starts rising and the thermal expansion process in the polymer takes place. The temperature rises up to processing temperature (between T_g of 58.39 °C and T_m of 161–170 °C). The melted particles of PLA make the bond with NaCl and HA. The compaction pressure of 0.02 MPa increases the bond strength in molecules. To

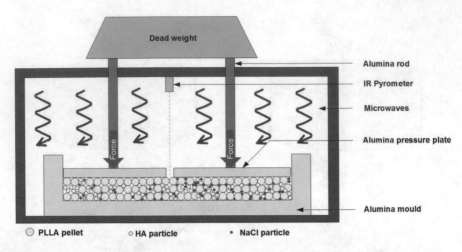

Fig. 4 Schematic of microwave applicator

obtain porous composite, leaching of the obtained composite is done using distilled water. While HA particles have negligible solubility in water and remain unaffected. The leached particles create porosity and interconnectivity in the porous biodegradable composite. The mechanism of development of PCL/HA biodegradable porous composite material is shown in Fig. 6.

The steps of fabrication may be discussed by six stages.

7.1   Initial Stage

In this stage, the PCL pellets mixed with nano-hydroxyapatite and NaCl in powder mixing equipment and loaded in alumina mould, which allows the microwave to pass from themselves up to 900 °C. Beyond 900 °C alumina will act as microwave absorbing material [16]. The average particle size of nHA was 100 nm. The particle size of NaCl was 350 μm. NaCl was acted as susceptor as well as leaching agent to induce the porosity in the fabricated composite.

7.2   Interaction Stage

In this stage, microwave interacts with loaded material (PCL, nHA and NaCl). The interaction of microwave was observed different for different material as per dielectric loss factor. The NaCl has the highest dielectric loss factor among all the materials (PCL, nHA and NaCl). The dielectric loss factor of nHA and NaCl are 0.014 and 0.20, respectively [10]. The exposed microwave directly couples with NaCl particle

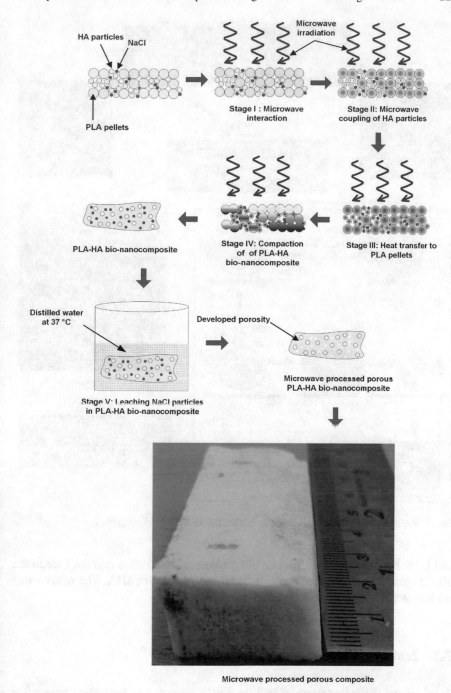

Fig. 5 Mechanism of fabrication process of PLA/HA porous bio-composite

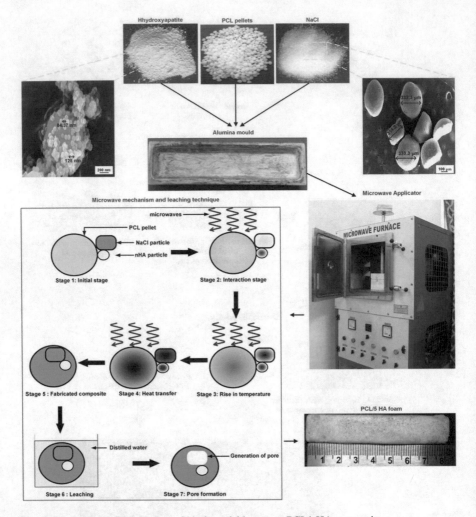

Fig. 6 Mechanism of fabrication of biodegradable porous PCL/nHA composite

and heats NaCl volumetrically. The temperature of NaCl rises rapidly. Due to the high thermal conductivity of NaCl, it transfers their heat to nHA. The microwaves are less interactive with nHA and PCL particles.

7.3 Temperature Rising Stage

In this stage, the temperature of PCL and nHA rises due to individual interaction of microwaves with nHA and PCL. nHA and NaCl were not affected significantly with the rise in temperature due to their higher thermal stability [9]. The pellets of affected by rise in temperature and start expanding and pellets of PCL make bonding in each other.

7.4 Heat Transfer Stage

In this stage heat transfer occurs from NaCl as well as nHA to PCL and temperature of PCL reach between glass transitions (T_g) and melting temperature (T_m), which is 42 °C and 60 °C, respectively.

7.5 Fabrication Stage

The PCL pellets expand on heating and make a stronger bond with nHA and NaCl. The bonding increases with the cooling of specimens.

7.6 Leaching Stage

Leaching for one week was done in distilled water by using an ultrasonicator. The particle of NaCl leached out and creates pores on the surface as well as throughout the depth of specimens. Particles of nHA were not affected by leaching due to its negligible solubility with water.

8 Microstructural Characterisation of Porous Composites

The micrographs of PCL/10HA and PLA/10HA can be observed in Fig. 7. The gold coating of 5 nm was done on the surface of specimens before microstructural observation on scanning electron microscopy (SEM). This is essential to avoid the accumulation of electrostatic charge on specimens, which created a barrier to the incoming electrons. The pore size varies between 142 μm and 261 μm. Nano-sized pores may be observed inside micro-sized pores which are known as interconnected pores. The dispersion of nHA particle may be observed in the highly magnified image.

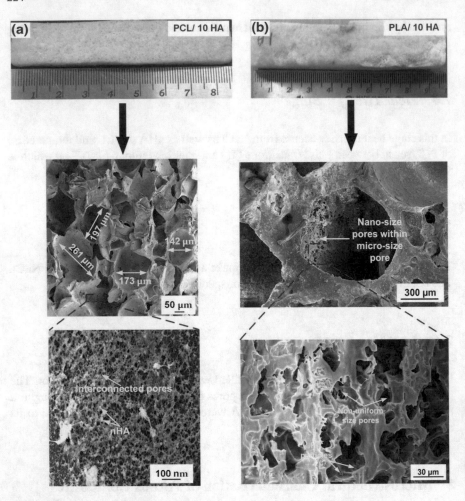

Fig. 7 Microstructure of hydroxyapatite reinforced porous PCL and PLA composite

9 Comparison of Microwave Processed HA Reinforced PCL and PLA Porous Composites

The pore size in both porous composites varies between 100 μm and 400 μm. This pore size is equivalent to the particle size of NaCl particles. The leaching time to induce the porosity within PLA/HA and PCL/HA composite was 24 h and 1 week, respectively. The flexural strength of PLA/HA porous composite is higher than composite PCL/HA composite. The lower flexural strength is due to higher porosity and high flexibility of PCL polymers. It can be concluded that porosity is a function of leaching time. From Table 4 it can be seen that the porosity obtained in PLA/HA porous composite is in the range of 23–25% and 70–72% in case of PCL/HA porous

Table 4 Comparison between PCL/HA porous composite and PLA/HA porous composite [9, 10]

Specimens	Power (W)	Exposure time (s)	Leaching time (s)	Flexural strength (MPa)	Pore size (μm)	Porosity (%)	Interconnectivity (%)
PLA/10 nHA	900	1220	24 h	88.7	180–340	23.1	96.67
PLA/15 nHA	900	1190	24 h	74.3	179–350	23.9	96.67
PLA/20 nHA	900	1140	24 h	66.5	182–345	24.9	96.67
PCL/10 nHA	540	412	1 week	6.26	142–261	70	51.7
PCL/15 nHA	540	393	1 week	6.99	98–169	70.8	54.8
PCL/20 nHA	540	368	1 week	7.2	98–261	72	60

composite. However, in terms of interconnectivity between pores, higher interconnectivity was obtained in PLA/HA porous composite. The less interconnectivity in the PCL/HA porous composite foam may be due to the hydrophobic nature of PCL. Table 4 shows the comparative data of PLA/HA and PCL/HA porous composite.

10 Effect of Microwave Power on Interfacial Bonding

The matrix of PCL is flexible in nature. The failure of the matrix occurs at low load due to its flexibility. The particle of nHA is stiff in nature. The uniform heating feature of microwave energy enhances the bonding between the PCL matrix and nHA particles. Thus the obtained mechanical property is greater at a lower power level. The extent of bodings between matrix and reinforcement decreases with increased power level. The enhanced bonding is due to the fact that load on the matrix material transferred nHA due to the higher extent of bonding. The interfacial bonding is a function of processing temperature of the specimen. Processing temperature depends on levels of microwave power. So, it can be seen that interfacial bonding decreases with an increase in power level of the microwave as shown in Fig. 8. This is due to decrease in relaxation time of oscillation dipolar moment of molecules [11].

Fig. 8 Effect of microwave power on temperature and interfacial bonding

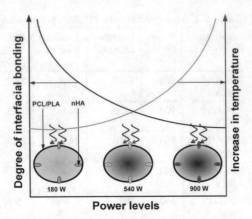

11 Effect of Dielectric Properties of Constituting Materials on Time–Temperature Curve

As discussed above, the major dependency of microwave interaction of material on dielectric properties of the materials. Therefore, the material having a higher value of dielectric constant has more interaction with the microwave. The dielectric constant values for PCL/PLLA, NaCl and HA are 1.2, 2.41 and 5.7, respectively. Since the dielectric constant of HA is maximum among all the materials used to fabricate porous composite. Therefore, the composite having a higher weight percentage of HA particles take lesser time to process. As seen in Fig. 9, the slope of neat PCL is minimum because of less microwave interaction. PCL/20 HA having the maximum slope. This is because of the higher value of the dielectric constant of HA.

12 Concluding Remarks

Microwave-assisted composite fabrication may be an alternative technique to fabricate biodegradable based polymer composite foam. The microwave technique of fabrication is rapid, cost-effective and eco-friendly technique to process polymers. The porosity of fabricated foam may be controlled by varying wt.% of NaCl and leaching time. The mechanical properties may be enhanced by varying wt.% of nano-HA.

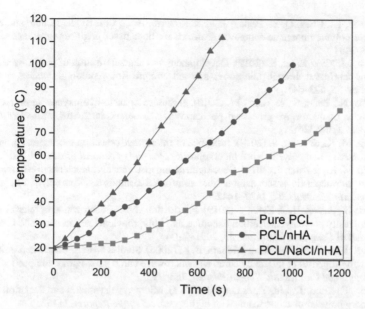

Fig. 9 Temperature time curve for HA reinforced PCL composite

Acknowledgement Authors would like to thank the editor and publisher for giving the opportunity to write this chapter. It has been a very educational experience. Again authors are very thankful to the authors around the world who have published their studies, which was very useful during the writing of this chapter.

References

1. Dursun, T., & Soutis, C. (2014). Recent developments in advanced aircraft aluminium alloys. *Materials and Design, 56*, 862–871.
2. Yanagimoto, J., & Ikeuchi, K. (2012). Sheet forming process of carbon fiber reinforced plastics for lightweight parts. *CIRP Annals-Manufacturing Technology, 61*, 247–250.
3. https://www.boeing.com/commercial/aeromagazine/articles/qtr_4_06/article_04_2.html (2019)
4. Tanner, K. E. (2010). Bioactive composite materials for bone augmentation. *Journal of the Royal Society, Interface, 7*, 541–557.
5. Coccorullo, I., Di Maio, L., Montesano, S., & Incarnato, L. (2009). Theoretical and experimental study of foaming process with chain extended recycled PET. *Express Polymer Letters, 2*, 84–96.
6. Zimmermann, M. V. G., da Silva, M. P., Zattera, A. J., & Santana, R. M. C. (2018). Poly(lactic acid) foams reinforced with cellulose micro and nanofibers and foamed by chemical blowing agents. *Journal of Cellular Plastics, 3*, 577–596.
7. Salerno, A., Di Maio, E., Iannace, S., & Netti, P. A. (2011). Solid-state supercritical CO_2 foaming of PCL and PCL-HA nano-composite: Effect of composition, thermal history and foaming process on foam pore structure. *Journal of Supercritical Fluids, 58*, 158–167.

8. Rezwan, K., Chen, Q. Z., Blaker, J. J., & Roberto, A. (2006). Biodegradable and bioactive porous polymer/inorganic composite scaffolds for bone tissue engineering. *Biomaterials, 18,* 3413–3431.

9. Singh, M. K., & Zafar, S. (2019). Development and characterisation of poly-L-lactide-based foams fabricated through microwave-assisted compression moulding. *Journal of Cellular Plastics, 55,* 523–541.

10. Verma, N., Zafar, S., & Talha, M. (2019). Influence of nano-hydroxyapatite on mechanical behavior of microwave processed polycaprolactone composite foams. *Materials Research Express, 8,* 085336.

11. Singh, M. K., & Zafar, S. (2018). Influence of microwave power on mechanical properties of microwave-cured polyethylene/coir composites. *Journal of Natural Fibers, 00,* 1–16.

12. Singh, M. K., & Zafar, S. (2019). Development and mechanical characterization of microwave-cured thermoplastic based natural fibre reinforced composites. *Journal of Thermoplastic Composite Materials, 32,* 1427–1442.

13. Arora, G., Pathak, H., & Zafar, S. (2019). Fabrication and characterization of microwave cured high-density polyethylene/carbon nanotube and polypropylene/carbon nanotube composites. *Journal of Composite Materials, 53,* 2091–2104.

14. Rao, R. M. V. G. K., Rao, S., & Sridhara, B. K. (2006). Studies on tensile and interlaminar shear strength properties of thermally cured and microwave cured glass-epoxy composites. *Journal of Reinforced Plastics and Composites, 25,* 783–795.

15. Zhang, F., Zhou, T., Liu, Y., & Leng, J. (2015). Microwave synthesis and actuation of shape memory polycaprolactone foams with high speed. *Scientific Reports,* 11152.

16. Bhattacharya, M., & Basak, T. (2017). Susceptor-assisted enhanced microwave processing of ceramics—A review. *Critical Reviews in Solid State and Materials Sciences, 42,* 433–469.

Cellulose Based Biomaterials: Benefits and Challenges

Faiza Sharif, Nawshad Muhammad and Tahera Zafar

Abstract Cellulose is amongst the most inexhaustible natural source of polymers available on the globe. It is present in trees, plants, fruits, barks and leaves in the form of key structural element of the cell wall of plant tissues. It contains lignin and hemicellulose as additional products when isolated, which need to be removed to obtain nanofibrous cellulose. It has numerous applications in paper, leather, cosmetic, pharmaceutical, food and packaging. Bacterial cellulose on the other hand is a micro fibrous membrane made by bacteria in low pH conditions at air liquid interphase. Bacterial cellulose (BC) is endowed with distinctive properties, for instance, ability to retain water, ability to mould, high rate of crystallinity, high tensile strength. These striking physical characteristics arise from its distinctive nanostructure, which consists of a three-dimensional network made of linear b-1, 4-glucan chains bonded together by hydrogen interactions. This structure is organized as twining ribbons made of microfibrillar bundles. These properties make BC an exceptional biomaterial which can be use in various ways in biomedical field. Although highly beneficial for biomedical applications cellulose does present some drawbacks. Basically, nanofibers of plant-based cellulose is isolated by acid hydrolysis and mechanical defibrillation, both processes have their own challenges similarly bacterial cellulose is naturally synthesized by bacteria which is a slow process and may make it difficult to commercially viable for biomedical application.

Keywords Cellulose · Bacterial cellulose · Cellulose based biomaterials · Biomedical cellulose

F. Sharif · N. Muhammad (✉)
Interdisciplinary Research Centre in Biomedical Materials, COMSATS University Islamabad, Lahore Campus, Lahore 54000, Pakistan
e-mail: nawshadmuhammad@cuilahore.edu.pk

T. Zafar
Department of Science of Dental Materials, de'Montmorency College of Dentistry, Lahore, Pakistan

1 Introduction

Polymeric materials and derivatives of petrochemicals are composed of high molecular mass and are non degradable [1]. Plastics being a useful and handy day to day material have various applications but it is hazardious due to its inertness and non-biodegradability [2]. It is quite obvious that the environment and its ecological diversity has suffred great damage due to the unrestrained and widespread use and disposition of such non-biodegradable materials [2]. The concequence of increase usage of plastics is now having an impact on the global environment since clearance and dispensation procedures are quite limited as compared to the rate of production. All the existing methods being used to treat the plastic goods and their waste have become inefficient and ineffective with passing time [3]. As a result of all these problems, revolutionary ideas are explored to substitute the conventional plastic material with the biodegradable materials which are the most favourable materials despite their high cost [4]. There are numerous biodegradable polymers which are being explored for multiple applications and cellulose is one of them (Fig. 1) [5–8].

Cellulose, a bountiful macromolecule, is universally identified as the key constituent of plant cell wall. This organic compound is literally the most abundant polysaccharide on the face of earth, yielding 1.5 trillion tons of yearly biomass production [9–11]. Chemically it is $(C_6H_{10}O_5)_n$, a linear polysaccharide (Fig. 2) held together by 1,4-β-glucosidic bonds containing highly functionalized hydroxyl group forming hydrogen bonds responsible for compact structure, formation of crystalline morphologies at different lengths and also for its low solubility [12–14]. The word, "cellulose", initially appeared in an article by A. Payen (1839), where it is defined as "a resistant fibrous solid that is left after treatment of various plant tissues with acids, ammonia, and after subsequent extraction with water, alcohol, and ether" [15, 16]. Plant extracted cellulose is useful where mass production is required [14, 17]. Plant cellulose is organized as a network of fibers nested in a matrix of lignin. A single fiber has a diameter of five nm and lengths in tens of micrometers [18].

Various types of cellulose exist based on their origin, anatomy and processing [16, 19]. Cellulose can yield derivatives using a two-way approach. It can be synthesized in a descending manner i.e. cellulose is exposed to shear forces to generate minute particle of cellulose in suspension [20–22]. Or the ascending manner also referred to as white biotechnology of cellulose [16, 23] which utilizes bacteria for the biogenesis of cellulose [24]. Nano cellulose is a unique merge of crucial properties of cellulose and nano-scale materials [16]. It is noteworthy that at nano-scale the cellulose fibres are entwined in the supramolecular cellulose complex [16]. This massive cellulose structure provides an appealing pattern for inorganic particle stabilization and different types of synthetic polymers [25, 26].

Cellulose is categorized further into three types of nanocellulose (NC). These are cellulose nanofiber (CNF), cellulose nanocrystal (CNC) and nanocellulose composites [9, 16]. NC, CNC and CNF have distinctive properties like dimensional stability, transparency, low thermal expansion coefficient, exceptional reinforcing potential. It

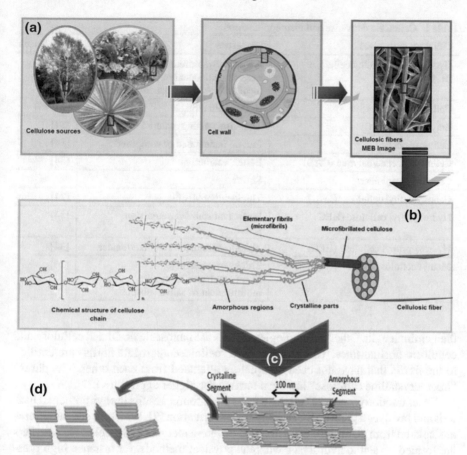

Fig. 1 Cellulosic fiber structural moieties [6]

Fig. 2 Cellobiose unit

is extremely crystalline with a high E ~150 GPa, which explains its strong mechanical properties [9].

Generally, nanofibrous structure is the basic aggregate of the discrete polymer units. The straight and longer cellulose crystals are named as whiskers, nanowires or rod-like microcrystals of cellulose. The term "microfibrillar cellulose" and "nanofibrillar cellulose" is used for the nanofibers that adhere together for at least a part of

Table 1 Cellulosic derivatise and their applications

Cellulose derivatives	Properties	References
Hydroxypropyl methylcellulose (HPMC)	Adhesive/water retention reagent/component in oral medication/salivary substitute	[32–34]
Cellulose nitrate	Highly inflammable	[34, 35]
Ethyl cellulose (EC)	Thin-film coating material	[34]
Cellulose acetate	Can be incinerated or composed	[34]
Cellulose acetate/butyrate (CAB)	Better weathering	[34]
Cellulose propionate	Stiffer	
Caboxymethyl cellulose (CMC)	Thickener/emulsifier	[34]
Hydroxyethyl cellulose (HEC)	Surfactant stabilizer/emulsion stabilizer	[34]
Hydroxypropyl cellulose (HPC)	Thickner/emulsion stabilizer/binder	[34]
Methyl cellulose (MC)	Thickner/emulsifier Lubricant/bacterial motility inhibitor/tear or saliva substitute	[36]

their entire length. Other terminologies for this are nanoscale fibrillated cellulose and cellulosic fibrillar fines. The nomenclature "cellulose aggregate fibrils" are entitled to the fibrils that have not been completely alienated from each other. The phrase "microcrystalline cellulose" has been termed for bigger aggregates [27].

The extraction process of cellulose from plant source results in environmental hazards and involves a great deal of energy for its formation [7]. Whiskers and nanofibres are isolated from plant cellulose using two different methods [28]. Cellulose whiskers are formed by acidic hydrolysis, whereas physical methods, for instance high pressure homogenization, are employed for nanosized fibers [29–31]. Other than this electrospinning is commonly being used to produce nanocellulose fibres from a cellulose solution or derivatives of cellulose. Further development is still necessary in order to achieve Nano sized fibre's length in the range of 1–100 nm.

Cellulose intended to be used in industries comes from cotton, wood and cotton (Table 1) [6]. The kraft process is used to separate cellulose from lignin, another major component of plant matter.

2 Bacterial Cellulose

Bacterial nanocellulose (BNC) is another division of cellulose also known as bio cellulose or microbial cellulose. Bacterial cellulose is beneficial in terms of purity as it is devoid of certain components like wax, lignin, pectin and hemicelluloses, which are the integral part of plant-derived cellulose [15, 37]. Additionally architectural parameters like molecular weight and repeating units of bacterial cellulose could be engineered by harnessing the process during biofabrication. BC basically has the porous fibrous framework of nanoscale cellulose [38]. The diameter of BC fibril is approximately 30 nm. This shows that it's diameter is roughly hundred times lesser than that of cellulose fibers. BC has ultrafine ribbon shaped nanofibers, approximately 100 nm in width, which constitutes a porous 3-D network (conductive polypyrole). The subfibrils having a width of ~1.5 nm tie-up in a ribbon form. Typically these ribbons are 70–80 nm wide, 3–4 nm thick and 1–9 nm long. The ribbons twirl and overlay to form a thick gel-like pellicle [39]. The surface area of the BC is further apmlified by the nanosize of these fibrils, this is manifested in its substaintial connection not only with the adjacent environment but also with other polymers, such as the extracellular matrices of the living tissue and the various types of nanoparticles [3].

BNC and plant cellulose has similar molecular formula but BNC is distinguished from plant cellulose on the basis of its nanofiber architecture [40]. The chief differentiating characteristics of BNC are: first; the constitution of cellulose with substrates having less molecular-weight in laboratory environment, second; The evolution of an exceptional material formed by uniting the attributes of cellulose with unique traits of nanoscale material and third, the advantage of in situ control of cellulose formation where certain traits of cellulose synthesis, for instance structure, shape and composite generation can be regulated directly at some stage in biosynthesis [41, 42]. The BC morphology can be regulated by using different kinds of bacterial strains, adding different constituents in cultural medium, providing various growing conditions, and post-processing drying methods [43].

2.1 Synthesis of BC

BC is produced by aerobic gram negative bacteria, for instance acetic acid bacteria of the genus Gluconacetobacter [44, 45]. Strains of Achrobacter, pseudomonas, Aerobacter, Azotobacter and Alcaligene can also be used to synthesize cellulose as an unadulterated constituent of their own biofilms (Table 2) [46]. Such bacteria are widespread in nature as a byproduct of sugar and plant carbohydrates fermentation. In comparison to MFC and NCC which are extracted from plants, biotechnological assembly is used for manufacture of BNC as a nanomaterial and polymer by using carbon sources which are low-molecular weight, for example bacterial d-glucose.

BC uses Gluconacetobacter strains for instance G. Xylinus which was reported by Brown first [14, 51]. It is assumed that bacteria make cellulose to guard itself from UV radiation and extremes in chemical environment and contact with oxygen [52–54]. The bacteria are cultured in glucose rich media to obtain cellulose while producing ethanol as a fermentation product [55]. The BNC (bacterial nanocellulose) is formed as exopolysaccharide concerning the outer and inner plasma membrane of bacterial cell and a cellulose-synthesizing complex, using biochemically activated dextrose catalyst, cellulose synthase [56, 57]. Cyclic diguanylmonophosphate (c-di-GMP) is the sutaible initiator of cellulose synthase [56]. Here a single cell has

Table 2 Bacterial strain used for synthesis of various types of cellulose

Organism	Cellulose type	References
Acetobacter	Extracellular pellicle cellulose ribbons	[47, 48]
Achromobacter	Fibrils	[49]
Agrobacterium	Short fibrils	[47, 49]
Alcaligenes	Fibrils	[49]
Pseudomonas	No distinct fibrils	[47, 49]
Rhizobium	Short fibrils	[44]
Sarcina	Amorphous cellulose	[44, 50]
Zoogloea	Not well defined	[48]

the potential to transform 100 plus dextrose molecules into cellulose every hour. This forms an enduring BNC hydrogel which is made of a nanofiber network with fiber 20–100 nm in diameter, having 99% water content. The pure cellulosic material exhibits high crytallinity, good mechanical strength and high molecular weight [14]. A temperature range of 25–29 °C is normally employed to meet the reqiorement of the bacteria [47]. Most common medium used for research work is made by Schramm and Hestrin, comprising 0.5% peptone and 0.5% yeast extract [58]. Apart from the nitrogen sources in culture media, a few amino acids are always considered essential: methionine and glutamate [56]. Certain vitamins, for instance nicotinic acid, pyridoxine, biotin and p-aminobenzoic acid; when added augments the process of cellulose synthesis and cell growth [59]. Cellulose production largely depends on the uninterrupted oxygen supply and carbon source. Other factors affecting the efficiency of production are kind of Gluconacetobacter strain used, the kind of material as well as its surface morphology, temperature, constituents of its culture medium (for instance additives) and pH. For cellulose synthesis by Acetobacter xylinum the optimum pH range is 4–7 [56, 58, 60].

2.2 BC Cultivation Methods

There are two methods used to produce the flat range of variable geometry i.e. static cultivation involving liquid culture medium or thin layer cultivation associated with solid phases for example agar. Certain parameters of BNC foils and fleeces for instance thickness and size can vary, depending on the kind of strain used, volume of culture medium and the time elapsed for cultivation.

2.2.1 Static Culture

By using matrix in the static culture, a thick gelatinous membrane can be formed on the surface of BC [61]. Agitated cultivation environment (stirred or shaken) provides different shapes like spheres, fibres and pallets [62, 63].

2.2.2 Agitated Culture

Agitated culture is found more appropriate for bulk production of such cellulose specifically because of their higher production rates [61].

Cellulose manufacture in horizontal fermentors, which is basically a mix of both stationary and agitated cultures, has also shown promising end-products [64].

2.3 Composite Formation

Pure bacterial cellulose is deficient in certain properties which limit its applications in industrial and medical field, therefore synthesis of BC composites is carried out to overcome these limitations [65]. Composites can be prepared by in situ incorporation of composite partners to the BC synthetic medium, or by post processing i.e. ex situ infiltration of reinforcement materials in BNC conventionally. Numerous materials are used for the synthesis of BC composite such as organic compounds, bioactive agents, polymerizable monomers, polymers such as polyacrylates, resins, polysaccharides, proteins, inorganic substances like metals and metal oxides [16, 66].

2.4 BNC Coating

The beneficial traits of BC, like high Young's modulus and increased fibre–matrix and fibre–fibre stress transfer, can used in coating sisal fibres to generate hierarchical composites. Short sisal fibres coated on Nano sized BC during fermentation create fibre-reinforced nanocomposites with enhanced adhesion properties. This also allows the extensive use of natural fibres in sustainable composites [67].

2.5 Properties of BC

A precise nano-structured BNC has remarkable intrinsic properties, and is utilized as a reinforcement material for nanocomposites, for actuation systems, and for biomedical purposes [9]. A supramolecular structure of choice can be formed just by controlling certain cultivation conditions and factors like type of bacterial strain, carbon source in culture medium, and types of additives used during biosynthesis. Further properties of BNC is as follows:

1. The Bacterial bio cellulose is highly pure in nature [68].
2. Direct formation of cellulose bodies for instance hydrogels and aerogels (after drying), can be carried out by using BC.
3. Surface cellulozation occurs if cultivated in a thin layer for coating different materials.
4. Ability to form In situ composite by using dispersed particles or water soluble additives.
5. It can be modified/processed effectively.
6. Natural fibers exhibit poor mechanical properties. Therefore BC plays a vital role as a nano scale reinforcement material due to its high Young's modulus >138 Gpa [69] and tensile strength >2 Gpa (conductive—polypyrrole) and high density of 1600 kg m^{-3} [70]. It gets attached in situ to the surface of fibers (hemp and sisal) through strong hydrogen bonding. Yano et al. and Nogi et al. found that the nanofiber network potential of BNC can be further exploited to achieve optically transparent and low thermal expansion material [71]. Guhados et al. worked out the elastic modulus of a single bacterial cellulose fiber [72]. BC exhibits remarkable mechanical strength in the dry state because of its 3D crystalline nano fiber structure. The strength of its single fiber are comparable to that of Kevlar and steel. Thus, it is a reputable source of reinforcing agent for polymer composites.
7. Macroporosity of BC can be effectively controlled by introducing porogens in the culture medium.
8. BC has a high molecular weight, characterized by lengthy polymer chains of roughly 3000–9000 repeating units. Other characteristics are high crystallinity (80–90%), high water content upto 99%, and superior fibre stability as well as proven biocompatibility [16]. BC holds a huge amount of water not less than 200 times the dry weight of cellulose [73] which performs like a spacer and stabilizes the nanofibril network.
9. Synthesis controlled shape of bacterial cellulose. A suitable reactor form and function (agitated or static cultivation) is the deciding factor in choosing the shape of BNC hydrogel [74]. Shaping of polymer-based materials is a vital process in polymer manufacture and application. The key methods for shaping are thermoplastic techniques, shaping from solution, in situ shaping and sintering. Cellulosic materials are shaped from solutions usually due to lack of thermoplastic behaviour. The BC formation by fermentation has offered a new perspective in the field of in situ shaping of cellulose.

2.6   BNC Cultivation Methods

To produce flat range of variable geometry either static cultivation in liquid culture medium or thin layer cultivation on solid phases like agar etc. are used. The parameters like size and thickness of the BNC fleeces and foils can vary and depends on the type of strain used, culture medium volume, and duration of cultivation. Static Culture.

By using matrix in the static culture thick gelatinous membrane of BC on the surface can be formed [61]. Spheres, fibres, pallets are formed under agitated cultivation conditions (shaking, stirring) [62, 63].

2.7   Agitated Culture

Agitated culture is more appropriate for mass production of BC specifically because of their higher production rates [61]. Cellulose manufacture in horizontal fermentors, which is basically a mix of both stationary and agitated cultures, has also shown promising end-products [64]. This bio-shaping is highly effective for the construction of flat materials (fleeces, foils), spheres, hollow bodies, fibers, and coatings during bacterial biosynthesis.

The primary step is selecting the appropriate bacterial culture in order to construct a specific fibre-network of BNC materials [14]. By managing the biological and physiological conditions of bacterial growth envioronment for instance pH, temperature and composition of the culture media, and oxygen tension, pellicle of different morphology is acquired [75]. Few polysaccharides complexes for instance cationic starch and CMC, brings variety to the supramolecular alignment and remain somewhat integrated in the hydrogen-bond of BNC [76].

2.8   Effects of Drying Methods on Morphology of Membranes

BC hydrogel materials can be dried by various techniques such as

(1) Dewatering by organic solvents such as ethanol or acetone. In slow dewatering, the fibril network remains sturdy and aero gels are created.aero gels are characterized by high re-swelling capacity.
(2) Critical point drying.
(3) Hot-press drying.
(4) Air drying under normal or higher pressure. Drying under pressure creates flat foils with 99% loss of water. Flat foils have high density and are highly stable mechanically. Such foils have technical application such as wound dressing and membranes for audio technologies.
(5) Freeze-drying.

2.9 In situ Modifications of Preformed BC

In situ or post modification done to design the BC composite is a step forward in development of BC materials. In case of a hydroxyl group as an additive, the in-situ insertion is altered by hydrogen bridge formation; in the case of nanoparticles, the BC fibers can act as nanofiber templates. A recent example is the impregnation of BC with silver nitrate and the subsequent reduction with water-soluble sodium borohydride (a technical reducing agent), which forms silver metal deposited as nanoparticles on the supporting BC nanofibers. Nanofibers of BC material, when modified with carboxylic functionalities, have shown the unique ability to induce biomimetic crystallization of calcium-deficient hydroxyapatite (28). The presence of nano-sized hydroxyapatite crystals on the surfaces of cellulose nanofibrils promotes osteoprogenitor cell adhesion and differentiation (29). The ability to control macro porosity combined with surface decoration with hydroxyapatite makes BC materials promising candidates for bone grafts.

2.10 Uses of BC

Because of its peculiarity, BC has gained reasonable importance as a worthy raw material for various products related to biomedical field. Its use is considered in microsurgery as an artificial blood vessels [77], scaffolds for tissue engineering [78], dental implants [56], and drug delivery systems [79]. Other uses for the bacterial cellulose includes its use as a fortification in very fine papers [80, 81], as coatings, paint additives, cosmetics and pharmaceuticals [46, 82], diaphragms for electro-acoustic transducers [83], and reinforcement for optically transparent films [71]. Bacterial cellulose is also utilized as a dietary fiber source (nata-de-coco) [84], as thickening or binding agents [85].

Paper products: Cellulose is the major constituent of paper, paperboard, and card stock.

Fibers: Cellulose is the main ingredient of textiles made from cotton, linen, and other plant fibers. It can be turned into rayon, an important fiber that has been used for textiles since the beginning of the twentieth century. Both cellophane and rayon are known as "regenerated cellulose fibers"; they are identical to cellulose in chemical structure and are usually made from dissolving pulp via viscose. A more recent and environmentally friendly method to produce a form of rayon is the Lyocell process.

Consumables: Microcrystalline cellulose (E460i) and powdered cellulose (E460ii) are used as inactive fillers in drug tablets [40] and a wide range of soluble cellulose derivatives, E numbers E461 to E469, are used as emulsifiers, thickeners and stabilizers in processed foods. Cellulose powder is, for example, used in processed cheese to prevent caking inside the package. Cellulose occurs naturally in some foods and is an additive in manufactured foods, contributing an indigestible component used for texture and bulk, potentially aiding in defecation [41].

Science: Cellulose is used in the laboratory as a stationary phase for thin layer chromatography. Cellulose fibers are also used in liquid filtration, sometimes in combination with diatomaceous earth or other filtration media, to create a filter bed of inert material.

Energy crops: The major combustible component of non-food energy crops is cellulose, with lignin second. Non-food energy crops produce more usable energy than edible energy crops (which have a large starch component), but still compete with food crops for agricultural land and water resources [42]. Typical non-food energy crops include industrial hemp (though outlawed in some countries), switchgrass, Miscanthus, Salix (willow), and Populus (poplar) species.

Biofuel: TU-103, a strain of Clostridium bacteria found in zebra waste, can convert nearly any form of cellulose into butanol fuel [43, 44].

Building material: Hydroxyl bonding of cellulose in water produces a sprayable, moldable material as an alternative to the use of plastics and resins. The recyclable material can be made water- and fire-resistant. It provides sufficient strength for use as a building material [45]. Cellulose insulation made from recycled paper is becoming popular as an environmentally preferable material for building insulation. It can be treated with boric acid as a fire retardant.

Miscellaneous: Cellulose can be converted into cellophane, a thin transparent film. It is the base material for the celluloid that was used for photographic and movie films until the mid-1930s. Cellulose is used to make water-soluble adhesives and binders such as methyl cellulose and carboxymethyl cellulose which are used in wallpaper paste. Cellulose is further used to make hydrophilic and highly absorbent sponges. Cellulose is the raw material in the manufacture of nitrocellulose (cellulose nitrate) which is used in smokeless gunpowder.

Pharmaceuticals: Cellulose derivatives, such as microcrystalline cellulose (MCC), have the advantages of retaining water, being a stabilizer and thickening agent, and in reinforcement of drug tablets [46].

3 Bacterial Cellulose for Biomedical Applications

3.1 Skin

The striking similarity existing between bacteria cellulose and human skin is evident, thats is why bacterial cellulose can be used as skin replacement in treating large-scale burns [86]. The fibril framework of bacterial cellulose is comparable to extracellular matrix of human skin [38]. Czaja et al. briefed the external usage of BNC in medicine: cure for chronic burns and wounds, transitory wound coverage, use in cosmetics [87]. Currently, Lohmann and Rauscher "Suprasorb" wound dressing is available in market [16]. Wang et al. explained the BNC dressings formation, medical gauze and artificial skins, by using a special culture medium for the membrane [16] through this method a suitable yield and thickness of BNC layer could be attained.

3.2　Vascular Grafts

The recent state of BC biomaterial development confirms that BC implants meet important demands on biomaterials, such as blood and tissue compatibility, ingrowths of cells, good surgical handling, endothelization and the use of common methods of sterilization. BC tubes for instance BASYC (Bacterial SYnthesized Cellulose) exemplifies a novel approach for outliving reconstructive issues by providing vascular grafts of small-caliber (Table 3) [14]. The BNC used most inside the body as grafts is to treat cardiovascular diseases—the illness common worldwide is the most

Table 3 Biomedical application of bacterial cellulose

Trade name	Application	Treatment	Properties	Effect	References
Suprasorb	Wound dressings				
Bionext	Wound dressing	Ulcers and burns, lacerations		Pain relief, reduced infection and inflammation, faster healing	[88]
BASYC	Bacterially synthesized cellulose tubes	Used as a micro vessel	Smooth surface		[14]
Biofill@, Bioprocess@	Temporary artificial skin	Therapy for burns and ulcers	High mechanical strength in wet state, low roughness of the inner surface, high water retention	Pain relief, quick healing, pain relief	[56, 77, 89]
Membracell	Temporary skin substitute	For burns lacerations and ulcers		Fast skin regeneration	[88]
XCell	Skin substitute		Promoting autolytic debridement, reducing pain, accelerating granulation	Proper and quick wound healing	[88]
Gengiflex	Dental implants, grafting material	Periodontal tissue recovery		Few surgical steps involved, reduced inflammation	

important purpose. Up till now, for these coronary bypasses (inner diameter <6 mm) only autografts (vena saphena and arteria mammaria) can be used. Suitable artificial implant material for this purpose has not been developed yet [16].

3.3 For Bone

Hutchens et al. carried out an investigation by combining hydroxyapatite and BNC to evaluate bone repair. It demonstrated that BNC can be used as a template to produce calcium-deficient hydroxyapatite (CDHAP), the natural mineral constituent of bone. Since it encourages bone colonization when inserted in bone defects and is also biodegardable [16].

3.4 Tissue Biocompatibility

Tissue biocompatibility is main evaluating factor of biomedical material. Currently no systemic study regarding cellulose absorbance and foreign body reaction on living tissue has been reported. The reason could be poor biodegradability of cellulose. Cellulose has poor biodegradability due to its higher-order arrangement. The distinguishing disparity in absorbance by living tissue chiefly recommends that the amorphous regions are absorbed and crystalline regions in BNC are poorly absorbed in tissue. This shows that absorption of cellulose (in vivo) relies on its degree of crystallinity [13].

3.5 Degradation of Cellulose

Biological solubilisation of cellulose probably takes place in at least two steps: (a) conversion of the native cellulose molecule into linear anhydroglucose chains and (b) hydrolysis of the 1,4-b-glucosidic linkage to form soluble sugars [90].

3.6 Cellulosic Composites

Various cellulosic composites are prepared and used in various biomedical applications. Some of the data are summarized in Table 4.

Table 4 Cellulose based composite materials for biomedical applications

Application	Nature of material	Effect	References
Wound dressing	BC membranes impregnated with Ag nanoparticles	Exhibit antibacterial activity against *S. aureus* and *E. coli*	[91]
Wound dressing	Chitosan-BC membranes	Faster epithelialization rates than BC or Tegaderm®	[92]
Burn treatment	BC membranes	Faster wound healing of second-degree burns	[93]
Wound dressing	BC-Hyaluronan (0.1%) composite films	Higher regeneration rates than gauze or BC	[94]
Cornea scaffold	BC-chitosan/CMC composites	A more suitable interface for retinal pigment epithelium adhesion and proliferation	[95]
Artificial cornea	BC/PVA composites	Artificial cornea and eye bioengineering	[96]
Bone healing	BC/hydroxyapatite nanocomposites	Improved cellular adhesion and differentiation	[97]

3.7 Drawbacks

In order to reduce the use of plant cellulose, greater emphasis on utilizing bacterial cellulose is advocated. However, the microbial cellulose is found to be 80 times more costly as compared to the plant cellulose.

Economical mass production necessitates designing a culture aeration and agitation process [15] another drawback of using bacteria as a source for cellulose production, is the aggregation of by-product during bacteria growth time. This retards the rate of production on industrial level [98].

In conclusion, cellulose especially bacterial cellulose has wide application in general applications as well as in biomedical applications.

Acknowledgement This work supported Interdisciplinary Research Centre in Biomedical Materials (IRCBM) COMSATS University Islamabad, Lahore Campus, Pakistan and NRPU research project 4146.

References

1. Wang, S., et al. (2016). Modification and potential application of short-chain-length polyhydroxyalkanoate (SCL-PHA). *Polymers, 8*(8), 273.
2. Nithin, B., & Goel, S. (2017). Degradation of plastics. In *Advances in solid and hazardous waste management* (pp. 235–247). Springer.
3. GQ, C. (2010). Plastics completely synthesized by bacteria: Polyhydroxyalkanoates. In C. GQ (Ed.), *Microbiology monographs*. Berlin, Heidelberg: Springer.

4. Song, J. H., et al. (2009). Biodegradable and compostable alternatives to conventional plastics. *Philosophical Transactions of the Royal Society of London. Series B, Biological Sciences, 364*(1526), 2127–2139.
5. Schröpfer, S. B., et al. (2015). Biodegradation evaluation of bacterial cellulose, vegetable cellulose and poly (3-hydroxybutyrate) in soil. *Polímeros, 25*, 154–160.
6. Ng, H.-M., et al. (2015). Extraction of cellulose nanocrystals from plant sources for application as reinforcing agent in polymers. *Composites Part B: Engineering, 75*, 176–200.
7. Man, Z., et al. (2011). Preparation of cellulose nanocrystals using an ionic liquid. *Journal of Polymers and the Environment, 19*(3), 726–731.
8. Sarwono, A., et al. (2017). A new approach of probe sonication assisted ionic liquid conversion of glucose, cellulose and biomass into 5-hydroxymethylfurfural. *Ultrasonics Sonochemistry, 37*, 310–319.
9. Kim, J.-H., et al. (2015). Review of nanocellulose for sustainable future materials. *International Journal of Precision Engineering and Manufacturing-Green Technology, 2*(2), 197–213.
10. Muhammad, N., et al. (2015). Dissolution and separation of wood biopolymers using ionic liquids. *ChemBioEng Reviews, 2*(4), 257–278.
11. Jamshaid, A., et al. (2019). Fabrication and evaluation of cellulose-alginate-hydroxyapatite beads for the removal of heavy metal ions from Aqueous Solutions. In *Zeitschrift für Physikalische Chemie* (p. 1351).
12. Iwamoto, S., Nakagaito, A., & Yano, H. (2007). Nano-fibrillation of pulp fibers for the processing of transparent nanocomposites. *Applied Physics A, 89*(2), 461–466.
13. Miyamoto, T., et al. (1989). Tissue biocompatibility of cellulose and its derivatives. *Journal of Biomedical Materials Research, 23*(1), 125–133.
14. Klemm, D., et al. (2011). Nanocelluloses: A new family of nature-based materials. *Angewandte Chemie International Edition, 50*(24), 5438–5466.
15. Mohite, B. V., & Patil, S. V. (2014). A novel biomaterial: Bacterial cellulose and its new era applications. *Biotechnology and Applied Biochemistry, 61*(2), 101–110.
16. Klemm, D., et al. (2009). Nanocellulose materials–different cellulose, different functionality. In *Macromolecular symposia*. Wiley Online Library.
17. Ummartyotin, S., & Manuspiya, H. (2015). A critical review on cellulose: From fundamental to an approach on sensor technology. *Renewable and Sustainable Energy Reviews, 41*, 402–412.
18. Hult, E.-L., Larsson, P., & Iversen, T. (2001). Cellulose fibril aggregation—An inherent property of kraft pulps. *Polymer, 42*(8), 3309–3314.
19. Jamshaid, A., et al. (2017). Cellulose-based materials for the removal of heavy metals from wastewater—An overview. *ChemBioEng Reviews, 4*(4), 240–256.
20. Nakagaito, A., & Yano, H. (2004). The effect of morphological changes from pulp fiber towards nano-scale fibrillated cellulose on the mechanical properties of high-strength plant fiber based composites. *Applied Physics A, 78*(4), 547–552.
21. Wu, C. (2010). *Production and characterization of optically transparent nanocomposite film*. Faculty of Forestry.
22. Sain, M. M., & Bhatnagar, A. (2008). *Manufacturing process of cellulose nanofibers from renewable feed stocks*. Google Patents.
23. Juntaro, J. (2009). *Environmentally friendly hierarchical composites*. Imperial College London.
24. Klemm, D., et al. (2005). Cellulose: Fascinating biopolymer and sustainable raw material. *Angewandte Chemie International Edition, 44*(22), 3358–3393.
25. Barud, H., et al. (2008). Bacterial cellulose–silica organic–inorganic hybrids. *Journal of Sol-Gel Science and Technology, 46*(3), 363–367.
26. Kramer, F., et al. (2006). Nanocellulose polymer composites as innovative pool for (bio) material development. In *Macromolecular symposia*. Wiley Online Library.
27. Islam, M. T., et al. (2014). Preparation of nanocellulose: A review. *AATCC Journal of Research, 1*(5), 17–23.
28. Azizi Samir, M. A. S., Alloin, F., & Dufresne, A. (2005). Review of recent research into cellulosic whiskers, their properties and their application in nanocomposite field. *Biomacromolecules, 6*(2), 612–626.

29. Revol, J.-F., et al. (1992). Helicoidal self-ordering of cellulose microfibrils in aqueous suspension. *International Journal of Biological Macromolecules, 14*(3), 170–172.
30. Bondeson, D., Mathew, A., & Oksman, K. (2006). Optimization of the isolation of nanocrystals from microcrystalline cellulose by acid hydrolysis. *Cellulose, 13*(2), 171.
31. Zimmermann, T., Bordeanu, N., & Strub, E. (2010). Properties of nanofibrillated cellulose from different raw materials and its reinforcement potential. *Carbohydrate Polymers, 79*(4), 1086–1093.
32. Macleod, G. S., Collett, J. H., & Fell, J. T. (1999). The potential use of mixed films of pectin, chitosan and HPMC for bimodal drug release. *Journal of Controlled Release, 58*(3), 303–310.
33. Nagy, G., et al. (1995). Use of hydroxy-propyl-methyl cellulose (methocel) and carboxy-methyl cellulose containing artificial saliva in the symptomatic treatment of xerostomia. *Fogorvosi Szemle, 88*(9), 299–304.
34. Khalil, H. A., et al. (2015). Cellulosic nanocomposites from natural fibers for medical applications: A review. In *Handbook of polymer nanocomposites. Processing, performance and application* (pp. 475–511). Springer.
35. Koob, S. P. (1982). The instability of cellulose nitrate adhesives. *The Conservator, 6*(1), 31–34.
36. Floury, J., et al. (2003). Effect of high pressure homogenisation on methylcellulose as food emulsifier. *Journal of Food Engineering, 58*(3), 227–238.
37. Colvin, J.R. (1980). The biosynthesis of cellulose. In *Carbohydrates: Structure and function* (pp. 543–570). Elsevier.
38. Kwak, M.H., et al. (2015). Bacterial cellulose membrane produced by Acetobacter sp. A10 for burn wound dressing applications. *Carbohydrate Polymers, 122*, 387–398.
39. Benson, R., et al. (2011). *Development of bacterial cellulose nanocomposites*. MRS Online Proceedings Library Archive, 1312.
40. Czaja, W., Romanovicz, D., & Malcolm Brown, R. (2004). Structural investigations of microbial cellulose produced in stationary and agitated culture. *Cellulose, 11*(3–4), 403–411.
41. Bodin, A., et al. (2007). Influence of cultivation conditions on mechanical and morphological properties of bacterial cellulose tubes. *Biotechnology and Bioengineering, 97*(2), 425–434.
42. Czaja, W. K., et al. (2007). The future prospects of microbial cellulose in biomedical applications. *Biomacromolecules, 8*(1), 1–12.
43. Gatenholm, P., & Klemm, D. (2010). Bacterial nanocellulose as a renewable material for biomedical applications. *MRS Bulletin, 35*(3), 208–213.
44. Nakagaito, A., Iwamoto, S., & Yano, H. (2005). Bacterial cellulose: The ultimate nano-scalar cellulose morphology for the production of high-strength composites. *Applied Physics A, 80*(1), 93–97.
45. Moosavi-Nasab, M., & Yousefi, A. (2011). Biotechnological production of cellulose by Gluconacetobacter xylinus from agricultural waste. *Iranian Journal of Biotechnology, 9*(2), 94–101.
46. Byrom, D. (1991). *Biomaterials: Novel materials from biological sources*. Springer.
47. Cannon, R. E., & Anderson, S. M. (1991). Biogenesis of bacterial cellulose. *Critical Reviews in Microbiology, 17*(6), 435–447.
48. Fiedler, S., Füssel, M., & Sattler, K. (1989). Gewinnung und Anwendung von Bakteriencellulose: I. Übersicht zum Stand der Forschung und Untersuchungen zur Fermentationskinetik. *Zentralblatt für Mikrobiologie, 144*(7), 473–484.
49. Matthysse, A. (1983). Role of bacterial cellulose fibrils in Agrobacterium tumefaciens infection. *Journal of Bacteriology, 154*(2), 906–915.
50. Canale-Parola, E. (1970). Biology of the sugar-fermenting Sarcinae. *Bacterioloical Reviews, 34*(1), 82.
51. Huang, Y., et al. (2014). Recent advances in bacterial cellulose. *Cellulose, 21*(1), 1–30.
52. Somerville, C. (2006). Cellulose synthesis in higher plants. *Annual Review of Cell and Developmental Biology, 22*, 53–78.
53. Putra, A., et al. (2008). Tubular bacterial cellulose gel with oriented fibrils on the curved surface. *Polymer, 49*(7), 1885–1891.

54. Brown, R. M., Jr. (2004). Cellulose structure and biosynthesis: What is in store for the 21st century? *Journal of Polymer Science Part A: Polymer Chemistry, 42*(3), 487–495.
55. Shah, J., & Brown, R. M. (2005). Towards electronic paper displays made from microbial cellulose. *Applied Microbiology and Biotechnology, 66*(4), 352–355.
56. Jonas, R., & Farah, L. F. (1998). Production and application of microbial cellulose. *Polymer Degradation and Stability, 59*(1–3), 101–106.
57. White, D. G., & Brown, R. M., Jr. (1989). Prospects for the commercialization of the biosynthesis of microbial cellulose. *Cellulose and Wood-Chemistry and Technology, 573,* 573–590.
58. Hestrin, S., & Schramm, M. (1954). Synthesis of cellulose by Acetobacter xylinum. 2. Preparation of freeze-dried cells capable of polymerizing glucose to cellulose. *Biochemical Journal, 58*(2), 345.
59. Ishikawa, A., et al. (1995). Increase in cellulose production by sulfaguanidine-resistant mutants derived from Acetobacter xylinum subsp. sucrofermentans. *Bioscience, Biotechnology, and Biochemistry, 59*(12), 2259–2262.
60. Masaoka, S., Ohe, T., & Sakota, N. (1993). Production of cellulose from glucose by Acetobacter xylinum. *Journal of Fermentation and Bioengineering, 75*(1), 18–22.
61. Krystynowicz, A., et al. (2002). Factors affecting the yield and properties of bacterial cellulose. *Journal of Industrial Microbiology and Biotechnology, 29*(4), 189–195.
62. Watanabe, K., et al. (1998). Structural features and properties of bacterial cellulose produced in agitated culture. *Cellulose, 5*(3), 187–200.
63. Dudman, W. (1960). Cellulose production by Acetobacter strains in submerged culture. *Microbiology, 22*(1), 25–39.
64. Bungay, III, H. R., & Serafica, G. C. (1999). *Production of microbial cellulose using a rotating disk film bioreactor.* Google Patents.
65. Moniri, M., et al. (2017). Production and status of bacterial cellulose in biomedical engineering. *Nanomaterials, 7*(9), 257.
66. Shah, N., et al. (2013). Overview of bacterial cellulose composites: A multipurpose advanced material. *Carbohydrate Polymers, 98*(2), 1585–1598.
67. Lee, K.-Y., et al. (2012). Hierarchical composites reinforced with robust short sisal fibre preforms utilising bacterial cellulose as binder. *Composites Science and Technology, 72*(13), 1479–1486.
68. Lestari, P., et al. (2014). Study on the production of bacterial cellulose from Acetobacter xylinum using agro-waste. *Jordan Journal of Biological Sciences, 147*(1570), 1–6.
69. Yamanaka, S., et al. (1989). The structure and mechanical properties of sheets prepared from bacterial cellulose. *Journal of Materials Science, 24*(9), 3141–3145.
70. Wan, Y., et al. (2006). Synthesis and characterization of hydroxyapatite–bacterial cellulose nanocomposites. *Composites Science and Technology, 66*(11–12), 1825–1832.
71. Nogi, M., et al. (2006). Fiber-content dependency of the optical transparency and thermal expansion of bacterial nanofiber reinforced composites. *Applied Physics Letters, 88*(13), 133124.
72. Guhados, G., Wan, W., & Hutter, J. L. (2005). Measurement of the elastic modulus of single bacterial cellulose fibers using atomic force microscopy. *Langmuir, 21*(14), 6642–6646.
73. Evans, B. R., et al. (2003). Palladium-bacterial cellulose membranes for fuel cells. *Biosensors and Bioelectronics, 18*(7), 917–923.
74. Yan, Z., et al. (2008). Biosynthesis of bacterial cellulose/multi-walled carbon nanotubes in agitated culture. *Carbohydrate Polymers, 74*(3), 659–665.
75. Pandey, L. K., Saxena, C., & Dubey, V. (2005). Studies on pervaporative characteristics of bacterial cellulose membrane. *Separation and Purification Technology, 42*(3), 213–218.
76. Ishida, T., et al. (2003). Role of water-soluble polysaccharides in bacterial cellulose production. *Biotechnology and Bioengineering, 83*(4), 474–478.
77. Klemm, D., et al. (2001). Bacterial synthesized cellulose—Artificial blood vessels for microsurgery. *Progress in Polymer Science, 26*(9), 1561–1603.

78. Helenius, G., et al. (2006). In vivo biocompatibility of bacterial cellulose. *Journal of Biomedical Materials Research Part A: Official Journal of the Society for Biomaterials, The Japanese Society for Biomaterials, and The Australian Society for Biomaterials and the Korean Society for Biomaterials, 76*(2), 431–438.
79. Sokolnicki, A. M., et al. (2006). Permeability of bacterial cellulose membranes. *Journal of Membrane Science, 272*(1–2), 15–27.
80. Iguchi, M., Yamanaka, S., & Budhiono, A. (2000). Bacterial cellulose—A masterpiece of nature's arts. *Journal of Materials Science, 35*(2), 261–270.
81. Sukjoon, Y., & Jeffery, S. (2010). Composites, enzyme-assisted preparation of fibrillated cellulose fibers and its effect on physical and mechanical properties of paper sheet composites. *Industrial and Engineering Chemistry Research, 49*, 2161–2168.
82. Turbak, A.F., Snyder, F. W., & Sandberg, K. R. (1983). Microfibrillated cellulose, a new cellulose product: properties, uses, and commercial potential. *Journal of Applied Polymer Science: Applied Polymer Symposium (United States)*. Shelton, WA: ITT Rayonier Inc.
83. Nishi, Y., et al. (1990). The structure and mechanical properties of sheets prepared from bacterial cellulose. *Journal of Materials Science, 25*(6), 2997–3001.
84. Stephens, R. S., Westland, J. A., Neogi, A. N. (1990). *Method of using bacterial cellulose as a dietary fiber component*. Google Patents.
85. Barud, H., et al. (2007). Thermal characterization of bacterial cellulose–phosphate composite membranes. *Journal of Thermal Analysis and Calorimetry, 87*(3), 815–818.
86. Czaja, W., et al. (2006). Microbial cellulose—The natural power to heal wounds. *Biomaterials, 27*(2), 145–151.
87. Czaja, W., et al. (2007). Biomedical applications of microbial cellulose in burn wound recovery. In *Cellulose: Molecular and structural biology* (pp. 307–321). Springer.
88. Lopes, T. D., et al. (2014). Bacterial cellulose and hyaluronic acid hybrid membranes: Production and characterization. *International Journal of Biological Macromolecules, 67*, 401–408.
89. Fontana, J., et al. (1990). Acetobacter cellulose pellicle as a temporary skin substitute. *Applied Biochemistry and Biotechnology, 24*(1), 253–264.
90. Reese, E. T., Siu, R. G., & Levinson, H. S. (1950). The biological degradation of soluble cellulose derivatives and its relationship to the mechanism of cellulose hydrolysis. *Journal of Bacteriology, 59*(4), 485.
91. Rejeb, S. B., et al. (1998). Functionalization of nitrocellulose membranes using ammonia plasma for the covalent attachment of antibodies for use in membrane-based immunoassays. *Analytica Chimica Acta, 376*(1), 133–138.
92. Tammelin, T., et al. (2006). Preparation of Langmuir/Blodgett-cellulose surfaces by using horizontal dipping procedure. Application for polyelectrolyte adsorption studies performed with QCM-D. *Cellulose, 13*(5), 519.
93. Lou, Z. (2016). Treatment of tympanic membrane perforation using bacterial cellulose: a randomized controlled trial. *Brazilian Journal of Otorhinolaryngology, 82*(5), 618–619.
94. Wei, B., Yang, G., & Hong, F. (2011). Preparation and evaluation of a kind of bacterial cellulose dry films with antibacterial properties. *Carbohydrate Polymers, 84*(1), 533–538.
95. Dutton, J. J. (1991). Coralline hydroxyapatite as an ocular implant. *Ophthalmology, 98*(3), 370–377.
96. Goncalves, S., et al. (2015). Bacterial cellulose as a support for the growth of retinal pigment epithelium. *Biomacromolecules, 16*(4), 1341–1351.
97. Mohan, T., et al. (2012). Exploring the rearrangement of amorphous cellulose model thin films upon heat treatment. *Soft Matter, 8*(38), 9807–9815.
98. Lee, K. Y., et al. (2014). More than meets the eye in bacterial cellulose: biosynthesis, bioprocessing, and applications in advanced fiber composites. *Macromolecular Bioscience, 14*(1), 10–32.

Cellulose Based Bio Polymers: Synthesis, Functionalization and Applications in Heavy Metal Adsorption

Vijaykiran N. Narwade, Yasir Beeran Pottathara, Sumayya Begum, Rajendra S. Khairnar and Kashinath A. Bogle

Abstract Water pollution due to tremendous increase in industrialization, urbanization and population become serious concerns since the last some decade and will be the major global nightmare. Various contaminations viz; dyes, heavy metals, pesticides, pharmaceutical effluents from industries are getting discharged into water bodies. Among these contaminants, Heavy metals are the main wastewater pollutants due to their ability to cause the nuisance to living beings and to persist in the environment. Hence lot of efforts are being taken for treating waste water contained with heavy metals. Materials scientist are trying to utilise various methods and materials for solving these problems. Cellulose the natural biopolymer is one of the materials gaining attention because of its extra ordinary physio-chemical, as well as mechanical properties compared to other natural biopolymer materials. The present book chapter deals with the preparations, modifications and its heavy metal adsorption studies.

1 Introduction

The separation of pollutants, those are speedily rising in our environs due to the massive comprehensive industrialization, has become essential interest. The pollutants are generally in the form of perilous organic or inorganic compounds they may be include metals, non-metals, chemicals, dyes etc. Specially, the heavy metals are coming towards us or towards the whole environment through the water streams from the

V. N. Narwade (✉) · S. Begum · R. S. Khairnar · K. A. Bogle
School of Physical Sciences, Swami Ramanand Teerth Marathwada University, Nanded 431606, Maharashtra, India
e-mail: vkiranphysics@gmail.com

V. N. Narwade
Faculty of Natural Sciences, Department of Inorganic Chemistry, Comenius University, Bratislava, Slovakia

Y. B. Pottathara
Faculty of Mechanical Engineering, University of Maribor, Smetanova ul. 17, 2000 Maribor, Slovenia

© Springer Nature Switzerland AG 2020
A. Khan et al. (eds.), *Biofibers and Biopolymers for Biocomposites*,
https://doi.org/10.1007/978-3-030-40301-0_12

different water resources [1–4]. The existence of such entities are posing a serious global problem to public health and leading to the destruction of our environment.

Today's scenario of the research fields is mostly relating with nanoworld. The nanoworld can define as the one dimensional tiny objects which are nanometric in size. Tiny objects can be tubular, spherical, wire like, sheet like or fibrous in shapes. Consequently, we can term as nanotubes, nanopowder, nanowires, nanosheets or nanofibers. This nanoworld is playing very important role in the cleansing of contaminated water.

The most plentiful biopolymer cellulose at its nano scale, having diameters in the range of some 1–100 nm and while its length varies from nanometres to several micrometres called as Nanocellulose, is highly participating in the waste water treatment area. Cellulose is one of the natural polymeric materials that consisting of hundreds—and sometimes even thousands—of carbon, hydrogen and oxygen atoms with the formula $(C_6H_{10}O_5)_n$. It is made from repeated units of the monomeric glucose; hence it is a linear organic polysaccharide. It's having specific chemical structure which depicted in Fig. 1. Nanocellulose show enhanced properties over bulk cellulose. Nanocellulose is having properties such as high strength and large surface area, which make cellulose as highly capable material for the engineering of high-performance filters as well as membranes. Among the synthesis methods of nanocellulose, Mechanical and chemical pre-treatments are the two different approaches generally employed for the transformation of cellulose materials into the,

(i)   Cellulose nanocrystals (CNCs),
(ii) Cellulose nanofibers or nanofibrils (CNFs),
(iii) Cellulose nanowhiskers (CNWs) and
(iv) Bacterial nanocellulose (BNC)

with an outstanding ability for treating industrial and drinking water system. The literature survey is also proofed the nanocellulose application in the form of various structures for water purification.

Recently, the modified as well as functionalized cellulose are giving more interest towards the water treatment. It is investigated that Cellulose as well as modified

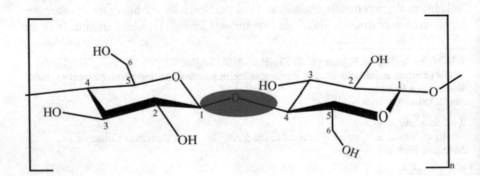

Fig. 1 The chemical structure of cellulose

cellulose are used as adsorbent for the heavy metal ions adsorption, like chromium Cr(IV), copper Cu(II), Zinc Zn(II), lead Pb(II), cadmium Cd(II), and cobalt Co etc. In addition to this, removal of bacteria from water through electrostatic interactions using paper filters modified which further modified with polyelectrolyte multilayers is also investigated [5]. Because of –OH group the cellulose can be successfully functionalise into anionic and cationic nature by functionalization procedures developed for the attaching functional groups on the cellulose [6].

This chapter mainly stress on the synthesis methods of cellulose, functionalization of cellulose based bio polymers, and particularly its applications in eradication by means of different methods for heavy metals.

2 Nanocellulose: Synthesis, Functionalization and Applications

2.1 Synthesis of Nanocellulose

Cellulose having the ample applications in almost every field and researcher are still working to improve its usability in all aspects and hence in order to use nanocellulose on globally, it is necessary to produce a sustainable and environmentally beneficial pre-treatment processes. The pre-treatment processes are necessary to produce nanocellulose product in the pure form. Ample research articles have been already reported the fabrication of nanocellulose using pre-treatments methodologies. Preparations of nanocellulose can be carried out by chemical as well as mechanical pre-treatment [7]. Cryocrushing and refining, are the mechanical pre-treatment technique which combines the steps like (i) combination of simple cropping in a refiner tank and then (ii) high-impact crushing under liquid nitrogen atmosphere to avoid impurities [8] (Fig. 2).

Nanocellulose can be further classified on the basis of characteristics, shapes and sources are presented below.

Cellulose Nanocrystals (CNCs)
TEMPO oxidation and subsequent cavitation is utilized for the acid-free preparation of CNCs [9]. Microcrystalline cellulose (MCC) subjected to NaOH/urea treatment which further followed by regeneration and sonication to get cellulose nanocrystals [10]. Industrial agricultural waste has been reported as the sources for the production of cellulose nanocrystals [11]. Moreover, various bio-wastes including wheat straw, rice straw, potato pulp, banana fibres, sugar beet, yellow pea cotton, algae, chitin, and waxy maize are also reported sources of the production of CNCs [12–19].

Cellulose Nanofibrils (CNFs)
CNFs are prepared from Softwood, Hardwood, and Tunicate [20]. Sludge materials from Paper mills industries including dried and never dried papers are utilized for the preparation of cellulose nanofiber [21]. High pressure homogenization

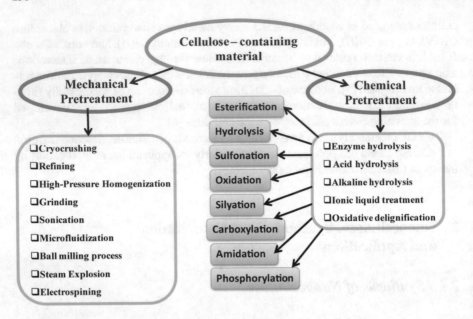

Fig. 2 Some Pre-treatment methodologies for the synthesis of nanocellulose

method were used for production of CNFs as network structure with widths varying from 10 to 50 nm. This method is not good in terms of energy consumption as it requires high amount of energy [22, 23]. Carboxymethylation [24, 25], phosphorylation [26], acetylation, and $NaClO_2$-oxidation followed by periodate oxidation [21], Pre-treatment of cellulose fibers derived from plants to efficiently prepare new nanocellulose using catalytic oxidation with 2,2,6,6-tetramethylpiperidine-1-oxyl radical which is considered as TEMPO reagent under aqueous conditions has been developed, this methodology is widely accepted for cellulose treatments to prepare nanocellulose [21, 27–29].

Microfibrillated Cellulose (MFC)

Microfibrillated cellulose (MFC) can be synthesized with the methods reported herewith. The micro cellulose fibrils and strong gel were produced by enzyme treatment collectively with mechanical shearing and high-pressure homogenization [30]. Wood source has been utilised for the synthesis of composite of nanocellulose polypyrrole and microfibrilillated cellulose [31]. Three different precursor materials such as China cotton, South African cotton and waste tissue papers are employed for the facile synthesis and characterization of conventional nano-cellulose like nano fibrils, nano-whiskers, microfibrillated cellulose by acid hydrolysis route [32]. An environmentally friendly method is applied for the production of enzyme-assisted preparation of microfibrillated cellulose (MFC) nanofibers [33]. High shear refining and cryocrushing methods are utilize for the synthesis of MFC [34].

Cellulose Nanowhiskers (CNWs)

Melt processed CNWs-filled polyactide-based nanocomposites are produced by interfacial ring opening polymerization [35]. CNWs are also obtained from waste recycling of paper industry [36] Coconut husk fibers [37], cotton fibres by controlled microbial hydrolysis [38] are employed for the preparation of CNWs.

Bacterial Nanocellulose (BNC)

Bacterial nanocellulose is produced from cheap stock of wheat straw [39]. BNC also have been produced from CI pre-treated waste cotton fabrics [40]. Enzymatic treatment methods like Gluconacetobacter xylinus for saccharification of dissolution pre-treated waste cellulosic fabrics have been effectively used for BNC synthesis [41].

2.2 Functionalization

The functional groups can be induced on the surface of the nanocellulose particularly –OH groups through some functionalization methods viz; phosphorylation, sulfonation, esterification, etherification, carboxylation, amination and oxidation which cause the surface of charge of cellulose. It is further examined that the introducing such functional groups to cellulose surface now will acts as binding functional groups such as carboxyl, sulfonate, and phosphonate groups, can provide the large number of capturing sites to adsorb the pollutant. The various functionalities attached with the nanocellulose biopolymer are schemed in Fig. 3.

The cellulose paper has been converted to the phosphorylated cellulose. The performance of these phosphorylated nanopaper is examined for copper as heavy metal ion. The result of reported nanopaper shows capability of eradicating copper ions during a process of filtration by means of adsorption [42].

Fig. 3 The scheme for different chemical functionalization for modification of cellulose surface

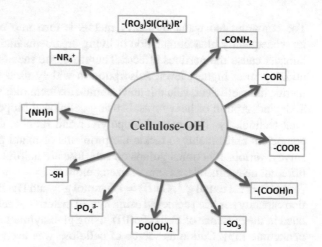

Tetraethoxysilane combined with cellulose acetate to form composite material for the adsorption purpose of Cr(VI) by alkalization process. The prepared composite membrane shows the good rejection activity for Cr(VI) [43].

Fe(III)-coordinated amino-functionalized poly(glycidylmethacrylate)-grafted TiO_2-densified cellulose is utilized for the removal of As(V) from polluted water [44]. The carboxylation of the waste pulp residue from paper industries carried out by means of TEMPO-mediated oxidation and modified carboxylated substrates has been used to adsorption of Cu(II), Ni(II) [45]. It is also examined the chemical modification of nanocellulose is achieved using carboxylic acids [46].

Liu et al. reported the comparative study of sulfonated CNC with sulfonated CFCs. They found that, Sulfonated CNC have greater adsorption capability (34 mg/g) for Ag(I) compared to its counterpart i.e. sulfonated CNC (14 mg/g) [45]. Sulfonated wheat pulp nanocellulose is suited for the Lead adsorption [47].

The removal of Cu(II), Ni(II) and Cd(II) ions from aqueous solution are successfully removed by using microfibrilillated cellulose which further modified by amino propyl triethoxy silane improved [48]. Gurgel et al. reported the adsorption of Cr(VI) using the cellulose functionalized methyl-iodide and triethylenetetramine [49]. P-aminobenzoic groups modified by cellulose is utilized for the removal of various heavy metals [50]. Amine functionalized cellulose which was further pretreated with microwave-H_2O_2 used for heavy metal ions adsorption [51].

Esterification cellulose has been performed to remove range of heavy metals including Zn, Pb, Cd, Cu etc. Various ratios of CNF and citric acid has been studies to get perfect esterification of CNFs. It's found that 1:1 ratio of CNF to citric acid shows greater adsorption capacities [52].

3 Functionalized Nanocellulose for the Adsorption of Heavy Metals from Contaminated Water

The contaminated water by heavy metals is turn into environmental disputes as its subsequently bioaccumulation in living organisms including plants, animals and humans cause the perilous effects. Therefore, the metal removal technologies are utilizing in at highest level. Adsorption is widely used techniques because of its merits, its an efficient, valuable and trouble-free technology for heavy metal removal. Today adsorption of heavy metals using cellulosed biopolymers has been become vital technology. This natural biopolymer and its chemically modified forms are competent constituents to tackle the problems of metal poisoning. From literature survey, various functional groups on cellulose are helpful for the remediation of the different heavy metals, as summarized below.

Ag(I) = 1.2 mmol g^{-1}, Cu(II) = 1.5 mmol g^{-1}, and Hg(II) = 2 mmol g^{-1} of adsorption capacity has been achieved using hybrid material of cellulose. CNF were precipitated in the presence of Fe(II)/Fe(III), using glycidylmethacrylate and tetraethylenepentamine [53]. Cotton as source of cellulose was used which first hydrolysed to

Table 1 Functional groups assisted different nano cellulose for Cu(II) adsorption

Sr. no	Adsorbent	Functional groups	References
1	Cellulose	Linked 8-hydroxy quinolone	[58]
2	Cellulose	TEMPO oxidation, PEI grafting	[59]
3	Cellulose nanofibrils	Mechanical disintegration	[60]
4	Cellulose Paper	PEI cross linked	[61]
5	Cellulose paper	Amidoxime	[62]
6	Cellulose	Triethylenetetramine	[63]
7	Cellulose from wood pulp	Citric acid (carboxyl)	[64]

achieve CNCs. CNCs were then termed as SCNCs after chemically modified with succinic anhydride. The sodic Nano adsorbent (NaSCNCs) was further prepared by treatment of SCNCs with saturated $NaHCO_3$ aqueous solution. The prepared adsorbent used for Cd and Pb removal [54].

The cellulose/chitin beads can adsorb efficiently Pb^{2+}, Cd^{2+} and Cu^{2+} ions, with uptake capacities are 0.33 mmol/g, 0.32 mmol/g and 0.30 mmol/g, respectively [55]. The succinic anhydride modified mercerized nanocellulose were used for removal of Zn(II), Ni(II), Cu(II), Co(II), and Cd(II) ions [56]. In-situ TEMPO surface functionalization of nanocellulose membranes are accountable for the improved removal capability of Ag(I), Cu(II)/Fe(III)/Fe(II) ions through contaminated waste water [57].

Moreover, concerning with different functionalities on nanocellulose surface are effective for the adsorption/removal of Cu(II) ions, as listed according to the literature in the Table 1.

4 Concluding Remarks

Cellulose, the most abundant biopolymer present in nature can be efficiently can be prepare and it can play a major role to treat waste water. Research herewith studied shows the cellulose can be prepared easily with mechanical as well as chemical processes. The prepared cellulose can be easily can modify for its specific adsorption of heavy metal from waste sources. The future research work will be the applicability of cellulose based materials in actual commercial way.

References

1. Blanchard, G., Maunaye, M., & Martin, G. (1984). Removal of heavy metals from waters by means of natural zeolites. *Water Research, 18*(12), 1501–1507.
2. Uslu, H., Yankov, D., Isewar, K. L., Azizian, S., Ullah, N., & Ahmad, W. (2015). Separation of organic and inorganic compounds for specific applications. *Journal of Chemistry, 2015.*

3. Järup, L. (2003). Hazards of heavy metal contamination. *British Medical Bulletin*, *68*(1), 167–182.
4. IARC Working Group on the Evaluation of Carcinogenic Risks to Humans. (2006). Inorganic and organic lead compounds. In *IARC Monographs on the Evaluation of Carcinogenic Risks to Humans* (Vol. 87, p. 1).
5. Ottenhall, A., Henschen, J., Illergård, J., & Ek, M. (2018). Cellulose-based water purification using paper filters modified with polyelectrolyte multilayers to remove bacteria from water through electrostatic interactions. *Environmental Science: Water Research & Technology*, *4*(12), 2070–2079.
6. Bethke, K., Palantöken, S., Andrei, V., Roß, M., Raghuwanshi, V. S., Kettemann, F., et al. (2018). Functionalized cellulose for water purification, antimicrobial applications, and sensors. *Advanced Functional Materials*, *28*(23), 1800409.
7. Islam, M. T., Alam, M. M., Patrucco, A., Montarsolo, A., & Zoccola, M. (2014). Preparation of nanocellulose: A review. *AATCC Journal of Research*, *1*(5), 17–23.
8. Chakraborty, A., Sain, M., & Kortschot, M. (2005). Cellulose microfibrils: A novel method of preparation using high shear refining and cryocrushing. *Holzforschung*, *59*(1), 102–107; Zhou, Y., Saito, T., Bergström, L., & Isogai, A. (2018). Acid-free preparation of cellulose nanocrystals by TEMPO oxidation and subsequent cavitation. *Biomacromolecules*, *19*(2), 633–639.
9. Shankar, S., & Rhim, J. W. (2016). Preparation of nanocellulose from micro-crystalline cellulose: The effect on the performance and properties of agar-based composite films. *Carbohydrate Polymers*, *135*, 18–26.
10. Mazlita, Y., Lee, H. V., & Hamid, S. B. A. (2016). Preparation of cellulose nanocrystals biopolymer from agro-industrial wastes: Separation and characterization. *Polymers and Polymer Composites*, *24*(9), 719–728.
11. Dufresne, A., Cavaille, J. Y., & Helbert, W. (1997). Thermoplastic nanocomposites filled with wheat straw cellulose whiskers. Part II: Effect of processing and modeling. *Polymer Composites*, *18*, 198–210. https://doi.org/10.1002/pc.10274.
12. Lu, P., & Hsieh, Y.-L. (2012). Preparation and characterization of cellulose nanocrystals from rice straw. *Carbohydrate Polymers*, *87*, 564–573. https://doi.org/10.1016/j.carbpol.2011.08.022.
13. Heux, L., Chauve, G., & Bonini, C. (2000). Nonflocculating and chiral-nematic self-ordering of cellulose microcrystals suspensions in nonpolar solvents. *Langmuir*, *16*(21), 8210–8212.
14. Imai, T., Boisset, C., Samejima, M., Igarashi, K., & Sugiyama, J. (1998). Unidirectional processive action of cellobiohydrolase Cel7A on Valonia cellulose microcrystals. *FEBS Letters*, *432*, 113–116. https://doi.org/10.1016/S0014-5793(98)00845-X.
15. Nair, K. G., Dufresne, A., Gandini, A., & Belgacem, M. N. (2003). Crab shell Chitin whiskers rein-forced natural rubber nanocomposites. 3. Effect of chemical modification of chitin whiskers. *Biomacromolecules*, *4*, 1835–1842. https://doi.org/10.1021/bm030058g.
16. Dufresne, A., & Cavaillé, J.-Y. (1998). Clustering and percolation effects in microcrystalline starch-reinforced thermoplastic. *Journal of Polymer Science: Part B: Polymer Physics*, *36*, 2211–2224. https://doi.org/10.1002/(SICI)1099-0488(19980915)36:12<2211:AIDPOLB18>3.0.CO;2-2.
17. Dubief, D., Samain, E., & Dufresne, A. (1999). Polysaccharide microcrystals reinforced amorphous poly(â-hydroxyoctanoate) nanocomposite materials. *Macromolecules*, *32*, 5765–5771. https://doi.org/10.1021/ma990274a.
18. Angellier, H., Molina-Boisseau, S., Lebrun, L., & Dufresne, A. (2005). Processing and structural properties of waxy maize starch nanocrystals reinforced natural rubber. *Macro-molecules*, *38*, 3783–3792. https://doi.org/10.1021/la047530j.
19. Zhao, Y., Moser, C., Lindström, M. E., Henriksson, G., & Li, J. (2017). Cellulose nanofibers from softwood, hardwood, and tunicate: Preparation–structure–film performance interrelation. *ACS Applied Materials & Interfaces*, *9*(15), 13508–13519.
20. Adu, C., Berglund, L., Oksman, K., Eichhorn, S. J., Jolly, M., & Zhu, C. (2018). Properties of cellulose nanofibre networks prepared from never-dried and dried paper mill sludge. *Journal of Cleaner Production*, *197*, 765–771.

21. Isogai, A., Saito, T., & Fukuzumi, H. (2011). TEMPO-oxidized cellulose nanofibers. *Nanoscale, 3*(1), 71–85.
22. Klemm, D., Kramer, F., Moritz, S., Lindström, T., Ankerfors, M., Gray, D., et al. (2011). Nanocelluloses: A new family of nature-based materials. *Angewandte Chemie International Edition, 50*(24), 5438–5466.
23. Isogai, A. (2013). Wood nanocelluloses: Fundamentals and applications as new bio-based nanomaterials. *Journal of Wood Science, 59*(6), 449–459.
24. Fall, A. B., Lindström, S. B., Sundman, O., Ödberg, L., & Wågberg, L. (2011). Colloidal stability of aqueous nanofibrillated cellulose dispersions. *Langmuir, 27*, 11332–11338.
25. Noguchi, Y., Homma, I., & Matsubara, Y. (2017). Complete nanofibrillation of cellulose prepared by phosphorylation. *Cellulose, 24*(3), 1295–1305.
26. Yang, H., Chen, D., & van de Ven, T. G. (2015). Preparation and characterization of sterically stabilized nanocrystalline cellulose obtained by periodate oxidation of cellulose fibers. *Cellulose, 22*(3), 1743–1752.
27. Saito, T., Nishiyama, Y., Putaux, J. L., Vignon, M., & Isogai, A. (2006). Homogeneous suspensions of individualized microfibrils from TEMPO-catalyzed oxidation of native cellulose. *Biomacromolecules, 7*, 1687–1691.
28. Saito, T., Kimura, S., Nishiyama, Y., & Isogai, A. (2007). Cellulose nanofibers prepared by TEMPO-mediated oxidation of native cellulose. *Biomacromolecules, 8*(8), 2485–2491.
29. Shinoda, R., Saito, T., Okita, Y., & Isogai, A. (2012). Relationship between length and degree of polymerization of TEMPO-oxidized cellulose nanofibrils. *Biomacromolecules, 13*(3), 842–849.
30. Pääkkö, M., Ankerfors, M., Kosonen, H., Nykänen, A., Ahola, S., Österberg, M., et al. (2007). Enzymatic hydrolysis combined with mechanical shearing and high-pressure homogenization for nanoscale cellulose fibrils and strong gels. *Biomacromolecules, 8*(6), 1934–1941.
31. Nyström, G., Mihranyan, A., Razaq, A., Lindström, T., Nyholm, L., & Strømme, M. (2010). A nanocellulose polypyrrole composite based on microfibrillated cellulose from wood. *The Journal of Physical Chemistry B, 114*(12), 4178–4182.
32. Maiti, S., Jayaramudu, J., Das, K., Reddy, S. M., Sadiku, R., Ray, S. S., et al. (2013). Preparation and characterization of nano-cellulose with new shape from different precursor. *Carbohydrate polymers, 98*(1), 562–567.
33. Henriksson, M., Henriksson, G., Berglund, L. A., & Lindström, T. (2007). An environmentally friendly method for enzyme-assisted preparation of microfibrillated cellulose (MFC) nanofibers. *European Polymer Journal, 43*(8), 3434–3441.
34. Chakraborty, A., Sain, M., & Kortschot, M. (2005). Cellulose microfibrils: A novel method of preparation using high shear refining and cryocrushing. *Holzforschung, 59*(1), 102–107.
35. Goffin, A. L., Raquez, J. M., Duquesne, E., Siqueira, G., Habibi, Y., Dufresne, A., et al. (2011). From interfacial ring-opening polymerization to melt processing of cellulose nanowhisker-filled polylactide-based nanocomposites. *Biomacromolecules, 12*(7), 2456–2465.
36. Souza, A. G., Rocha, D. B., & Rosa, D. S. (2017). Cellulose nanowhiskers obtained from waste recycling of paper industry. In *Materials design and applications* (pp. 101–111). Cham: Springer.
37. Rosa, M. F., Medeiros, E. S., Malmonge, J. A., Gregorski, K. S., Wood, D. F., Mattoso, L. H. C., et al. (2010). Cellulose nanowhiskers from coconut husk fibers: Effect of preparation conditions on their thermal and morphological behavior. *Carbohydrate Polymers, 81*(1), 83–92.
38. Satyamurthy, P., Jain, P., Balasubramanya, R. H., & Vigneshwaran, N. (2011). Preparation and characterization of cellulose nanowhiskers from cotton fibres by controlled microbial hydrolysis. *Carbohydrate Polymers, 83*(1), 122–129.
39. Hong, F., Zhu, Y. X., Yang, G., & Yang, X. X. (2011). Wheat straw acid hydrolysate as a potential cost-effective feedstock for production of bacterial cellulose. *Journal of Chemical Technology & Biotechnology, 86*(5), 675–680.
40. Guo, X., Chen, L., Tang, J., Jönsson, L. J., & Hong, F. F. (2016). Production of bacterial nanocellulose and enzyme from [AMIM] Cl-pretreated waste cotton fabrics: Effects of dyes on enzymatic saccharification and nanocellulose production. *Journal of Chemical Technology & Biotechnology, 91*(5), 1413–1421.

41. Kuo, C. H., Lin, P. J., & Lee, C. K. (2010). Enzymatic saccharification of dissolution pre-treated waste cellulosic fabrics for bacterial cellulose production by Gluconacetobacter xylinus. *Journal of Chemical Technology & Biotechnology, 85*(10), 1346–1352.
42. Mautner, A., Maples, H. A., Kobkeatthawin, T., Kokol, V., Karim, Z., Li, K., et al. (2016). Phosphorylated nanocellulose papers for copper adsorption from aqueous solutions. *International Journal of Environmental Science and Technology, 13*(8), 1861–1872.
43. Taha, A. A., Wu, Y. N., Wang, H., & Li, F. (2012). *Journal of Environmental Management, 112*, 10–16. https://doi.org/10.1016/j.jenvman.2012.05.031.
44. Anirudhan, T. S., Divya, L., & Parvathy, J. (2013). *Journal of Chemical Technology and Biotechnology, 88*(5), 878–886.
45. Liu, P., Sehaqui, H., Tingaut, P., Wichser, A., Oksman, K., & Mathew, A. P. (2014). *Cellulose, 21*(1), 449–461.
46. Espino-Pérez, E., Domenek, S., Belgacem, N., Sillard, C., & Bras, J. (2014). Green process for chemical functionalization of nanocellulose with carboxylic acids. *Biomacromolecules, 15*(12), 4551–4560.
47. Suopajärvi, T., Liimatainen, H., Karjalainen, M., Upola, H., & Niinimäki, J. (2015). Lead adsorption with sulfonated wheat pulp nanocelluloses. *Journal of Water Process Engineering, 5*, 136–142.
48. Hokkanen, S., Repo, E., Suopajärvi, T., Liimatainen, H., Niinimaa, J., & Sillanpää, M. (2014). Adsorption of Ni(II), Cu(II) and Cd(II) from aqueous solutions by amino modified nanostructured microfibrillated cellulose. *Cellulose, 21*(3), 1471–1487.
49. Gurgel, L. V., Perin de Melo, J. C., de Lena, J. C., & Gil, L. F. (2009). *Bioresource Technology, 100*(13), 3214–3220. https://doi.org/10.1016/j.biortech.2009.01.068.
50. Castro, G. R. d., Alcantara, I. L. d., Roldan, P. d. S., Bozano, D. d. F., Padilha, P. d. M., Florentino, A. d. O. et al. (2004). *Journal of Materials Research 7*(2), 329–334.
51. Zhang, C., Su, J., Zhu, H., Xiong, J., Liu, X., Li, D., et al. (2017). The removal of heavy metal ions from aqueous solutions by amine functionalized cellulose pretreated with microwave-H_2O_2. *RSC Advances, 7*(54), 34182–34191.
52. Madivoli, E.S., Kareru, P.G., Gachanja, A.N., Mugo, S., Murigi, M.K., Kairigo, P.K., Kipyegon, C., Mutembei, J.K. and Njonge, F.K., 2016. Adsorption of selected heavy metals on modified nano cellulose. *International Research Journal of Pure and Applied Chemistry*, pp.1–9.
53. Donia, A. M., Atia, A. A., & Abouzayed, F. I. (2012). *Chemical Engineering Journal, 191*, 22–30. https://doi.org/10.1016/j.cej.2011.08.034.
54. Yu, X., Tong, S., Ge, M., Wu, L., Zuo, J., Cao, C., et al. (2013). Adsorption of heavy metal ions from aqueous solution by carboxylated cellulose nanocrystals. *Journal of Environmental Sciences, 25*(5), 933–943.
55. Zhou, D., Zhang, L., Zhou, J., & Guo, S. (2004). Cellulose/chitin beads for adsorption of heavy metals in aqueous solution. *Water Research, 38*(11), 2643–2650.
56. Hokkanen, S., Repo, E., & Sillanpää, M. (2013). Removal of heavy metals from aqueous solutions by succinic anhydride modified mercerized nanocellulose. *Chemical Engineering Journal, 223*, 40–47.
57. Karim, Z., Hakalahti, M., Tammelin, T., & Mathew, A. P. (2017). In situ TEMPO surface functionalization of nanocellulose membranes for enhanced adsorption of metal ions from aqueous medium. *RSC Advances, 7*(9), 5232–5241.
58. Gurnani, V., Singh, A. K., & Venkataramani, B. (2003). Cellulose functionalized with 8-hydroxyquinoline: New method of synthesis and applications as a solid phase extractant in the determination of metal ions by flame atomic absorption spectrometry. *Analytica Chimica Acta, 485*(2), 221–232.
59. Zhang, N., Zang, G. L., Shi, C., Yu, H. Q., & Sheng, G. P. (2016). A novel adsorbent TEMPO-mediated oxidized cellulose nanofibrils modified with PEI: Preparation, characterization, and application for Cu(II) removal. *Journal of Hazardous Materials, 316*, 11–18.
60. Sehaqui, H., de Larraya, U. P., Liu, P., Pfenninger, N., Mathew, A. P., Zimmermann, T., & Tingaut, P. (2014). Enhancing adsorption of heavy metal ions onto biobased nanofibers from waste pulp residues for application in wastewater treatment. *Cellulose, 21*(4), 2831–2844.

61. Setyono, D., & Valiyaveettil, S. (2016). Functionalized paper—A readily accessible adsorbent for removal of dissolved heavy metal salts and nanoparticles from water. *Journal of Hazardous Materials, 302*, 120–128.
62. Saliba, R., Gauthier, H., & Gauthier, R. (2005). Adsorption of heavy metal ions on virgin and chemically-modified lignocellulosic materials. *Adsorption Science & Technology, 23*(4), 313–322.
63. Gurgel, L. V. A., & Gil, L. F. (2009). Adsorption of Cu(II), Cd(II) and Pb(II) from aqueous single metal solutions by succinylated twice-mercerized sugarcane bagasse functionalized with triethylenetetramine. *Water Research, 43*(18), 4479–4488.
64. Low, K. S., Lee, C. K., & Mak, S. M. (2004). Sorption of copper and lead by citric acid modified wood. *Wood Science and Technology, 38*(8), 629–640.

Arundo Donax Fibers as Green Materials for Oil Spill Recovery

Luigi Calabrese, Elpida Piperopoulos and Vincenzo Fiore

Abstract Oil spillage is considered one of the most devastating forms of pollution, for its effect on the environment, particularly on aquatic life. This kind of disaster can impact in two ways, directly caused by the polluting spilled oil or due to the cleanup process. In fact, oil floating on water does not allow sunlight to pass through and its toxicity puts the life of aquatic animals at risk. Furthermore, other factors can also contribute to this damage. In fact, a wrong oil recovery system can add a further pollution level. Polymer sorbents used for the oil spill recovery, if not properly treated, increase the level of marine and ground pollution. For this reason, in the last years, green materials are increasingly studied and used for this purpose. Green adsorbents (such as lignocellulosic, fruits fibers) are recently employed with excellent results. Aim of this book chapter is the evaluation of the oil sorption properties of natural fibers extracted from the stem of the giant reed *Arundo Donax* L., a perennial rhizomatous grass belonging to the Poaceae family that grows naturally all around the world thanks to its ability to tolerate different climatic conditions.

1 Introduction

Oil represents one of the most important raw material sources for synthetic polymers and chemicals worldwide. Consequently, there is a high risk of oil spillage whenever oil is explored, transported and stored or its derivatives are used, thus leading to significant environmental impact [1]. Oil spills have dramatic concern cause of the adverse impact on economic, social and ecological systems. In the last years several efforts were focused in order to identify an optimal cleaning-up technology able to give high effectiveness and reliability.

L. Calabrese (✉) · E. Piperopoulos
Department of Engineering, University of Messina, Contrada di Dio, Sant'Agata, 98166 Messina, Italy
e-mail: lcalabrese@unime.it

V. Fiore
Department of Engineering, University of Palermo, Viale delle Scienze, Edificio 6, 90128 Palermo, Italy

© Springer Nature Switzerland AG 2020
A. Khan et al. (eds.), *Biofibers and Biopolymers for Biocomposites*,
https://doi.org/10.1007/978-3-030-40301-0_13

Aim of this book chapter is the evaluation of the oil sorption properties of natural fibers extracted from the stem of the giant reed *Arundo Donax* L., a perennial rhizomatous grass belonging to the *Poaceae* family that grows naturally all around the world thanks to its ability to tolerate different climatic conditions. It was widely introduced by humans as ornamental plant to many countries surrounding the Mediterranean basin (even in the urban areas) but it has since became an invasive and aggressive species that displaces autochthonous plant communities in a number of riparian habitats [2, 3]. Its ability of adapting a wide variety of ecological conditions can be mainly related to the development of coarse, drought-resistant rhizomes, and deeply penetrating roots [4]. In particular, its deep and permanent roots, able to extend as much as 1.2 m downwards, play a key role in the prevention of soil erosion in addition to perform their primary task of insuring efficient water absorption [4].

A further feature of *Arundo Donax* giant reed is the high growth rate even if it can be slow down by water shortage [5]. In more detail, established rhizomes show growth rates of about 6.25 cm/day in the first 40 days and about 2.67 cm/day in the first 150 days. Furthermore, under optimal conditions (i.e., cultivation) *Arundo Donax* L. plant is reported to grow in the range from 4 to 10 cm/day [6].

Arundo Donax L. plant shows a wide range of applications. Its stem (or culm) is still today used in the production of musical woodwind instruments: i.e., no satisfactory substitutes have been developed. In particular, optimum playing conditions of woodwind instruments using *Arundo Donax* appear after some time of usage due to the high content of water-soluble extractives, mainly of glucose, fructose, and sucrose that influence their acoustic properties [7, 8]. Furthermore, the giant reed stem is often used to make fences, trellises, stakes for plants, windbreaks, sun shelters [9], as source of fibers for printing paper [10], as a diuretic and as a source of biomass for chemical feedstocks and for energy production. In particular, Papazoglou et al. [11] evaluated the grown of *Arundo Donax* L. on surface soil and irrigated with solutions of mixed heavy metals (i.e., cadmium and nickel) to study the impact of these heavy metals on its growth and photosynthesis. From the experimental results, it was evidenced that *Arundo Donax* L. can be cultivated in contaminated soils to provide biomass for energy production purposes, thus representing very promising energy plants.

Moreover, this non-wood plant has been also considered for the production of chipboard panels alternative to the wood-based ones. In this context, it was shown by Flores et al. [12] that it is possible to produce panels based on *Arundo Donax* particles (using urea formaldehyde resin as matrix) suitable for indoor use, since they reach the minimum structural parameters for that use (i.e., acceptable resistance compared to the wood panels available on the market). Fiore et al. [13] extracted fibers from the outer part of the plant stem in order to evaluate their usefulness as reinforcement in polymer based composites. To this aim, the morphology, the chemical composition, the physical and mechanical properties of extracted fibers were investigated. The promising results allowed to state that *Arundo Donax* fibers represent a valid alternative to other common natural fibers as reinforcing phase of composite materials. Starting from these preliminary study, the same Authors used the extracted fiber as reinforcement of both epoxy resin [14] and poly (lactic acid) [15]. Ismail and Jaeel [16] evaluated the partial replacement of sand in concrete mixes by using giant reed ash and air-dried giant reed fibers, thus evidencing that increments of the compressive strength (up to 7.96% by replacing 7.5 wt% of sand with giant reed ash) were achieved. In a most recent paper [17], the effects of plasma treatment both on the properties of fibers extracted from the leaves of *Arundo Donax* L. plant and on their compatibility with a bio-based epoxy resin were evaluated. The mechanical characterization showed remarkable enhancements of the performances of the resulting composites, thus indicating that plasma treatment was able to improve the compatibility between *Arundo Donax* fibers and the epoxy matrix. However, the considerable industrial flexibility of this class of materials does not limit its application fields to structural composite materials, but suggests its potential use in more diversified contexts where the chemical-physical characteristics of the fibers are integrated to the mechanical properties. In such a context, the use of these fibers as oil spillage systems represents a stimulating and innovative challenge in the panorama of the current literature.

2 Advances on Green Materials for Oil Spill Recovery Technologies

An oil spill is the accidental loss or intentional release of petroleum hydrocarbons into an ecosystem, particularly the marine environment. An oil spill in water represents a significant environmental risk because it cannot be easily confined to the area where the contamination occurred. The oil spill can be carried by the wind, current or sea waves increasing oil transportation and weathering speed. Furthermore the spilled oil evaporates, it forms a large surface layer that widely disperses in the water, or submerges and accumulates in the seabed, implying contamination both in the marine fauna and flora.

The research of new high performing adsorbents for oil spill recovery is a hot topic today. Three main approaches can be identified for the clean-up oil spill: physical, chemical and biological methods [18]. Skimmers are stationary or mobile devices, that were used to remove floating and/or emulsified oil from the water surface [19]. In adverse environmental conditions, such as rough seawater, the chemical dispersant application is preferred. However, using chemical methods, seawater ecosystem may be affected [20]. Biological methods are based on addition of microbes, nutrients or oxygen to favor bacterial growth able to biodegrade the spilled oil [21].

The reliability of a technology compared to another one is related to the specific conditions of use that are identified where the spill oil occurred (position, type and volume of spilled oil, environmental conditions). Among the several techniques that can be identified in oil spill recovery, the physical method applying sorbent materials is considered to be one of the most efficient and high performing methods [22].

In such a context, the main aim is to find a material with high adsorbent capacity, superhydrophobic, reusable, low cost and recyclable, with a very low environmental impact, to simplify disposal procedures at the end of its life and optimize the costs-benefits ratio.

Three main categories of oil sorbents can be considered: i.e., synthetic polymers, mineral materials, and natural biomass-based oil sorbents. Synthetic polymers (i.e., mainly polypropylene, polyester and polyurethane fibers) are typically used as sorbents, thanks to their great efficiency in the oil removal from oil-water mixtures. Although, synthetic polymers as polypropylene [23], polyurethane [24], methacrylate [25] are considered ideal materials for oil spillage clean-up, due to their low density, good hydrophobicity and optimal physical and chemical resistance. Unfortunately, they are made from non-biodegradable and expensive oil by-products. Furthermore, often, polymers are employed to make hydrophobic some waste materials, to enhance their properties and give them a second life and consequently an added value. On the other hand, mineral materials (i.e., vermiculite, exfoliated graphite and fly ash) can be also used even if most of them are difficult to handle onsite with poor buoyancy and oil sorption capacity [26].

Due to the above drawbacks shown by synthetic polymers and mineral materials, an increasing attention was focused in the last decades toward biomass-based oil sorbents such as lignocellulosic fibers or particles. As a consequence, a quite wide literature describing oil removal by natural biomass-based oil sorbents can be found.

In this context, Annunciado et al. [27] investigated the use of vegetable fibers such as mixed leaves residues, mixed sawdust, sisal (Agave sisalana), coir fiber (Cocos nucifera), sponge-gourd (Luffa cylindrica) and silk-floss as sorbent materials of crude oil, showing that the latter evidenced highest hydrophobicity and oil sorption capacity. Adhithya et al. [28] evaluated the feasibility of bamboo fibers for separation of oil from water showing the great advantage in treating oil-water mixture with this kind of natural fibers: i.e., very high adsorption capacities were obtained for both vegetable and synthetic oils. Moreover, platanus fruit fibers having unique hollow tubular structures were successfully utilized in the preparation of an efficient oil sorbents by chemical modification with acetic anhydride by Yang et al. [26]. They showed that the hydrophobic modification of platanus fruit fibers greatly enhances their sorption capacity for various oils and organic solvents.

Liu et al. developed a cellulose-based oil adsorbent using the waste paper through a simple modification process [29]. They sprayed with polyethylene wax and alkyl ketene dimer the pulverized paper to increase the hydrophobicity of the raw material. With this process they reached a contact angle of 125.6° in water ad an adsorbing capacity of 16–28 times their own weight in oil. But these modifications may result in decreased biocompatibility and the resulting materials may be not renewable and biodegradable. Green materials are environmentally friendly and low cost. Furthermore they exhibit a low density, similar to polymeric materials and a competitive adsorption capacity. Wang et al. assessed a simple and green approach to produce a cost effective, ultralight, elastic, and highly recyclable superadsorbent by using renewable cellulose fibers through a simple and eco-friendly microfibrillation treatment and freeze-drying [30]. The as realized adsorbent reached 88–228 gg^{-1} oil

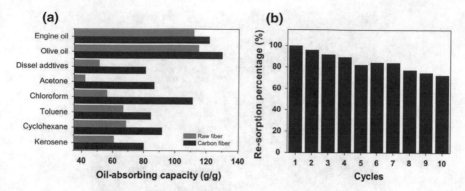

Fig. 1 Oil adsorption capacity (**a**) and reusability (**b**) of pyrolysed Calotropis gigantea fibers [31]

adsorption capacity, but what is worthy to note is that the material showed an excellent flexibility and elasticity, being reused for at least 30 adsorbing/squeezing cycles without changing its oil sorption capability. Tu et al. used carbon fibers derived from Calotropis gigantea pyrolysis process to obtain an eco-friendly and sustainable oil adsorbent [31]. The obtained material was able to adsorb up to $130 \, gg^{-1}$ of oil in few seconds and could be recovered by extracting the tested oils with absolute ethanol and could be reused for more than 10 cycles (Fig. 1).

Structured cattail fibers were employed by Cao et al. [32] to remove oils from runoff. The inner structural properties of the fibers were the reason of their good oil cyclic cleanup ($10.62 \, gg^{-1}$ of diesel oil). Superoleophilic and superhydrophobic carbon fibers obtained from sisal leaves alkalization, bleaching, freeze-drying and carbonization, were utilized by Liu et al. [33]. The sorbent material exhibited an excellent oil sorption ability (from 90 to 188 times of its own weight in different organic oils). Recently, porous carbon spheres derived by a simple carbonization of the Liquidambar formosana (Fig. 2a–c) were used for oil and organic solvent adsorption [34]. Due to their porous structure and hydrophobic nature the organic spheres could float (Fig. 2d) and showed a high rate of adsorption kinetics.

Furthermore the adsorbed oil could be removed by ethanol treatment and the spheres could be reused for several times. Further studies were focalized on the investigation of oil recovery properties of raw barley straw [35], kapok fiber [36] and raw bagasse [37].

In such a context, Wang et al. realized magnetically superhydrophobic oil sorbent by directly immobilization of Fe_2O_3 nanoparticles on kapok fibers surface [38] (Fig. 3).

This led to a water contact angle of 143° (Fig. 3b) also after several adsorbing cycles. The as modified kapok could increase by 30.5% its sorption capacity in diesel oil. In general, based on the low-cost and abundance of raw material obtained from nature, green materials result very attractive to be applied in oil spillage recovery field.

Fig. 2 Photos of Liquidambar formosana (**a**), before (**b**) and after (**c**) carbonization. The sphere can float on water (**d**) [34]

Fig. 3 Schematic representation of magnetically superhydrophobic kapok fibers realization process (**a**), wettability of water and toluene (**b**), mirror effect in water (**c**), magnetic properties (**d**) [38]

3 Fibers Preparation and Characterization

3.1 Arundo Donax *Fibers Preparation*

The *Arundo Donax* L., used in this reference experimental campaign on oil sorption characterization, was collected after flowering in a plantation in the area of Palermo (Sicily). After collecting the fresh plant, the stem, having outer and inner average diameters equal to 25 and 17 mm, was separated from the foliage. The stem was cut into small parts and dried at 103 °C for 24 h in an oven to remove all the moisture content, according to ASABE S358.3 (2012). After this phase, the fibers with a length between 100 and 160 mm were extracted from the stem by mechanical separation. In particular, the outer part of the culm was manually decorticated with the help of blades obtaining thin strips from which fibers were easily separated with the aid of a scalpel and a Leica optical microscope model MS5. Long fibers were chopped at different length in order to evaluate the effect of fiber geometry on the sorption performances on different oil pollutants. Therefore, three batches were prepared as function of fiber length of *Arundo Donax* fiber in the range of 2000–300 μm, coding them AD-L, AD-M and AD-S for large, medium and small fiber size, respectively.

3.2 *Sorption Capacity Experiment*

Sorption experiments were carried out by equilibrating AD fibers sample bags in 250 mL of different commercial oils at room temperature and under slow stirring. In particular, four commercial oils were used for the sorption experiments (i.e., with kerosene, virgin naphtha, pump oil and crude oil). For comparison sorption of distilled water was also performed. Some characteristics of selected oils are summarized in Table 1 (where ρ and μ are density and dynamic viscosity of the pollutants, respectively).

At increasing immersion time, the AD samples were accurately removed from the beaker. The impregnated bag was places in a watch glass, open to air for 30 s in order to release the residual pollutant and subsequently weighed. Furthermore, sorption measurements on empty bags were performed as reference and used to extrapolate the AD fiber sorption performances. The mass adsorption capacity was calculated as:

Table 1 Density (ρ) and the dynamic viscosity (μ) of selected oils

	Water	Virgin naphtha	Kerosene	Crude oil	Pump oil
ρ (kg/m^3)	1000	630	780	890	858
μ (Pa s)	0.0010	0.0012	0.0019	0.2710	0.1231

$$Q_t = \frac{m_t - m_0}{m_0} \tag{1}$$

where Q_t is sorption capacity of the fiber at sorption time t, m_0 and m_t are the initial weight (at instant $t = 0$ s) and the weight of AD sample after sorption time t (reduced by the bag contribute), respectively. The sorption measurements were performed up to weight stabilization. This value was identified as saturation sorption capacity.

4 Performance Evaluation and Characterization of *Arundo Donax* (AD) Fibers for Oil Spill Recovery Applications

4.1 *Morphology of* Arundo Donax *Fibers*

As other lignocellulosic fibers, the surface morphology of the *Arundo Donax* fibers is characterized by several elementary fibers (known as fibrils or fiber-cells) bonded each other along their length by pectin and other non-cellulosic compounds [13, 39], thus forming a fiber bundle.

Usually, their surface is characterized by impurities, typical of raw natural fibers. In order to eliminate these impurities and to increase the surface functional groups, natural fibers are usually chemically treated [13]. In Fig. 4a, the cross section of an Arundo fiber is shown. Vascular bundles and fiber-cells, with polygonal shape and a central hole, named lumen, can be clearly identified [13]. In particular, by analyzing the detail reported in Fig. 4b, it is possible to identify the morphology of fiber-cells structure. The AD fiber has a large porous structure with very wide channels with a hexagonal-type geometry along the preferential orientation of fibrils. However, the channel size is not regular with defects and heterogeneities commonly highlighted on the structure. The variability in diameter of fiber-cells and lumen has a great influence on the chemo-physical and mechanical properties of *Arundo Donax* fibers [39]. For oil spill recovery applications, a fundamental key factor is therefore to evaluate the morphology of the structure at the fiber size variable in order to be able to uniquely define the relationship between structure and sorption performances. The morphological peculiarity of a wide and regular internal porosity in AD fibers plays a fundamental role in the adsorbent performance with the different contaminating oils.

Arundo Donax fibers show a porous orthotropic morphology with lamellar structure that identifies a preferential orientation and several interconnected pores, as observed in Fig. 5. The analyzed samples belong to three different sizes. The AD-L has mean size higher than 500 μm and, as shown in Fig. 5a, the morphology is characterized by circular channels side by side, each formed by many communicating cells, like a beehive. AD-M mean size is in the range of 300–500 μm and the sample is constituted by smaller *Arundo Donax* particles, where decreased channels size and amount but still maintaining the characteristics of the base material (Fig. 5b).

Fig. 4 SEM micrographs of cross section of Arundo fibers at two different magnifications: **a** lower and **b** higher [13]

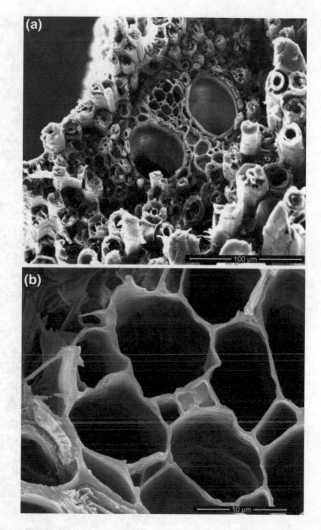

AD-S sample presents a mean size between 150 and 300 μm and the original beehive structure is completely destroyed (Fig. 5c). Owing to their channels morphology and subsequently to their low density, the *Arundo Donax* fibers float.

4.2 Sorption Performances

The sorption capacity, q_t (mg/g), evolution at increasing sorption time for all AD fiber batches at varying pollutants is shown in Fig. 6.

Fig. 5 SEM images of
a large **b** medium and **c** small
shaped *Arundo Donax* fibers

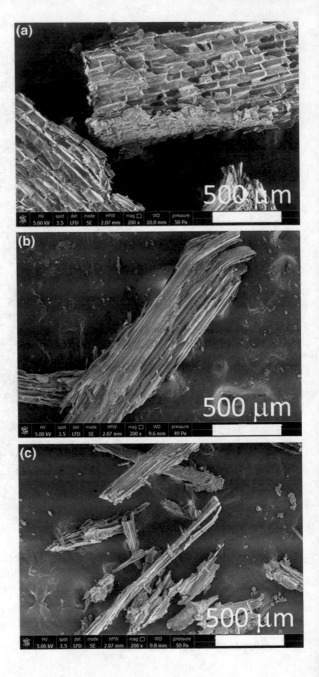

Fig. 6 Sorption capacity at increasing time for all pollutants of AD-L **a** AD-M **b** AD-S **c** samples

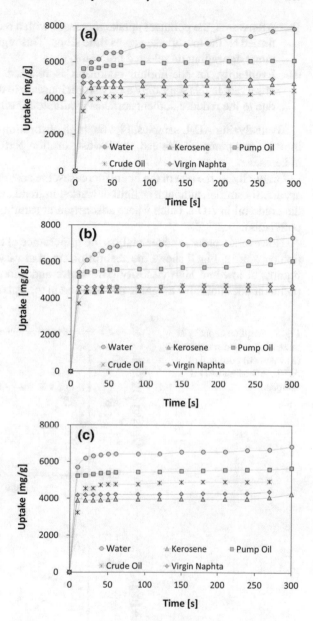

All curves highlight monotone trend with a progressive increase at increasing adsorption time until reaching a constant weight. In particular, three stages can be identified:

I.   At first, at low sorption time (range 0–30 s), the sorption capacity increases very quickly, indicating a high sorption rate.

II. Afterwards, the pollutant uptake increases with a reduced sorption rate as confirmed by the low uptake versus time slope. This region occurs at different time range depending on selected pollutant.

III. Eventually, an equilibrium condition is reached and an uptake plateau is observed. In this stage only a very slight increase in the sorption capacity occurs due to the reduced sorbent surfaces available for oil entrapment [40].

By analyzing AD-L samples, Fig. 6a, high absorption capabilities were observed in water and pump oil. Instead, the lowest sorption performances were highlighted in kerosene.

Gradually, as the size of the fibers increases, the adsorbent properties appear to be apparently similar, although a slight deviation in trend can be found (e.g., analyzing the crude oil in AD-L batch where adsorption at saturation is the lowest than other pollutants).

However, in order to better analyze the importance of fiber dimensions on absorption capacities, Fig. 7 shows the absorption trend at varying time for two reference liquids, at low and high viscosity (i.e., water and pump oil). While the pump oil pollutant does not show a significant variation in the uptake trend as a function of the

Fig. 7 Sorption capacity at increasing sorption time in (**a**) water (**b**) pump oil of AD-L, AD-M and AD-S samples

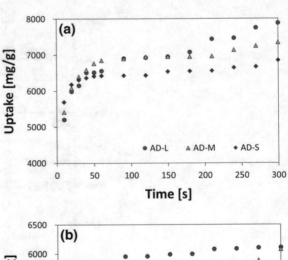

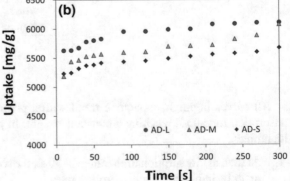

particle size (only a difference in uptake at saturation can be evidenced), it is worth noting that that in water several phenomena participate in the sorption process.

In particular, the initial phase is kinetically favored in small size *Arundo Donax* fibers. On the other hand, the uptake at saturation is very high for the large size fibers. In fact, for AD-M and mainly AD-L batches, show a very extensive phase II, in which a slow but progressive increase in weight can be observed. This trend cannot be identified for AD-S specimens.

By analyzing the pump oil trend there is a bilinear trend. The initial phase at short times is associated with the rapid absorption phenomena that take place due to a combined action of mass transfer (i.e., transport from the solution to the adsorbent surface) and intra-particle diffusion (i.e., diffusion toward surface pores). The second phase can be identified in correspondence of uptake plateau when the absorption equilibrium is reached. This phase is identifiable by a quite horizontal straight line indicating the independence of the pollutant uptake percentage from time. In fact, as confirmed by Wahi et al. [41] in its study on sago bark fiber waste for removal of oil from palm oil mill effluent (POME), a saturation step is reached when the adsorption-desorption process that occurs at saturated sorbent surfaces takes place (Fig. 8). After 30–40 min, adsorption capacity remains constant between 60 and 90% for the examined sago bark fibers samples. In this case, the oil removal efficiency is reduced or constant, because only small increase is observed due to the limited availability of sorbent surfaces, for oil entrapment. It is worth noting that AD fibers highlight a faster adsorption kinetics and almost compatible adsorption values as saturation, assessing a better costs-benefits ratio.

Conversely, using water as adsorbed there is a further intermediate step. This step extends over a relatively wide range (range 50–250 s). It may be due to the combined

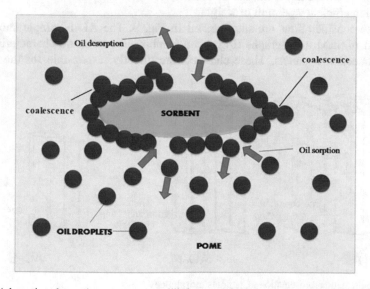

Fig. 8 Adsorption-desorption process at equilibrium step [41]

action of two competing mechanisms: film diffusion (diffusion of the liquid film from the sorbent surface toward the fiber bulk) and the surface interactions on the active sites.

The first stage is relatively fast due to the superficial absorption of the water and its diffusion on the superficial pores. The intermediate stage, instead, can be induced by the diffusion of the adsorbed in the internal and interconnected porosities of the adsorbent [42].

The first phase, linked to the interaction of the pollutant with the adsorbent, is strongly influenced by the surface area. In fact, the slope of the uptake curve versus time is steeper as smaller are the *Arundo Donax* fiber size. At the same time, phase II, which is the intermediate stage, can be associated with the diffusion of adsorbed within the micro-porosity of the *Arundo Donax* structure. The tortuosity of the fiber surface acts as active site for oil adsorption enhancing sorption efficiency [43]. As evidenced by the SEM images (Fig. 5) these interconnections are evident in large shaped fibers. The AD-S samples showed a very fine structure, comparable to the micro-fibrils size. Consequently, a large part of the porous channels in the fibrillar structure were damaged or destroyed.

The sorption mechanism is influenced by fibers morphology and the presence of channels structure, especially during the sorption of high viscose oils. Al-Din et al. [44] evidenced that the pores can make oil entrance into the sorbent internal parts easier and helpful in the sorption process, conducting a significant role in the sorption performance in oil spilling. Moreover, high mean size samples (and consequently low density samples for the channels presence) favor a higher and faster sorption of oils. As a consequence, the stage II for small fibers (AD-S) is very short and very rapidly the saturation is reached with a lower maximum adsorption level than AD-L batch for which the liquid has spread inside the porous structure, thus increasing the adsorption efficiency per unit of weight.

These considerations are summarized in Fig. 9. The AD-L sample shows, as reported in SEM micrographs (Fig. 5), a bamboo like structure characterized by interconnection channels. These channels are directly responsible for the higher

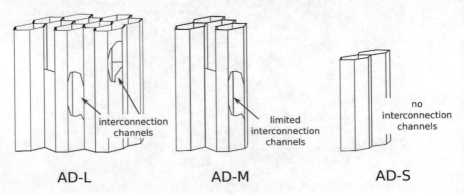

interconnection channels

limited interconnection channels

no interconnection channels

AD-L   AD-M   AD-S

Fig. 9 Schematic representation of samples morphology

Table 2 Sorption capacity and time at equilibrium parameters for all liquids and all investigated AD fiber batches

		Water	Kerosene	Virgin Naphta	Pump oil	Crude oil
AD-L	S_e (mg g^{-1})	7880 ± 394	4851 ± 243	5058 ± 253	6109 ± 305	4434 ± 222
	t_e (s)	290.0 ± 17.3	33.3 ± 5.8	20.0 ± 10	80.0 ± 17.3	46.7 ± 5.8
AD-M	S_e (mg g^{-1})	7343 ± 367	4611 ± 231	4697 ± 235	6076 ± 304	4844 ± 242
	t_e (s)	280.0 ± 17.3	23.3 ± 5.8	16.7 ± 5.8	53.3 ± 5.8	53.3 ± 5.8
AD-S	S_e (mg g^{-1})	6847 ± 342	4253 ± 213	4379 ± 219	5672 ± 284	5179 ± 259
	t_e (s)	270.0 ± 30.0	16.7 ± 5.8	13.3 ± 5.8	43.3 ± 5.8	56.7 ± 5.8

sorption capacity of the material. In fact, they may be filled by the adsorbed oil. In AD-M sample, the number of interconnection channels is limited and the adsorption is mainly conducted on the fibers surface. This totally occurs in the AD-S sample where the interconnection channels are completely absent.

Table 2 summarizes the equilibrium sorption capacity, S_e, at equilibrium time, t_e, for AD fiber batches at varying pollutants. It is worth of note that in water solution all fibers are able to obtain high sorption capacity with S_e value in the range 7880–6847 mg g^{-1} with the highest value for AD-L batch. At same time, very high equilibrium time can be highlighted. This result indicates a good interaction of *Arundo Donax* fibers with water, thanks to a surface hydrophilic behavior, but at the same time, as previously discussed, the intra-particellar diffusion of low viscosity liquid takes place at long sorption time. With oil pollutants the sorption capacity is lower than with water liquid even if a very fast saturation condition is reached. All samples showed an equilibrium time, t_e, in the range 13–80 s, significantly lower than that observed in water. In particular, all the investigated samples reveal a good affinity with pump oil, showing, among the analyzed oils, the highest sorption capacity. AD-L in particular reaches 6109 mg g^{-1} of saturation sorption capacity, approaching the result obtained in water even if in less than a third of the time. Conversely, AD-S, shows the highest equilibrium sorption value, among the other samples, in crude oil (5179 mg g^{-1}). Probably this behavior is due to the crude oil characteristics, i.e. highest density and viscosity, 890 kg m^{-3} and 0.27 Pa s, respectively [45], to respect the other polluting oils. For this reason, the diffusion of crude oil in interconnection channels results more difficult and the adsorption is mainly conducted on the surface of the fibers, thus favoring the sample with smaller size and fewer channels (AD-S) than those with more channels and larger size (AD-M and AD-L). On the contrary, for lighter and less viscous oils (i.e., virgin naphtha and kerosene) the diffusion into the interconnection channels is easier and the AD-L sample shows a higher equilibrium sorption capability. However, as the diffusion in entering in the samples channels is easy, its release is equally simple. As reported by Dong et al. [46] in the study on a highly efficient and recyclable depth filtrating system using structured kapok filters for oil removal and recovery from wastewater, three mechanism are described to understand the sorption capacity of kapok fibers (Fig. 10). The bad wettability and

Fig. 10 Separating process scheme of oil from water **a** bad wettability and adhesiveness between oil and fiber, **b** good wettability but low adhesiveness, **c** excellent wettability and adhesiveness [46]

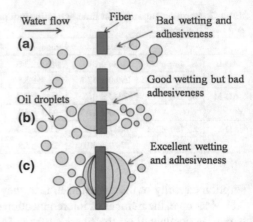

adhesiveness oil-fiber can lead to easy loss of oil drops from pores of fibrous filter (Fig. 10a). As schematized in Fig. 10b, if the oil wettability is good, oil droplets are attached on fiber, but easily distort and detach again due to the weak adhesiveness with fiber. Nevertheless, if the fiber present high wettability and excellent adhesion properties, a large number of oil droplets are captured and retained by the fiber (Fig. 10c). The interactions of diesel and vegetable oil with kapok fibers in Dong's research were just associated with cases in Fig. 10b, c, respectively. The low diesel adhesion with kapok fiber and its low viscous property cause it easily to deform, detach and transfer from kapok filter body to the water.

Further information can be acquired defining a parameter that relates the sorption performances with fiber size. In particular, an effect of sorbent size (ESS) parameter was defined as:

$$ESS_i[\%] = \frac{m_i - m_L}{m_L} \cdot 100$$

where m_L is the sorption capacity (mg g^{-1}) of the AD-L sample at a specific pollutant. m_i is the sorption capacity (mg g^{-1}) in the same pollutant of AD-M or AD-S samples, obtaining ESS$_M$ and ESS$_S$ index respectively. A negative ESS index indicates that the sorption capacity of large size of *Arundo Donax* fiber is higher than smaller one. At the same time, the higher or lower this parameter is, the greater the effect of the fiber size on the adsorption properties in the specific pollutant.

By analyzing the Fig. 11, that shows the evolution of the ESS$_M$ and ESS$_S$ indices at varying the pollutant viscosity, there is a monotonous trend with a gradual growth as the liquid viscosity increases. In particular for low viscosity pollutants, the index has a negative value and progressively increases until it becomes positive for liquids with viscosity higher than about 0.2 Pa s. This result indicates that in low viscosity pollutants the larger fibers have higher efficiency in terms of absorption capacity than the smaller one. Conversely, the higher the viscosity of the liquid, more small sized fibers play a fundamental role in having effective adsorbent performance. This

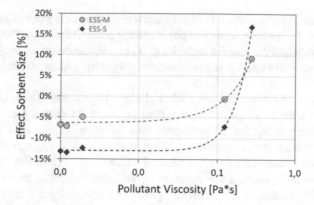

Fig. 11 Evolution of effect sorbent size index at varying pollutant viscosity

behavior can be associated with the microstructure of the fibers and the relative surface interaction with the liquid under examination. The presence of porosity and interconnected channels in the surface and bulk fibers, as in the case of AD-L batch, allows better sorbent properties in liquids with low viscosity. In such configuration, the mass flow towards the fiber bulk is favored and therefore, thanks to the low viscosity of the pollutant, all the internal channels can be imbibed increasing the sorption efficiency. In fact, the ESS_S index is ~15% compared to ESS_M that is approximately one-third lower (ESS_M ~ 5%) indicating a greater effectiveness in terms of adsorption the larger the fiber size.

For high pollutant viscosity, this liquid diffusion toward the inner part of the fiber is significantly limited. Therefore, the surface interaction is the driving force in the sorption phenomena. AD-S is characterized by small sized fiber with subsequent large surface area that allows a higher sorption capacity. Considering that micro channel diffusion of pollutant is avoided in high viscous liquids, AD-L samples showed not effective capabilities in these conditions. This is due, as previously discussed, to the low ability of large sized fibers to host the adsorbed pollutant oil in their porosity, reducing their sorption efficiency. In Fig. 12, the high selectivity towards oil of

Fig. 12 Oil adsorption steps in water of *Arundo Donax* fibers. **a** crude oil drops are poured into water, **b** sorbent fibers are put into the beaker, **c** sorbent fibers are taken out and the crude oil drops disappear, **d** the oil remain trapped into the sorbent material, **e** the water is cleaned up

Arundo Donax fibers is shown. In Fig. 12a, crude oil drops are poured into the water. Subsequently, the sorbent fibers are put into the beaker (Fig. 12b). What is worthy to note is that, when the sorbent material is taken out, the crude oil drops disappear (Fig. 12c), remaining trapped in the adsorbent (Fig. 12d) and the water is cleaned up (Fig. 12e).

So, although the *Arundo Donax* fibers have a high water adsorption capacity, the diffusion kinetics in oils is higher than in water, as shown from the shorter equilibrium time found for oil pollutants than for water (Table 2). Consequently, when fibers interact with a water-oil solution, there is a rapid impregnation of oil by fibers that inhibit the possible adsorption of water. As shown in the figures above, the AD fibers have enabled the water to be decontaminated in few seconds. The results indicate efficiency (in terms of speed) and selectivity (towards oil). These two conditions can represent an effective precondition for the use of AD fibers on industrial scale, thanks also to the cheapness of the product which can offset the adsorbing performance with respect to some new emerging materials [45, 47, 48].

4.3   Morphological and Structural Aspects of Oil Spill AD Materials

Morphological characteristics are fundamental to understand the adsorption process of adsorbent materials. Surface irregularities and porosity are required to promote oil adsorption and trapping. Baker et al. [49] confirm that the selectivity of a membrane for water treatment is related to the microstructure and chemistry of the membrane material. They discriminate the membranes class according to the pore size.

Figure 13 summarizes these considerations indicating for each specific fiber pore size, in the range from 1 to 3000 Å, the optimal filtration approach.

It is well known that porosity has a main role in the sorbent adsorption capacity. Piperopoulos et al. [47] examining carbon nanotubes sponges for oil recovery applications describe a sorption mechanism, represented in Fig. 14a, and comparing it in water. In Fig. 14a1, oil molecules diffuse from bulk liquid and wet the sorbent. At the same time, physical (van der Waals forces between adsorbate molecules and adsorbent surface) and/or chemical interactions take place (Fig. 14a2). Increasing sorption time (Fig. 14a3), the sorption rate due to the oil diffusion into sorbent micropores decreases. In water (Fig. 14b), the hydrophobic behavior of carbon nanotubes foam inhibits the sorption mechanism. After the mass transfer and the film diffusion (Fig. 14b1), water diffusion occurs with difficulty due to few hydrophilic active sites (Fig. 14b2), not filling micropores (Fig. 14b3).

Also Alaa El-Din et al. [44] attributes to irregular morphology and the porous surface of banana peels (Fig. 15) the good result obtained during adsorption tests. In fact, the pores can make able oil to enter into the internal parts of the material easier, thus favoring the sorption process.

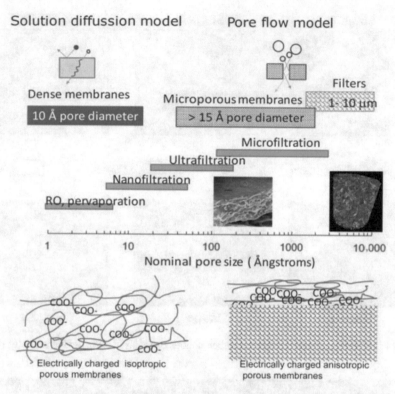

Fig. 13 Schematic illustration of membranes classification according to pore size [50]

Natural surface roughness of green adsorbent materials is an added value for this kind of application. Cui et al. [51] examine adsorption mechanism on cattail fiber assembly. They observe that oil sorption by cattail fibers can be explained by a combination of more mechanisms: i.e., oil adsorption by interactions with waxes on fiber surface and adsorption by physical trapping on the fiber irregular surface. Indeed, the rough surface and the open morphology of the cattail fibers provide the high surface area thus guarantying the oil trapping.

Based on the previous remarks and on the supposed approaches, more competing adsorption stages can be recognized:

I.   mass flow towards the fiber bulk (external mass transfer and film diffusion);
II. liquid diffusion toward the inner part (intraparticle diffusion).

For water adsorption there is a further intermediate stage:

III. surface interaction on active sites and film diffusion.

In this scheme, analyzing oil adsorption process in AD-L, intra-particle diffusion mechanism can be considered as the rate limiting step of the process instead external mass transfer is generally rapid. So in the stage 1, as represented in Fig. 16, oil molecules diffuse from bulk liquid and soak the AD sorbent. Subsequently, in the

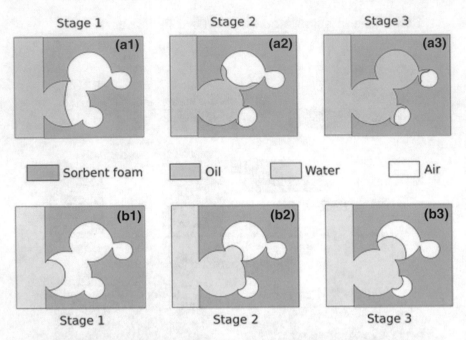

Fig. 14 Sorption mechanism stages for carbon nanotubes foams in oil (**a**) and in water (**b**) [47]

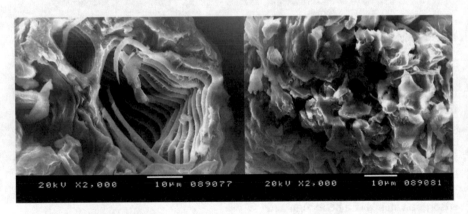

Fig. 15 SEM analysis of banana peel surface [44]

stage 2 the oil diffusion into sorbent micropores takes place (Fig. 16, Stage 2) increasing the oil sorption capacity of the fibers [42]. For AD-S sample the interconnection channels are almost absent, so the adsorption process is conducted only in stage 1 (Fig. 16) and saturation occurs more quickly (Table 2). In water, the first stage is fast, due to the superficial adsorption and the subsequent diffusion on the superficial pores. The intermediate stage is described by the water diffusion from the sorbent

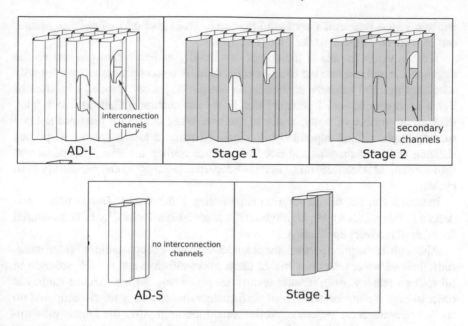

Fig. 16 Sorption mechanism for AD-L and AD-S samples

surface to the fiber bulk and the surface interactions on the active sites. These preliminary results evidenced good potential applicability of this type of green adsorbents in oil recovery application, indicating a suitable interaction with large common oil pollutants.

5 Conclusions and Future Trends

Recently most of the researchers working on oil recovery systems are moving to the employment of natural fibers. Zhang et al. [41] synthesized anisotropic bamboo-derived cellulose nanofibrils, improving shape recovery ratio of 92% after 100 cycles at 80% compression. The obtained material, due to its low density, high hydrophobicity and high compressive recoverability, can be used several times without any damaging. Whang et al. [52] remodeled raw cotton fibers into flexible and resistant macropourous cellulose aerogels with excellent oil trapping capability. The studied material can adsorb several oils and organic solvents, reaching up to 19.8–41.5 times its initial weight. Phantong et al. [53] fabricated a nanocellulose sponge by freeze-drying method, obtaining a superhydrophobic and superoleophilic material, that can be reused at least 10 cycles without modifying and can easily disposed of for its biodegradable characteristics. Furthermore, the use of waste materials to be used and revaluated for this purpose can be an added value. Tesfaye et al. [54] used chicken feathers for adsorption of liquid oils, being the disposal of waste chicken feathers a huge problem. They found that up to 16.21 g of oil/g of chicken feather can be recovered at fast uptake time (i.e., 10 min).

What is worth to note is the use of ecofriendly materials that can reduce the disposal cost and increase the costs-benefits ratio. A biocompatible and ecofriendly adsorbent material has been studied by Lee et al. [55]. They realized a mixture of lignin nanoparticles and 1-pentanol that is able to confine spilled oil into a thick slick on the surface of water, easily to remove. Wang et al. [56] prepared a novel superhydrophobic and superoleophilic material using a loofah sponge as porous skeleton and carnauba wax and rice bran wax as coating material. This adsorbent shows excellent selectively oil separation properties (>9.5 gg^{-1}) and reusability (>10 cycles).

In such a context, the oil sorption capabilities of the *Arundo Donax* fibers, evidenced in this book chapter, highlighted the acceptable affordability of this material for oil spill recovery application.

Although the sorption performances are lower than other conventional green materials, their oil/water selectivity makes them a cost-effective and reliable solution of oil spillage remedy. Even if these results are promising, further focused studies in order to improve the knowledge on performances/morphology relationship and on surface treatments are welcome. At the same time to improve the kinetic oil sorption capabilities and to reduce the water interaction of the vegetable fiber surface is another relevant issue for future activities.

References

1. Ariharasudhan, S., & Dhurai, B. (2019). Adsorption of oil from water surfaces using fibrous material—An overview. *Man-Made Textiles in India, 47*(4), 124–126.
2. Aguiar, F. C. F., & Ferreira, M. T. (2013). Plant invasions in the rivers of the Iberian Peninsula, south-western Europe: A review. *Plant Biosystems—An International Journal Dealing with All Aspects of Plant Biology, 147*(4), 1107–1119.
3. Celesti-Grapow, L., Capotorti, G., Del Vico, E., Lattanzi, E., Tilia, A., & Blasi, C. (2013). The vascular flora of Rome. *Plant Biosystems—An International Journal Dealing with All Aspects of Plant Biology, 147*(4), 1059–1087.
4. Sharma, K., Kushwaha, S. P. S., & Gopal, B. (1998). A comparative study of stand structure and standing crops of two wetland species, *Arundo Donax* and Phragmites karka, and primary production in *Arundo Donax* with observations on the effect of clipping. *Tropical Ecology, 39*, 3–14.
5. Pompeiano, A., Remorini, D., Vita, F., Guglielminetti, L., Miele, S., & Morini, S. (2017). Growth and physiological response of *Arundo donax* L. to controlled drought stress and recovery. *Plant Biosystems—An International Journal Dealing with All Aspects of Plant Biology, 151* (5), 906–914.
6. Perdue, R. E. (1958). *Arundo Donax*—Source of musical reeds and industrial cellulose. *Economic Botany, 12*(4), 368–404.
7. Weidenfeller, B., Lambri, O.A., Bonifacich, F.G., Arlic, U., Gargicevich, D., Weidenfeller, B., et al. (2018). Damping of the woodwind instrument reed material *Arundo Donax* L. *Materials Research, 21* (suppl 2).
8. Obataya, E., Umezawa, T., Nakatsubo, F., & Norimoto, M. (1999). The effects of water soluble extractives on the acoustic properties of reed (*Arundo Donax* L.). *Holzforschung, 53* (1), 63–67.

9. Pilu, R., Bucci, A., Cerino Badone, F., & Landoni, M. (2012). Giant reed (*Arundo Donax* L.): A weed plant or a promising energy crop? *African Journal of Biotechnology, 11*(38), 9163–9174.

10. Ververis, C., Georghiou, K., Christodoulakis, N., Santas, P., & Santas, R. (2004). Fiber dimensions, lignin and cellulose content of various plant materials and their suitability for paper production. *Industrial Crops and Products, 19*(3), 245–254.

11. Papazoglou, E. G., Karantounias, G. A., Vemmos, S. N., & Bouranis, D. L. (2005). Photosynthesis and growth responses of giant reed (*Arundo Donax* L.) to the heavy metals Cd and Ni. *Environment International, 31* (2), 243–249.

12. Flores, J. A., Pastor, J. J., Martinez-Gabarron, A., Gimeno-Blanes, F. J., Rodríguez-Guisado, I., & Frutos, M. J. (2011). *Arundo Donax* chipboard based on urea-formaldehyde resin using under 4 mm particles size meets the standard criteria for indoor use. *Industrial Crops and Products, 34*(3), 1538–1542.

13. Fiore, V., Scalici, T., & Valenza, A. (2014). Characterization of a new natural fiber from *Arundo Donax* L. as potential reinforcement of polymer composites. *Carbohydrate Polymers, 106*, 77–83.

14. Fiore, V., Scalici, T., Vitale, G., & Valenza, A. (2014). Static and dynamic mechanical properties of *Arundo Donax* fillers-epoxy composites. *Materials & Design, 57*, 456–464.

15. Fiore, V., Botta, L., Scaffaro, R., Valenza, A., & Pirrotta, A. (2014). PLA based biocomposites reinforced with *Arundo Donax* fillers. *Composites Science and Technology, 105*, 110–117.

16. Ismail, Z. Z., & Jaeel, A. J. (2014). A novel use of undesirable wild giant reed biomass to replace aggregate in concrete. *Construction and Building Materials, 67*, 68–73.

17. Scalici, T., Fiore, V., & Valenza, A. (2016). Effect of plasma treatment on the properties of *Arundo Donax* L. leaf fibres and its bio-based epoxy composites: A preliminary study. *Composites Part B: Engineering, 94*, 167–175.

18. Sun, R. (2010). *Cereal straw as a resource for sustainable biomaterials and biofuels : Chemistry, extractives, lignins, hemicelluloses and cellulose*. Elsevier.

19. Chen, B., Ye, X., Zhang, B., Jing, L., & Lee, K. (2019). Marine oil spills—Preparedness and countermeasures. *World Seas: An Environmental Evaluation*, 407–426.

20. DeLeo, D. M., Ruiz-Ramos, D. V., Baums, I. B., & Cordes, E. E. (2016). Response of deepwater corals to oil and chemical dispersant exposure. *Deep Sea Research Part II: Topical Studies in Oceanography, 129*, 137–147.

21. Azubuike, C. C., Chikere, C. B., & Okpokwasili, G. C. (2016). Bioremediation techniques—classification based on site of application: Principles, advantages, limitations and prospects. *World Journal of Microbiology and Biotechnology, 32*(11), 180.

22. Ceylan, D., Dogu, S., Karacik, B., Yakan, S. D., Okay, O. S., & Okay, O. (2009). Evaluation of butyl rubber as sorbent material for the removal of oil and polycyclic aromatic hydrocarbons from seawater. *Environmental Science & Technology, 43*(10), 3846–3852.

23. Wei, Q. F., Mather, R. R., Fotheringham, A. F., & Yang, R. D. (2003). Evaluation of nonwoven polypropylene oil sorbents in marine oil-spill recovery. *Marine Pollution Bulletin, 46*(6), 780–783.

24. Nikkhah, A. A., Zilouei, H., Asadinezhad, A., & Keshavarz, A. (2015). Removal of oil from water using polyurethane foam modified with nanoclay. *Chemical Engineering Journal, 262*, 278–285.

25. Feng, Y., & Xiao, C. F. (2006). Research on butyl methacrylate–lauryl methacrylate copolymeric fibers for oil absorbency. *Journal of Applied Polymer Science, 101*(3), 1248–1251.

26. Yang, L., Wang, Z., Li, X., Yang, L., Lu, C., & Zhao, S. (2016). Hydrophobic modification of platanus fruit fibers as natural hollow fibrous sorbents for oil spill cleanup. *Water, Air, & Soil Pollution, 227*(9), 346.

27. Annunciado, T. R., Sydenstricker, T. H. D., & Amico, S. C. (2005). Experimental investigation of various vegetable fibers as sorbent materials for oil spills. *Marine Pollution Bulletin, 50*(11), 1340–1346.

28. Adhithya, N., Goel, M., & Das, A. (2017). Use of bamboo fiber in oil water separation. *International Journal of Civil Engineering and Technology, 8*(6), 925–931.

29. Liu, J., & Wang, X. (2019). A new method to prepare oil adsorbent utilizing waste paper and its application for oil spill clean-ups. *BioResources, 14*(2), 3886–3898.

30. Wang, S., Peng, X., Zhong, L., Tan, J., Jing, S., Cao, X., et al. (2015). An ultralight, elastic, cost-effective, and highly recyclable superabsorbent from microfibrillated cellulose fibers for oil spillage cleanup. *Journal of Materials Chemistry A, 3*(16), 8772–8781.

31. Tu, L., Duan, W., Xiao, W., Fu, C., Wang, A., & Zheng, Y. (2018). Calotropis gigantea fiber derived carbon fiber enables fast and efficient absorption of oils and organic solvents. *Separation and Purification Technology, 192*, 30–35.

32. Cao, S., Dong, T., Xu, G., & Wang, F. (2018). Cyclic filtration behavior of structured cattail fiber assembly for oils removal from wastewater. *Environmental Technology, 39*(14), 1833–1840.

33. Liu, Y., Peng, Y., Zhang, T., Qiu, F., & Yuan, D. (2018). Superhydrophobic, ultralight and flexible biomass carbon aerogels derived from sisal fibers for highly efficient oil–water separation. *Cellulose, 25*(5), 3067–3078.

34. Feng, Y., Liu, S., Liu, G., & Yao, J. (2017). Facile and fast removal of oil through porous carbon spheres derived from the fruit of Liquidambar formosana. *Chemosphere, 170*, 68–74.

35. Husseien, M., Amer, A. A., El-Maghraby, A., & Taha, N. A. (2009). Availability of barley straw application on oil spill clean up. *International Journal of Environmental Science & Technology, 6*(1), 123–130.

36. Dong, T., Xu, G., & Wang, F. (2015). Adsorption and adhesiveness of kapok fiber to different oils. *Journal of Hazardous Materials, 296*, 101–111.

37. Said, A. E.-A. A., Ludwick, A. G., & Aglan, H. A. (2009). Usefulness of raw bagasse for oil absorption: A comparison of raw and acylated bagasse and their components. *Bioresource Technology, 100*(7), 2219–2222.

38. Wang, J., Geng, G., Liu, X., Han, F., & Xu, J. (2016). Magnetically superhydrophobic kapok fiber for selective sorption and continuous separation of oil from water. *Chemical Engineering Research and Design, 115*, 122–130.

39. De Rosa, I. M., Kenny, J. M., Puglia, D., Santulli, C., & Sarasini, F. (2010). Morphological, thermal and mechanical characterization of okra (Abelmoschus esculentus) fibres as potential reinforcement in polymer composites. *Composites Science and Technology, 70*(1), 116–122.

40. Cheu, S. C., Kong, H., Song, S. T., Johari, K., Saman, N., Che Yunus, M. A., et al. (2016). Separation of dissolved oil from aqueous solution by sorption onto acetylated lignocellulosic biomass—Equilibrium, kinetics and mechanism studies. *Journal of Environmental Chemical Engineering, 4*(1), 864–881.

41. Wahi, R., Chuah Abdullah, L., Nourouzi Mobarekeh, M., Ngaini, Z., & Choong Shean Yaw, T. (2017). Utilization of esterified sago bark fibre waste for removal of oil from palm oil mill effluent. *Journal of Environmental Chemical Engineering, 5*(1), 170–177.

42. Aloulou, F., Boufi, S., & Labidi, J. (2006). Modified cellulose fibres for adsorption of organic compound in aqueous solution. *Separation and Purification Technology, 52*(2), 332–342.

43. Nishi, Y., Iwashita, N., Sawada, Y., & Inagaki, M. (2002). Sorption kinetics of heavy oil into porous carbons. *Water Research, 36*(20), 5029–5036.

44. Alaa El-Din, G., Amer, A. A., Malsh, G., & Hussein, M. (2018). Study on the use of banana peels for oil spill removal. *Alexandria Engineering Journal, 57*(3), 2061–2068.

45. Piperopoulos, E., Calabrese, L., Mastronardo, E., Proverbio, E., & Milone, C. (2018). Synthesis of reusable silicone foam containing carbon nanotubes for oil spill remediation. *Journal of Applied Polymer Science, 135*(14), 46067.

46. Dong, T., Cao, S., & Xu, G. (2017). Highly efficient and recyclable depth filtrating system using structured kapok filters for oil removal and recovery from wastewater. *Journal of Hazardous Materials, 321*, 859–867.

47. Piperopoulos, E., Calabrese, L., Mastronardo, E., Abdul Rahim, S.H., Proverbio, E., & Milone, C. (2019). Assessment of sorption kinetics of carbon nanotube-based composite foams for oil recovery application. *Journal of Applied Polymer Science, 136* (14).

48. Wang, N., & Deng, Z. (2019). Synthesis of magnetic, durable and superhydrophobic carbon sponges for oil/water separation. *Materials Research Bulletin, 115*, 19–26.

49. Baker, R. W. (2004). *Membrane technology and applications*. Chichester, UK: Wiley.

50. Voisin, H., Bergström, L., Liu, P., Mathew, A., Voisin, H., Bergström, L., et al. (2017). Nanocellulose-based materials for water purification. *Nanomaterials, 7*(3), 57.
51. Cui, Y., Xu, G., & Liu, Y. (2014). Oil sorption mechanism and capability of cattail fiber assembly. *Journal of Industrial Textiles, 43*(3), 330–337.
52. Wang, J., & Liu, S. (2019). Remodeling of raw cotton fiber into flexible, squeezing-resistant macroporous cellulose aerogel with high oil retention capability for oil/water separation. *Separation and Purification Technology, 221*, 303–310.
53. Phanthong, P., Reubroycharoen, P., Kongparakul, S., Samart, C., Wang, Z., Hao, X., et al. (2018). Fabrication and evaluation of nanocellulose sponge for oil/water separation. *Carbohydrate Polymers, 190*, 184–189.
54. Tesfaye, T., Sithole, B., & Ramjugernath, D. (2018). Valorisation of waste chicken feathers: Green oil sorbent. *International Journal of Chemical Sciences, 16*(3), 1–13.
55. Lee, J. G., Larive, L. L., Valsaraj, K. T., & Bharti, B. (2018). Binding of lignin nanoparticles at oil-water interfaces: An ecofriendly alternative to oil spill recovery. *ACS Applied Materials & Interfaces, 10*(49), 43282–43289.
56. Wang, F., Xie, T., Zhong, W., Ou, J., Xue, M., & Li, W. (2019). A renewable and biodegradable all-biomass material for the separation of oil from water surface. *Surface and Coatings Technology, 372*, 84–92.

Effect of Surface Modification on Characteristics of Naturally Woven Coconut Leaf Sheath Fabric as Potential Reinforcement of Composites

K. N. Bharath, S. Basavarajappa, S. Indran and J. S. Binoj

Abstract Increase in biological and ecological concern has made to find new natural fibers which are biodegradable, from renewable sources and flexibility to chemical modification. Naturally woven Coconut leaf sheath (CLS) fabric were investigated. Removal of impurities on alkali treatment was carried to check the performance of these CLS fabric. On alkali treatment tensile and thermal properties were enhanced due to increase in crystallinity. The eco-friendly coconut leaf sheath fabric was found to be a suitable reinforcement material in composite structure for biocomposite applications.

1 Introduction

Natural fibers from agricultural crop residues, such as oil palm, coir, sisal, banana, pineapple leaf, jute, etc., are generated worldwide in abundance, representing an affordable, huge and easily accessible source. Among these enormous quantities of residues, only some are reserved for animal feed or household purposes and much of the residue is burned in the field, causing environmental pollution [1]. A lot of technological innovations have taken place to satisfy the requirements of consumers

K. N. Bharath (✉)
Composite Materials and Engineering Center, Washington State University, Pullman, Washington State, United States of America
e-mail: kn.bharath@gmail.com

Department of Mechanical Engineering, G.M. Institute of Technology, Davangere, Karnataka, India

S. Basavarajappa
Indian Institute of Information Technology, Dharwad, Dharwad, Karnataka, India

S. Indran
Rohini College of Engineering & Technology, Palkulam, Tamil Nadu, India

J. S. Binoj
Sree Vidyanikethan Engineering College (Autonomous), Tirupati, Andhra Pradesh, India

© Springer Nature Switzerland AG 2020
A. Khan et al. (eds.), *Biofibers and Biopolymers for Biocomposites*,
https://doi.org/10.1007/978-3-030-40301-0_14

and their expectations, which has increased demand for global resources. A remarkable transformation has taken place in the field of biofibre in the last few decades [2]. They offer more importance to the efficient use of agricultural residues in developing nations, and their importance depends on accessibility and use [3, 4]. Many scientists are working to replace synthetic fibers for multiple apps with naturally accessible fibers [5, 6]. Natural fibers are nonabrasive to mixing and molding equipment, which reduces the cost significantly and has positive environmental impact [7].

Some natural fibers' characteristics are described in the literature. To evaluate the quality, the chemical and morphological features of rape are used. The proportions of hemicellulose, cellulose and lignin in hardwood and prevalent nonwood were comparable to those observed in stalks of rapeseed [8]. The fibers obtained from the Cynara cardunculus were defined by electron microscopy and FT-IR scanning. These fibers heat degradation behavior was investigated through TGA and DTG curves. The mechanical characteristics were also assessed using single fiber tensile tests [9]. In the production of excellent results, natural fiber composites, the thermo-mechanical conduct of hemp fibers was regarded [10].

A part of the palm family (arecaceae) is the coconut palm (nucifera of coco). Coconut palms grow abundantly in coastal regions of all tropical nations and are grown in around 100,000 km2 worldwide [11, 12]. Coconut leaf sheath is obtained from coconut tree in mat form that acts as a supporting framework between coconut tree and leaves. For different purposes, the fibers from many parts of the coconut trees are used, but the sheath fabrics are left to waste. No research on evaluating the structure, morphology and thermal characteristics of CLS fabric has been revealed to the best of the author's data.

In the view of the above an attempt has been made in the present investigation to study the effect of Sodium hydroxide on coconut leaf sheath fabric. The structural and thermal properties were investigated on these fabrics verifying the possibility to use them as reinforcement in a composite structure.

2 Materials

Figure 1 is shown on a coconut leaf sheath with fabrics. The leaf sheath gathered from the trees was carefully cleaned with tap water in the current inquiry and then dried for a week in the sunshine. The cleaned coconut leaf sheath is divided into the internal sheath mat and the fibers of the outer layer. Tissues are obtained from the inside sheath mat and handled with 5% aqueous Sodium hydroxide (NaOH) solution for 24 h at room temperature. The treated and untreated fabrics were used for further experiments.

(a) **(b)**

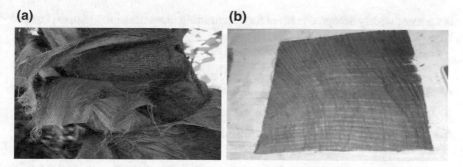

Fig. 1 Photographs of **a** coconut tree with leaf sheath; **b** leaf sheath inner mat

3 Experimentation

3.1 CLS Fabric Tensile Test

CLS fabric tensile test was performed for alkali treated and untreated fabric sample to determine tensile characteristics such as tensile strength, Young's modulus and percentage elongation at fabric break. Five samples have been used in each and average values have been recorded. Scanning electron microscope (JEOL JSM 820 SEM) was used to study the morphology of the treated and untreated fabrics after the test. The micrographs of the cross section of the fabrics were recorded.

4 Thermogravimetric Analysis (TGA)

To determine the thermal stability of the fabrics, TGA of the powdered alkali treated and untreated CLS fabric samples were carried out. Then these specimens were registered using a Perkin–Elmer TGA-7 instrument to undergo TGA at a temperature increase of 100 °C per minute. The range of temperatures was 250–5000 °C. In terms of worldwide mass loss, thermal decomposition was noted as well as, the initial degradation temperature and the final degradation temperature were calculated by using these thermograms.

5 Differential Scanning Calorimetry (DSC)

Measurements of differential calorimetry scanning (DSC) were performed by heating alkali treated and untreated samples of CLS fabric. These samples were produced to undergo DSC in an inert atmosphere from 0 to 450 °C at a temperature increase of 10oc per minute. DSC is performed to determine the melting temperature and it is the most widely accepted method for determining glass transition temperatures of the treated and untreated CLS fabrics.

6 Results and Discussion

6.1 Naturally Woven CLS Fabric Tensile Test

Table 1 presents the tensile properties of the untreated and alkali treated CLS fabric. The effects of chemical treatment on the properties of natural fabrics depend on the type and concentration of the chemical solution. It was proved from the previous chapter that, cellulose content increases on alkali treatment of natural fabrics and as a result the tensile properties also increases. Similarly, we found the same results in the naturally woven CLS fabric before and after alkali treatment of woven CLS fabrics.

Table 1 shows the results of tensile strength, young's modulus and percentage of elongation for the untreated and alkali treated CLS fabrics. Due to the treatment of fabrics there is increase of 40–60% in the tensile strength when compared to untreated fabrics. There was also increase of 30–50% in Young's modulus and 40–50% in the percentage of elongation on alkali treated CLS fabrics. Unwanted substances like wax and moisture were removed during the chemical treatments which led to an increase in the ratio of cellulose in the material. In the alkali treated fabrics, the increased quantity of cellulose showed a more significant increase in tensile strength [13].

Similar observations were made by Jayaramudu et al. in case of Sterculia urecs fabrics, in which increase in crystallinity was observed due to the removal of hemicellulose from the fabric resulting in thinning of fabric on alkali treatment which results an improvement in tensile properties [14]. It is evidenced from the Fig. 2a, b showed the micrographs of both the untreated and alkali treated fabrics respectively. The fabric has roughly uniaxial parallel fibers and these fibers are ligno-cellulosic in nature which are completely or partially soluble in alkali solutions. Due to alkali treatment, the amorphous hemicellulose and lignin were removed. From the above discussion, it is marked that the tensile properties increases on alkali treatment.

Table 1 Tensile properties of untreated and alkali treated CLS fabrics	Fabrics	Young's modulus (GPa)	Tensile strength (MPa)	Elongation at break (%)
	Untreated	5–8	150–250	6–13
	Alkali treated	10–14	220–320	17–22

(a)   (b)

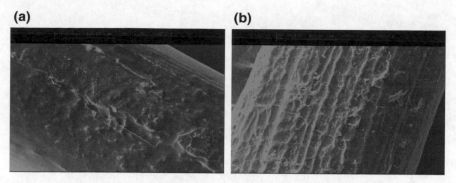

Fig. 2 Scanning electron micrographs of **a** untreated CLS fiber **b** alkali treated CLS fiber

7 Thermogravimetric Analysis (TGA)

Figure 4a, b shows TGA results of the untreated and alkali treated CLS fabrics respectively. Initially, weight change occurred is due to the release of moisture and further shoulder peak at about 226 °C of untreated CLS fabric and 218 °C of alkali treated CLS fabric corresponds to the removal of waxy layer and hemicellulose degradation. The peak at 248 °C (untreated) and 233 °C (alkali treated) corresponds to weight loss of the given material and the last peak at 371 °C (untreated) and 308 °C (alkali treated) may be attributed to thermal degradation. As the temperature goes on increasing, it is witnessed that thermal stability, decrease and further decomposition of fabric takes place.

It is observed from the Fig. 3 that, weight loss in the untreated CLS fabric is 77%, while weight loss in alkali treated CLS fabric is 51%. Lastly, residual char (mass) left in untreated fabric is 14.38% at 492.62 °C and in alkali treated CLS fabric is about 37.42% at 505.2 °C. Treated CLS fabric does not show a minor second weight change and peak discloses the removal of waxy layers and the nonexistence of hemicelluloses in the alkali treated CLS fabric [15, 16]. Similar observations were made in the TGA analysis of Manicaria fabric [17]. Three major weight loss steps were studied. The first step was associated with the evaporation of moisture in the fibers and the second step was related with the decomposition of hemicelluloses. The third step was due to lignin and alpha cellulose degradation [18]. There is very less amount of variation in moisture content of the untreated and alkali treated CLS fabrics. However, the percentage of char content was drastically reduced in case of alkali treated fabrics. Finally, due to reduction in amorphous hemicellulose content in the alkali treated CLS fabric have better thermal stability when compared with untreated CLS fabric which was confirmed after TGA analysis.

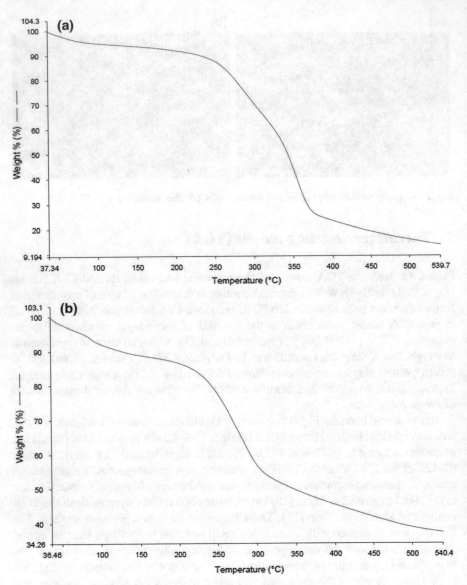

Fig. 3 Thermograms of **a** untreated CLS fabric **b** alkali treated CLS fabric

8 Differential Scanning Calorimetry (DSC)

Differential scanning calorimetric (DSC) results of the untreated and alkali treated CLS fabrics to examine the thermal behavior of the fabric is shown in Fig. 4a, b. Table 2 gives the DSC data for untreated and alkali treated CLS fabrics in an inert atmosphere (Heating rate 10 °C per minute).

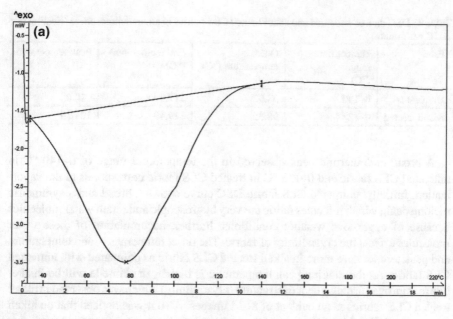

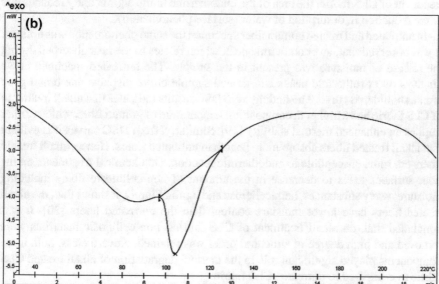

Fig. 4 DSC curves of **a** untreated CLS fabric **b** alkali treated CLS fabric

Table 2 DSC data of untreated and alkali treated CLS fabrics in an inert atmosphere (Heating rate 10 °C per minute)

Fabric	Temperature range (°C)	Onset temperature (°C)	End temperature (°C)	Peak temperature (°C)
Untreated	0–140	38.92	117.4	78.05
Alkali treated	0–128	98.35	119.49	104.15

A broad endothermic peak observed in the temperature range of 0–140 °C in untreated CLS fabric and 0–128 °C in treated CLS fabric corresponds to the vaporization. Initially, untreated CLS fabric DSC curve shows a broad and a symmetric melting peak, which indicates fabric are very hygroscopic and retain water molecules because of aggressive weather conditions. Further, accumulation of these water molecules affects the crystallinity of fabric. The onset temperature, end temperature and peak temperature more in alkali treated CLS fabric as compared with untreated CLS fabric. As the melting peak temperature is higher, the lamella will be thicker due to improved arrangement of structure (crystallinity) which was observed in alkali treated CLS fabrics as an evident of SEM images. Also it was noticed that on alkali treatment of CLS fabric, the area of the endothermic hump significantly reduces due to the reduction in desorption of water (surface phenomenon).

In untreated and treated gomuti fiber specimen the same phenomenon was noticed. It was observed that, an endothermic peak which relates to the heat absorbed during the release of moisture was present in the sample. The untreated specimen curve displays two exothermic peaks and treated sample curve displays one broad peak after a shouldered curve. The finding from DSC results indicates that alkali treatment of CLS fabric had a higher decomposition temperature of gomuti fibre, which further implies an enhanced thermal stability [19]. Similarly, from DSC curves it is evident that alkali treated fibers absorb more heat than untreated fibers. Hence, alkali treated fibers are more susceptible to endothermic reaction. On chemical treatments on the fiber surfaces leads to decrease in the amount of non-cellulosic fibers including moisture, waxy substances, hemicellulose and lignin. Finally, it shows that, the alkali treated fibers have lower moisture content than the untreated fibers [20]. It was concluded that, on alkali treatment of CLS fabrics, non-cellulosic materials were removed and high degree of structural order was retained. Nevertheless, both these components played significant role in the thermal degradation of alkali treated CLS fabric as compared to untreated CLS fabric.

9   Conclusions

The influence of alkali treatment on structural, morphological and thermal properties of CLS fibres and naturally woven fabric were investigated. Removal of non-cellulosic materials due to alkali treatment was the main reason for enhancement of fabric tensile properties. Fiber was found to be less thick and rough, due to the removal of amorphous on the surface of the fibers which was observed by Scanning Electron Micrographs. Increase in thermal stability and decrease in the moisture content in the fiber was due to the increase in crystallinity as indicated in TGA test. In DSC, the results obtained after alkali treatment showed increase in the surface area through D-condition which was due to water desorption. The obtained curves strengthen the assumption that moisture plays a predominant role in the fiber behavior during the glass transition region. Hence this work will provide a new approach for effective utilization of coconut leaf sheath fabric. The renewable and environment friendly coconut leaf sheath fibers were found to be a suitable reinforcement material in composite structure for biocomposite applications.

References

1. Abdul Khalil, H. P. S., Siti Alwani, M., Ridzuan, R., Kamarudin, H., & Khairul, A. (2008). Chemical composition, morphological characteristics, and cell wall structure of malaysian oil palm fibers. *Polymer-Plastics Technology and Engineering, 47*, 273–280.
2. Faruk, O., Bledzkia, A. K., Fink, H.-P., & Sain, M. (2012). Biocomposites reinforced with natural fibers: 2000–2010. *Progress in Polymer Science, 37*, 1552–1596.
3. Reddy, N., & Yang, Y. (2005). Biofibers from agricultural by products for industrial applications. *TRENDS in Biotechnology, 23*, 22–27.
4. Singha, A. S., & Thakur, V. K. (2009). Morphological, thermal, and physicochemical characterization of surface modified pinus fibers. *International Journal of Polymer Analysis and Characterization, 14*, 271–289.
5. John, M. J., & Thomas, S. (2008). Biofibres and Biocomposites. *Carbohydrate Polymers, 71*, 343–364.
6. Bharath, K. N., & Basavarajappa, S. (2016). Applications of biocomposite materials based on natural fibers from renewable resources: A review. *Science and Engineering of Composite Materials, 23*, 123–133.
7. Fiore, V., Valenza, A., & Di Bella, G. (2011). Artichoke (Cynara cardunculus L.) fibers as potential reinforcement of composite structures. *Composites Science and Technology, 71*, 1138–1144.
8. Placet, V. (2009). Characterization of the thermo-mechanical behaviour of Hemp fibres intended for the manufacturing of high performance composites. *Composites: Part A, 40*, 1111–1118.
9. Bharath, K. N., & Basavarajappa, S. (2014). Flammability characteristics of chemical treated woven natural fabric reinforced phenol formaldehyde composites. *Procedia Materials Science, 5*, 1880–1886.
10. Indran, R. E., & Raj, V. S. S. (2014). Characterization of new natural cellulosic fiber from Cissusquadrangularis root. *Carbohydrate Polymers, 110*, 423–429.
11. Jayaramudu, J., Guduri, B. R., & Rajulu, A. V. (2009). Tensile properties of polymethyl methacrylate coated natural fabric Sterculia urens. *Materials Letters, 63*, 812–814.
12. Ramamoorthy, S. K., Skrifvars, M., & Rissanen, M. (2015). Effect of alkali and silane surface treatments on regenerated cellulose fibre type (Lyocell) intended for composites. *Cellulose, 22*, 637–654.
13. Jayaramudu, J., Reddy, G. S. M., Varaprasad, K., Sadiku, E. R., Sinha Ray, S., & Rajulu, A. V. (2013) Structure and properties of poly (lactic acid)/Sterculia urens uniaxial fabric biocomposites, *Carbohydrate Polymers, 94*, 822–828.
14. Bennet, C., Rajini, N., Winowlin Jappes, J. T., Siva, I., Sreenivasan, V. S., & Amico, S. C. (2015). Effect of the stacking sequence on vibrational behavior of Sansevieria cylindrica/coconut sheath polyester hybrid composites. *Journal of Reinforced Plastics and Composites, 34*, 293–306.

15. Mulinari, D. R., Baptist, C. A. R. P., Souza, J. V. C., & Voorwald, H. J. C. (2011). Mechanical properties of coconut fibers reinforced polyester composites. *Procedia Engineering, 10*, 2074–2079.
16. Nitta, Y., Goda, K., Noda, J., Lee, & W.-I. (2013). Cross-sectional area evaluation and tensile properties of alkali-treated kenaf fibres. *Composites: Part A, 49*, 132–138.
17. Ticoalu, A., Aravinthan, T., & Cardona, F. (2013). A study into the characteristics of gomuti (Arenga pinnata) fibre for usage as natural fibre composites. *Journal of Reinforced Plastics and Composites, 33*, 179–192.
18. Porras, A., Maranon, A., & Ashcroft, I. A. (2015). Characterization of a novel natural cellulose fabric from Manicaria saccifera palm as possible reinforcement of composite materials. *Composites Part B, 74*, 66–73.
19. Sreenivasan, V.S., Rajini, N., Alavudeen, A., & Arumugaprabu, V. (2015). Dynamic mechanical and thermo-gravimetric analysis of Sansevieria cylindrica/polyester composite: Effect of fiber length, fiber loading and chemical treatment. *Composites: Part B, 69*, 76–86.
20. Hossain, M. K., Karim, M. R., Chowdhury, M. R., Imam, M. A., Hosur, M., Jeelani, S., et al. (2014). Comparative mechanical and thermal study of chemically treated and untreated single sugarcane fiber bundle. *Industrial Crops and Products, 58*, 78–90.

Effect of Glass and Banana Fiber Mat Orientation and Number of Layers on Mechanical Properties of Hybrid Composites

T. P. Sathishkumar, S. Ramakrishnan and P. Navaneethakrishnan

Abstract In this work, the effects of fiber mat orientation and number of layers on the tensile, flexural and impact properties of glass fiber random (SGFR) and banana fiber woven (BFW) mat reinforced epoxy laminated hybrid composites are investigated experimentally based on ASTM standards. The hybrid composites are prepared by compression molding process and results are compared with pure glass and banana fiber mat epoxy laminated composites. Results shows that introducing of SGFR mat in-between the BFW mats in the epoxy laminated composites reduces the overall weight of the composites and the mechanical properties of the hybrid composites are varied with BFW mat orientation. Moreover, the mechanical properties are varied by varying the number of layers in hybrid composites. The hybrid composites with four layers of glass and three layers of banana (i.e. G4B3) are showing higher tensile, flexural and impact properties compared to G3B2 composites. Also, by varying orientation of banana fiber woven mat, the maximum mechanical properties obtained for composites containing G4B3 layering pattern at 0° and 30° orientations.

1 Introduction

The composites are usually multiphase materials which are obtained by artificial combination of two or more materials with different properties. Hybrid composites are obtained by reinforcing more than one discontinuous phases in the continuous phase which are used for applications like automotive parts, airplanes interior parts, household appliances and infrastructure materials. For improving the mechanical properties of the laminated composites, the two fiber mats are usually used to prepare the laminated composites. Hybrid composites have prepared with two fibers, one is natural fiber and another one is synthetic fiber, or both fibers are natural or synthetic. Jun Hee Song et al. (2015) developed hybrid composites by lamination pairing of carbon/aramid fibers and carbon/glass fibers by vacuum assisted resin

T. P. Sathishkumar (✉) · S. Ramakrishnan · P. Navaneethakrishnan
Faculty of Mechanical Engineering, School of Building and Mechanical Sciences, Kongu
Engineering College, Perundurai, Erode, Tamilnadu 638060, India
e-mail: tpsathish@kongu.ac.in

© Springer Nature Switzerland AG 2020
A. Khan et al. (eds.), *Biofibers and Biopolymers for Biocomposites*,
https://doi.org/10.1007/978-3-030-40301-0_15

transfer molding technique. The tensile and bending behaviors of hybrid composites were evaluated according to ASTM standard. It was observed that the tensile strength increased with accumulation of central carbon layers and the mechanical behavior was dependent on the pairing sequence [1]. Arthur V. N. A. Lima et al. (2019) have reviewed the researches on sisal/glass hybrid composites with special emphasis on the specifications and parameters used. They have concluded that transverse orientation of fibers and higher glass content (%) significantly augmented the mechanical properties such as modulus, tensile, flexural and impact strengths. However, surface modification of sisal fiber and higher glass content resulted in improved the hardness, short-beam, shear strength, tear strength, and compression properties [2]. Fabrizio Sarasini et al. (2013) studied the low velocity impact behavior of E-glass/basalt reinforced hybrid epoxy composites, fabricated by resin transfer moulding method with diverse stacking sequences. It was found that higher impact energy absorption and capacity damage tolerance capabilities were exhibited by basalt and hybrid laminates having an intercalated configuration [3]. Silvio Leonardo Valença et al. (2015) investigated the tensile, bending and impact properties of Kevlar fiber plain fabric and Kevlar/glass hybrid fabric reinforced epoxy laminated composites. The composites plates were fabricated using hand lay-up technique. The results indicated that epoxy composites reinforced with Kevlar/glass hybrid structure exhibited higher tensile strength, bending and impact energy [4]. The water uptake behaviour of jute/glass/carbon fiber reinforced epoxy composites and its consequent impact on the in-plane shear performance of the laminates were evaluated by M. A. Abd El-baky and M. A. Attia (2018). The water absorption behaviour of both jute fiber reinforced epoxy composites and the hybrid composites with glass/carbon fibers adhered to Fickian pattern. The shear strength of the laminates decreased with water uptake. However, hybrid epoxy composites reinforced with jute/glass/carbon fibers showed better in-plane shear properties in both dry and wet conditions [5]. Athith et al. (2017) investigated the mechanical and tribological properties of natural rubber and epoxy matrices reinforced with jute/sisal/E-glass fabrics as a function of tungsten carbide (WC) loading. Results shows that incorporation of WC powder significantly improved the mechanical properties and decreased the wear rate of the hybrid polymeric composites [6]. Farah Hanan et al. (2018) prepared epoxy matrix composites reinforced with oil palm empty fruit bunch and kenaf fiber mats by hand-lay-up technique and evaluated the impact of hybridization on the mechanical properties of the prepared composites. The results demonstrated that the hybridization of kenaf fiber reinforced into oil palm EFB composites augmented the tensile and flexural properties. On the other hand, pure oil palm empty fruit bunch composites had better impact properties compared to those of hybrid composites [7]. Jothibasu et al. (2018) investigated the mechanical properties of four different composites obtained with different laminate stacking sequence involving areca sheath fiber/jute fiber/glass-woven fabric. The epoxy composites fabricated with intermittent layers of jute fiber, core layer of areca sheath fiber and skin layer with glass fabrics showed substantial improvement in mechanical properties owing to hybridization effect [8]. Hande Sezgin and Omer B. Berkalp (2016) studied the dual effects of hybridization and stacking sequence of the E-glass fabric and carbon fabric layers on the mechanical properties (tensile strength, impact strength) of jute fabric-reinforced polyester composite laminates. It was concluded that higher impact values can be achieved by incorporating high impact resistant fibers to the outer layers of the composites; in this connection, higher tensile strength of the composite laminates can be obtained by placing high tensile strength fibers at the inner layer [9]. Chensong Dong (2016) studied the flexural properties of epoxy hybrid composites reinforced with glass and carbon fibers. The composite laminates were prepared using two different kinds of glass fibres and two different types of carbon fibers, respectively. From the results, it was noted that partial replacement of glass fibres by carbon fibers in high strength carbon fibre reinforced composites improved the flexural strength significantly [10].

Praveen Kumar and Nalla Mohamed (2017) have assessed the tensile and moisture absorption behaviors of epoxy matrix composites reinforced with hybrid kenaf/glass composites with and without fly ash particulate filler. The results showed that the tensile properties of the kenaf/epoxy composites improved with addition of fly ash filler whereas their water absorption behavior decreased due to hybridization with glass fibers [11]. Erdem Selver et al. (2017) evaluated the tensile, flexural, and dynamic mechanical properties of epoxy laminates reinforced with flax/glass and jute/glass fibers. The laminates were prepared with different stacking sequences by vacuum infusion method. The composite laminates made from natural fibers exhibited higher specific strength values compared to glass fiber reinforced composites with normalization. These composites also showed enhanced visco- elastic behavior owing to synergistic effect of jute/flax fibers [12]. Yuqiu Yang et al. (2017) evaluated the optical properties, such as light transmission property and luminance distribution, of glass/silk fiber hybrid reinforced plastics. From the results, it was noted that the glass/silk fiber hybrid reinforced plastics diffused light effectively compared glass fiber reinforced plastics and light diffusion was positively correlated with the crepe degree of silk fabric [13]. P. N. B. Reis et al. (2007) investigated the flexural property of hybrid laminated composites having hemp fiber reinforced with polypropylene core and two glass fibers/polypropylene surface layers at either side of the sample. The hybrid composites showed economic and environmental benefits along with augmented mechanical behavior [14]. Sathishkumar et al. (2014) reviewed the static and dynamic mechanical, tribological, thermal and water absorption properties of natural fibers and glass fiber reinforced hybrid polymer composites as a function of different volume fraction or weight fraction, different fiber length and frequency. It was concluded that better thermo-mechanical properties were shown by hybrid polymer composites reinforced with treated natural fibers [15]. Morye and Wool (2005) evaluated the mechanical properties of thermoset matrix composite laminates reinforced with glass/flax fibers as a function of fiber arrangement. It was determined that the mechanical behavior of the hybrid composites was significantly influenced by glass/flax ratio and the arrangement of fibers [16]. Carlo Santulli (2005) explored an intermediary approach for fabricating hybrid laminates reinforced with glass/natural fiber with adequate impact properties. It has been inferred that better hybrid laminates can be developed by using larger volume of fibers and better interfacial adhesion between the matrix and reinforcements [17]. Amico et al. (2005) investigated the mechanical properties of compression-molded composites prepared with pure sisal, pure glass, and hybrid sisal/glass fibers with different stacking sequences. It was found that excellent properties were obtained for hybrid composites prepared with glass fibers on the top and bottom surfaces [18]. John and Venkata Naidu (2007) developed unsaturated polyester based hybrid composites reinforced with sisal/glass fibers using hand lay-up process to investigate their chemical resistance property. The results indicated that the sisal/glass hybrid composites were strongly resistant to most of the chemicals excluding chlorinated hydrocarbons [19]. Venkateshwaran et al. (2011) investigated the tensile, flexural, impact and water absorption properties of banana/sisal reinforced hybrid composites. From the results, it was noted that the mechanical properties improved with addition of sisal fiber in banana/epoxy composites up to 50% by weight. Furthermore, the water uptake was also reduced due to this hybridization [20]. Yahaya et al. (2016) investigated the impact and water absorption properties of hybrid epoxy composite using woven kenaf-Kevlar fibers. The results indicated that the physical properties of the composites considerably influenced their water absorption and thickness swelling behavior [21]. Vieira et al. (2018) explored the mechanical properties of epoxy composites reinforced with hybrid fabric with 70% Malva and 30% Jute fibers. The impact tests (Charpy and Izod) were carried out on the polymer matrix specimens prepared with 10, 20 and 30% in volume of hybrid fabric. It was observed that the samples incorporated with 30% volume fraction of fibers showed superior results [22]. Siddika et al. (2013) evaluated the mechanical properties of jute-coir fiber reinforced hybrid polypropylene composite as a function of filler loading. The results revealed that 20 wt% of jute and coir fibers (1:1 ratio) imparted better mechanical properties to the polymeric composites [23].

The more than one fiber reinforced with laminated composites have been increasing the mechanical, thermal and dynamic properties. There are different types of hybrid fibers laminated polymer composites were prepared such as natural-natural, natural-synthetic and synthetic-synthetic by varying the stacking sequence and number of layers. During synthetic-natural fiber composites preparation, the synthetic fiber mats were kept as outer layers for improving the mechanical properties and reducing the water absorption. However, the bidirectional banana fiber woven mat with random glass fiber mat is not yet discussed. The main aim of this work is to prepare the bidirectional banana fiber woven mat with random glass fiber mat reinforced laminated epoxy composites by varying the number of layers and banana fiber woven mat orientation, and studied these effects on tensile, flexural and impact properties. Then this analysis will be concluded the number of layer required for hybrid composites and fiber mat orientation in the composites.

2   Materials

Figure 1a shows the randomly oriented E-glass fiber mat. This mat is purchased From Covai, Senu Industry, Coimbatore, Tamilnadu, India. The density of glass mat is 2600 g/m^3. The special types bidirectional banana fiber woven mat (Fig. 1b) was

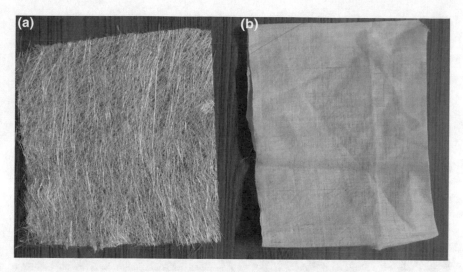

Fig. 1 **a** Random glass fiber mat, **b** bidirectional woven banana fiber mat

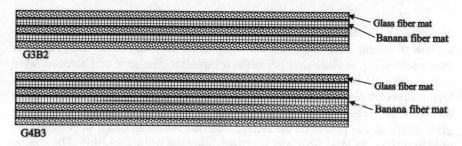

Fig. 2 Number of layers in the laminated composites **a** G3B2, and **b** G4B3

prepared in Erode Fiber Net India Pvt. Tamilnadu, India. The linear density of mat is 6 fiber/cm. The handloom machine was used to prepare the banana fiber mat with 1 m wide and 3 m long. This mat was dried under sun light for about 8 h to remove the moisture content (Fig. 2).

3 Preparation of Composites

The E-glass random mats and banana fiber woven mats were cut from the long mat of 240 mm (length) × 200 mm (width). The thickness of glass fiber mat and banana fiber woven mat are 0.45 mm and 0.25 mm respectively. The dried banana fiber mats were cut from the mat at various angles 0°, 30°, 45°, 60°, and 75° respectively. The laminated composites were prepared by varying the number of layers namely G3B2 and G4B3. The G3B2 is having three laminas of glass fiber mats and two laminas

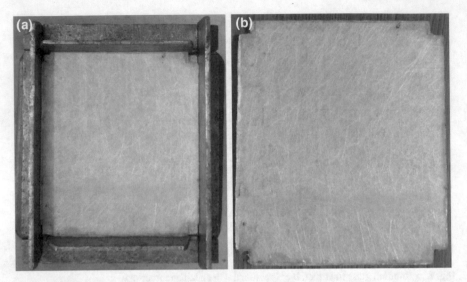

Fig. 3 a Die for preparing the composites, **b** composite plate

of banana fiber mats. The G4B3 is having four laminas of glass fiber mats and three laminas of banana fiber mats. In both composites, the outer layers are glass fiber mats which will give more strength to the composites. The resin and hardener were mixed with ratio of 200:20 mL. In composite preparation, the releasing agent of liquid polyvinylchloride was coated on the die set which could help to easily remove the composite plate. the small amount of resin was poured in the female die and then glass fiber mat was placed on the reins in the die. Followed that a steel roller was used to roll on the glass fiber mat for distributing the resin uniformly. The similar process was used to place the next laminas like banana and glass fiber mats. Finally, the male die was closed and the die set was kept under hydraulic press with 5 bar pressure for 4 h. The solidified composite (Fig. 3a) was taken from the die set and then post cured in hot air oven at 45 °C for 1 h. The pure banana fiber mat composite with seven layers (B7) and pure glass fiber mat composite with five layers (G5) were prepared for the analysis. The size of composite plates was 240 mm (length) × 200 mm (width) × 3 mm (thickness) size shown in Fig. 3b.

3.1 Characterization

The tensile properties of hybrid laminated composites were studied according to ASTM D638 standard [15]. The dog-bone shape specimens were drawn (Fig. 4a) and cut the specimens (Fig. 4b) from the prepared composite plate. The length, width, and thickness of specimen are 165 mm, 19 mm and 3 mm respectively. The gauge length of specimen is 50 mm. The testing speed is 5 mm/min. The ten identical specimens

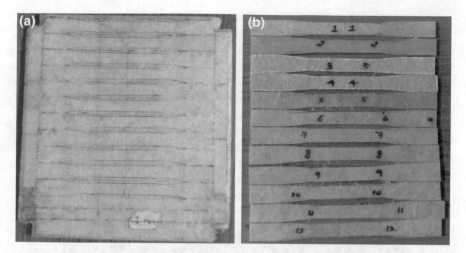

Fig. 4 Specimens preparation **a** dimension, and **b** dog-bone shape for tensile test specimens

were tested for all composites, and average values are taken for analysis. Figure 5a shows the tensile testing of dog-bone shape specimen using UTM. Figure 5b shows the fracture specimens after tensile test.

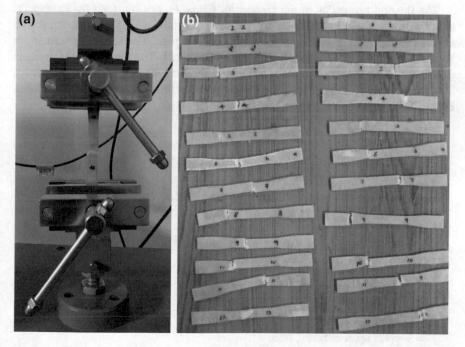

Fig. 5 **a** UTM machine for tensile testing, and **b** tensile fractured samples

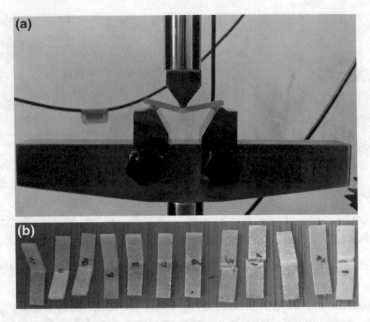

Fig. 6 **a** Three-point bending test, and **b** bending fracture samples

The flexural properties of the composite specimen are analyzed by three-point bending mode (Fig. 6a) according to ASTM D790 standard [15]. The flexural load versus displacement curves is obtained for all samples. The size of specimens is 48 mm length and 13 mm width with 3 mm thickness. The length (span) to depth (thickness) ratio of specimen is 16:1. The testing cross-head speed was 2 mm/min. The following Eqs. 1 and 2 are used to calculate the flexural strength and modulus. Figure 6a shows the bending fractured samples.

$$\sigma = \frac{3PL}{2bd^3} \tag{1}$$

σ = Flexural stress, MPa,
P = load at a given point on the load–deflection curve, N
L = support span of beam, mm,
b = width of beam, mm,
d = depth of beam, mm.

$$E_B = \frac{L^3 m}{4bd^3} \tag{2}$$

E_B = modulus of elasticity in bending, GPa,
L = support span of beam, mm,
b = width of beam, mm,

d = depth of beam, mm.

m = slope of the tangent to the initial straight-line portion of the load–deflection curve, N/mm of deflection.

The impact strength of all composites was measure by drop tower impact hammer (Model: IZOD TESTC-R, Make: Deepak poly-Plast Pvt. Ltd) according to ASTM D-256 [15]. The specimen length and width are 64 and 12.7 mm. The accuracy of impact machine is 0.01 J and maximum measuring range is up to 0–25 J. Figure 7a shows the impact testing of specimen and Fig. 7b shown the fractured samples from impact test.

Fig. 7 **a** Impact test, and **b** impact fractured samples

4   Results and Discussion

4.1   Effect of Number and Orientation of Layers on Tensile Properties

Figure 8 tensile stress versus tensile strain of glass and banana fiber mat laminated epoxy composites. The tensile stress is gradually increased with increasing tensile strain and all curves are seemed to be linear. The tensile stress curve of pure banana fiber mat composite (B7) is seemed lower compared to all other composites and also the pure glass fiber mat composite (G5) is seemed higher compared to all other composites. The tensile stress curves of all hybrid fibers laminated composites are seemed between the B7 and G5 composites. By varying the banana fiber mat orientation, the tensile stress curves are varying. The maximum tensile stress versus curve is found for hybrid composites containing 0° orientation of banana fiber mat at G4B3. Figure 9 shows the tensile strength of all laminated composites. Seeing that the tensile strength of hybrid composites is found higher compared to banana fiber mat composites and lower compared to glass fiber mat. The tensile strength of G4B3 composites is found higher compared to G3B2 composites. The increasing number of layers increases the tensile strength of the composites because the more layers can bear higher tensile loads. Five types of banana fiber mat orientations are taken for the discussion link 0°, 30°, 45°, 60°, and 75°. The tensile strength is varying by the orientation of the banana fiber woven mats. The maximum tensile strength of hybrid composite is found for composites containing 0° banana fiber mat orientation

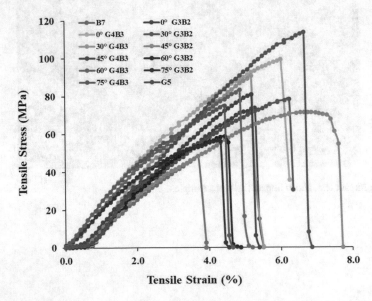

Fig. 8 Tensile stress versus strain of glass and banana fiber woven mat reinforced epoxy composites

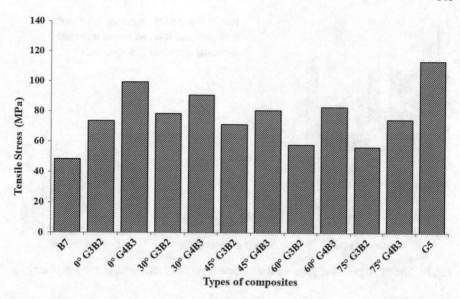

Fig. 9 Tensile strength of glass and banana fiber woven mat reinforced epoxy composites

and G4B3 layering pattern. Increasing the orientation of fiber mat decreases tensile strength of the hybrid composites. Fiber orientation is also playing am important roll on tensile strength.

The tensile strength of the G5 and B7 composites is 113.54 and 48.27 MPa. The tensile strength of G4B3 and G3B2 hybrid composites is 99.18 MPa and 73.6 MPa (at 0°), 90.55 MPa and 78.18 MPa (30°), 80.3 MPa and 71.24 MPa (at 45°), 83.1 MPa and 58.08 MPa (at 60°), 74.76 MPa and 59.42 MPa (at 75°) respectively. Based on the number of layers, the percentage difference of tensile strength between G4B3 and G3B2 is 25.79% (at 0°), 13.66% (30°), 11.83% (at 45°), 30.11% (at 60°), and 24.53% (at 75°). Based on orientation of banana fiber mat (G4B3 composites, the percentage difference between maximum tensile strength (at 0°, G4B3) and other strength is 8.7% (at 30°), 18.53% (at 45°), 16.2% (at 60°), and 24.62% (at 75°). The percentage difference of tensile strength between B7 and G4B3 is 51.33% (at 0°). Figure 10 shows the tensile properties of laminated composites. The maximum tensile modulus is found for composites containing G4B3 layers at 30° and 0° fiber mat orientations, 1.74 and 1.64 GPa. The elongation at breaks and tensile strain are varied for all composites based on the number of layers and fiber mat orientations.

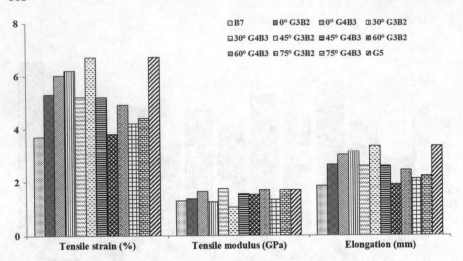

Fig. 10 Tensile properties of glass and banana fiber woven mat reinforced epoxy composites

5 Effect of Number and Orientation of Layers on Flexural Properties

The three-point bending test was used to measure the flexural properties of the hybrid and non-hybrid composites. Figure 11 shows the flexural stress versus flexural strain of glass and banana fiber mat laminated epoxy composites. The load cell of 500 N strain gauge was used to carry out the experiments. The flexural stress is gradually increased with increasing tensile strain and all curves are seemed to be linear. The

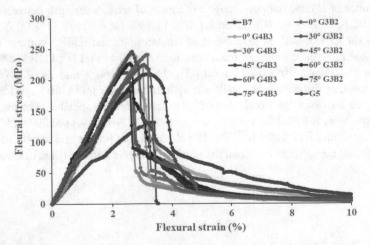

Fig. 11 Flexural stress versus strain of glass and banana fiber woven mat reinforced epoxy composites

flexural stress curve of pure banana fiber mat reinforced epoxy composite (B7) is seemed low compared to all other composites and the pure glass fiber random mat reinforced epoxy composite (G5) is seemed equal to hybrid fibers laminated composites. During flexural test, the loading surface (top surface) is subjected to compressive stress and unloading surface (bottom surface) is subjected to tensile stress. The failure mechanism is clearly shown in Fig. 6b. The top and bottom surfaces are seemed to be compressive and tensile region. The flexural stress curves of all hybrid fibers laminated composites are seemed between the B7 and G5 composites. By varying the banana fiber mat orientation, the flexural stress curves are varying. The maximum flexural stress versus curve is found for hybrid composites containing 0° and 60° orientation of banana fiber mat at G4B3. Figure 12 shows the flexural strength of all laminated composites. Seeing that the flexural strength of hybrid fibers laminated composites is found higher compared to banana fiber mat composites and little lower compared to glass fiber mat. The flexural strength of G4B3 composites is found higher compared to G3B2 composites. The increasing number of layers increases the flexural strength of the composites because the more layers can bear higher flexural loads, the glass fiber mat can bear more flexural strength than banana fiber mat due higher fiber strength. Five types of banana fiber mat orientations are taken for the discussion of flexural properties link 0°, 30°, 45°, 60°, and 75°. The various of flexural strength is associated with the change orientation of banana fiber woven mats. Increasing the orientation of fiber mat decreases flexural strength of the hybrid composites. The maximum flexural strength of hybrid composite is found for composites containing 0° and 60° banana fiber mat orientation at G4B3 layering pattern. Fiber orientation is also playing am important roll on flexural strength.

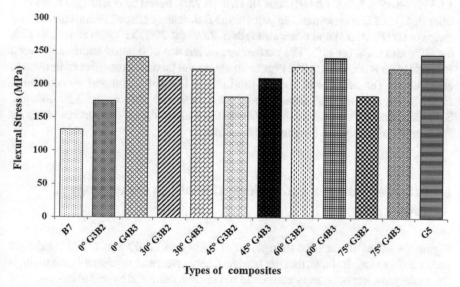

Fig. 12 Tensile strength of glass and banana fiber woven mat reinforced epoxy composites

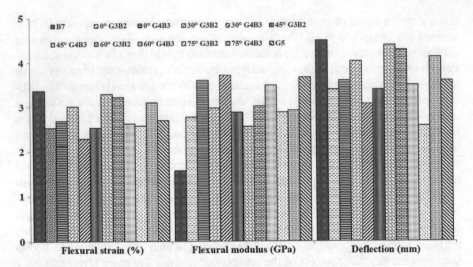

Fig. 13 Flexural properties of glass and banana fiber woven mat reinforced epoxy composites

The flexural strength of the B7 and G5 composites is 132.02 MPa and 246.23 MPa. The flexural strength of G4B3 and G3B2 hybrid fiber laminated epoxy composites is 241.57 MPa and 174.88 MPa (at 0°), 223.41 MPa and 212.38 MPa (30°), 209.86 MPa and 181.84 MPa (at 45°), 241.16 MPa and 227.16 MPa (at 60°), 224.89 MPa and 184 MPa (at 75°) respectively. Based on the number of layers, the percentage difference of flexural strength between G4B3 and G3B2 is 27.61 (at 0°), 4.9% (30°), 13.35% (at 45°), 5.18% (at 60°), and 18.18% (at 75°). Based on orientation of banana fiber mat (G4B3) composites, the percentage difference between maximum flexural strength (at 0°, G3B3) and other strength is 7.6% (at 30°), 13.13% (at 45°), 0.17% (at 60°), and 6.9% (at 75°). The percentage difference of flexural strength between B7 and G4B3 is 45.35% (at 0°). Figure 13 shows the flexural properties of laminated composites. The maximum flexural modulus is found for composites containing G4B3 layers at 0° and 30° banana fiber mat orientations, it is around 3.59 GPa and 3.70 GPa. The flexural strain and deflection are varied for all composites based on the number of layers and fiber mat orientations.

6 Effect of Number and Orientation of Layers on Impact Properties

Figure 14 shows the impact strength of B7, G4 and G3B2 and G4G3 laminated epoxy composites. It shows that the banana fiber woven mat reinforced composite is found very low impact energy compared to G5 composite and hybrid fiber laminated

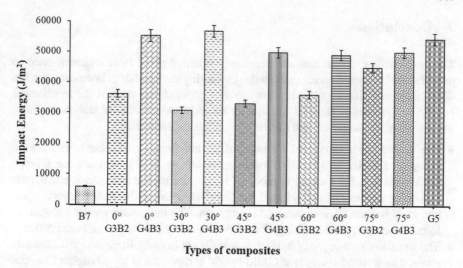

Fig. 14 Impact energy of glass and banana fiber woven mat reinforced epoxy composites

composites. Incorporation of glass fiber random mat in banana fiber woven mat laminated composites is increasing the impact energy absorption. The glass fiber mat composite (G5) is found higher impact energy compares to B7 composites. The Impact energy of B7 and G5 is 5905.51 and 54,383.3 J/m^2. In hybrid composites, the G4B3 composites is found high compared to G3B2 composites. Increasing the number of both layers increases the impact energy absorption. It shows that the G4B3 laminated composite at 0° banana fiber woven mat orientation is found equal impact energy compared G5 composites. The composite containing G4B3 layering pattern at 30° banana fiber woven mat orientation shows the maximum impact energy compared to G5 composites due that the orientation plays an important role on impact energy absorption. The impact energy transfers layers are uniformly occurred and resisted between the layers. This figure that the impact energy of G3B2 and G4B3 composites is 36,115.49 and 54,120.73 J/m^2 (at 0°), 30,695.54 and **56,732.28** J/m^2 (at 30°), 33,044.62 and 49,842.52 J/m^2 (at 45°), 35,667.38 and 49,094.49 J/m^2 (at 60°), and 44,960.63 and 4994.51 J/m^2 (at 75°). The maximum impact energy is found for composites containing G4B3 layering patterns at 0° banana fiber woven mat. The percentage difference between maximum impact energy (G4B3, at 30°) to other G4B3 hybrid composites is 4.6% (at 0°), 12.14% (at 45°), 13.46% (at 60°) and 11.96% (at 75°), and also the glass fiber mat composite (G5) is 4.14% low compared to G4B3 hybrid composite. So, the banana fiber woven mat can improve the impact energy of the hybrid fiber laminated composites.

7 Conclusion

The banana fiber woven mat and randomly oriented glass fiber mat was used to prepared the laminated epoxy composites by varying the number of layers and banana fiber mat orientation by compression molding process and studied the mechanical properties like tensile, flexural and impact according to the ASTM standards. From the extensive investigation, the following conclusion are obtained.

- The tensile stress curves of all composites are linearly increased. The tensile strength of G3B2 and G4B3 hybrid composites is found between the B7 and G5 composites. By varying number of layers, the tensile strength is varied, the maximum tensile strength is found for composites containing G4B3 layering pattern at 0° banana fiber woven mat orientation. The maximum tensile modulus is found for composite containing 0° and 30° banana fiber woven mat orientation.
- The flexural stress curve of hybrid composites is linearly increased with flexural strain. The flexural strength of G4B3 hybrid composites is found higher than G5, B7 and G3B2 hybrid composites. Incorporation of banana fiber mat is increasing the flexural strength of composites. By varying number of layers, the flexural strength is varied, the maximum flexural strength is found for composite containing G4B3 layering pattern at 0° banana fiber woven mat orientation. The maximum flexural modulus is found for composite containing 0° banana fiber woven mat orientation.
- The absorption of impact energy is varying for all hybrid composites. The impact energy of banana fiber woven mat laminated epoxy composite is very low compared to all hybrid composites and G5 composites. Incorporation of glass fiber random mat is increasing the impact energy absorption. The maximum impact energy absorption is found for composite containing 0° and 30° banana fiber woven mat orientation. It is higher compared to G5 laminated composites.

Increasing the banana fiber mat orientation more than 30° is decreasing the tensile, flexural and impact strength of the hybrid fibers laminated composites. Overall the maximum mechanical strength is found for laminated composites containing 0° banana fiber woven mat. Also, the weight of the banana fiber mat is low compared to glass fiber mat, it is reducing the total weight of the composites. This G4B3 laminated composites at 0° banana fiber woven mat orientation is a suitable composites material for various structural applications.

References

1. Song, J. H. (2015). Pairing effect and tensile properties of laminated high-performance hybrid composites prepared using carbon/glass and carbon/aramid fibers. *Composites Part B, 79,* 61–66.
2. Lima, A. V. N. A., Cardoso, J. L., & Lobo, C. J. S. (2019). Research on hybrid sisal/glass composites: A review. *Journal of Reinforced Plastics and Composites.* https://doi.org/10.1177/0731684419847272.

3. Sarasini, F., Tirillo, J., Valente, M., Valente, T., Cioffi, S., Iannace, S., & Sorrentino, L. (2013). Effect of basalt fiber hybridization on the impact behavior under low impact velocity of glass/basalt woven fabric/epoxy resin composites. *Composites: Part A, 47*, 109–123.

4. Valenca, S.L., Griza, S., Gomes de Oliveira, V., Sussuchi, E. M., & Carvalho de Cunha, F. G. (2015). Evaluation of the mechanical behavior of epoxy composite reinforced with Kevlar plain fabric and glass/Kevlar hybrid fabric. *Composites: Part B*: 70, 1–8.

5. Abd El-baky, M. A., & Attia, M. A. (2018). Water absorption effect on the in-plane shear properties of jute–glass–carbon reinforced composites using Iosipescu test. *Journal of Composite Materials*. https://doi.org/10.1177/0021998318809525.

6. Athith, D., Sanjay, M. R., Yashas Gowda, T. G., Madhu, P., Arpitha, G. R., Yogesha, B., & Omri, M. A. (2017). Effect of tungsten carbide on mechanical and tribological properties of jute/sisal/E-glass fabrics reinforced natural rubber/epoxy composites. *Journal of Industrial Textiles, 48*(4), 713–737.

7. Hanan, F., Jawaid, M., & Tahir, P. M. (2018). Mechanical performance of oil palm/kenaf fiber-reinforced epoxy-based bilayer hybrid composites. *Journal of Natural Fibers*. https://doi.org/10.1080/15440478.2018.1477083

8. Jothibasu, S., Mohanamurugan, S., Vijay, R., Lenin Singaravelu, D., Vinod, A., & Sanjay, M. R. (2018). Investigation on the mechanical behavior of areca sheath fibers/jute fibers/glass fabrics reinforced hybrid composite for light weight applications. *Journal of Industrial Textiles*. https://doi.org/10.1177/1528083718804207

9. Sezgin, H., & Berkalp, O. B. (2016). The effect of hybridization on significant characteristics of jute/glass and jute/carbon-reinforced composites. *Journal of Industrial Textiles, 47*(3), 283–296.

10. Dong, C. (2016). Uncertainties in flexural strength of carbon/glass fibre reinforced hybrid epoxy composites. *Composites Part B, 98*, 176–181.

11. Praveen Kumar, A., & Nalla, M. M. (2018). A comparative analysis on tensile strength of dry and moisture absorbed hybrid kenaf/glass polymer composites. *Journal of Industrial Textiles, 47*(8), 2050–2073.

12. Selver, E., Ucar, N., & Gulmez, T. (2017). Effect of stacking sequence on tensile, flexural and thermomechanical properties of hybrid flax/glass and jute/glass thermoset composites. *Journal of Industrial Textiles, 48*(2), 494–520.

13. Yang, Y., Zhao, D., Xu, J., Dong, Y., Ma, Y., Qin, X., et al. (2017). Mechanical and optical properties of silk fabric/glass fiber mat composites: An artistic application of composites. *Textile Research Journal, 88*(8), 932–945.

14. Reis, P. N. B., Ferreira, J. A. M., Antunes, F. V., & Costa, J. D. M. (2007). Flexural behaviour of hybrid laminated composites. *Composites: Part A, 38*, 1612–1620.

15. Sathishkumar, T. P., Naveen, J., & Satheeshkumar, S. (2014). Hybrid fiber reinforced polymer composites—A review. *Journal of Reinforced Plastics and Composites, 33*(5), 454–471.

16. Morye, S. S., & Wool, R. P. (2005). Mechanical properties of glass/flax hybrid composites based on a novel modified soybean oil matrix material. *Polymer Composites, 26*(4), 407–416.

17. Santulli, C. (2005). Impact properties of glass/plant fibre hybrid laminates. *Journal of Material Science, 42*, 3699–3707.

18. Amico, S. C., Angrizani, C. C., & Drummon, M. L. (2010). Influence of the stacking sequence on the mechanical properties of glass/sisal hybrid composites. *Journal of Reinforced Plastics and Composites, 29*(2), 179–189.

19. John, K., & Venkata, N. S. (2007). Chemical resistance of sisal/glass reinforced unsaturated polyester hybrid composites. *Journal of Reinforced Plastics and Composites, 26*(4), 373–376.

20. Venkateshwaran, N., ElayaPerumal, A., Alavudeen, A., & Thiruchitrambalam, M. (2011). Mechanical and water absorption behaviour of banana/sisal reinforced hybrid composites. *Materials & Design, 32*(7), 4017–4021.

21. Yahaya, R., Sapuan, S., Jawaid, M., Leman, Z., & Zainudin, E. (2016). Water absorption behaviour and impact strength of kenaf-kevlar reinforced epoxy hybrid composites. *Advanced Composites Letters, 25*(4), 98.

22. Vieira, J. D. S., Lopes, F. P., de Moraes, Y. M., Monteiro, S. N., Margem, F. M., Margem, J. I., & Souza, D. (2018). Comparative mechanical analysis of epoxy composite reinforced with malva/jute hybrid fabric by izod and charpy impact test. In *Proceedings of the TMS Annual Meeting & Exhibition*, pp. 177–183.
23. Siddika, S., Mansura, F., & Hasan, M. (2013). Physico-mechanical properties of jute-coir fiber reinforced hybrid polypropylene composites. *International Journal of Chemical, Molecular, Nuclear, Materials and Metallurgical Engineering, 7*(1), 60–64.

Printed in the United States
by Baker & Taylor Publisher Services